英和 舶用機関用語辞典

商船高専機関英語研究会 編

海文堂

編集委員（学校別・五十音順（所属は初版発行時））

学校	委員
富山商船高等専門学校	石井 秋吉、見上 宗博、佐藤 芳男
鳥羽商船高等専門学校	嶋岡 弘、多田 邦勝
弓削商船高等専門学校	松下 直幸、松永 英也、古賀 伊司
大島商船高等専門学校	三原 鉛文、上杉 博一
広島商船高等専門学校	大山 敏史、茶園 邦文、中島 朋廣、濱田 起、藤井 誠

序　文

　5商船高専15名の教員が，長年の経験を生かし舶用機関全般と関連する学問分野における専門英語を集めて，ここに『英和　舶用機関用語辞典』刊行の運びとなりました。本辞典の特徴として名詞以外に動詞，形容詞や副詞なども掲載しました。専門英語の辞典は名詞のみを対象としたものがほとんどです。これは専門英語の分野が「機械」「電気・電子」「情報処理」「金属」「化学・工業化学」「自動車」「地学・天文」「ビジネス・経済」「軍事」「輸送」など多方面にわたるため，名詞以外の品詞を加えると，用語の数が増え，また製作上煩雑になることから，敬遠されてきたと考えられます。しかし，動詞などの用語が掲載されていればとても便利な辞典になることは間違いありません。そこで，便利で使いやすい辞典を目指して，名詞以外に専門英語において使用頻度の高い動詞，形容詞，副詞なども掲載しました。また，日本語の訳は，極力カタカナ表示を避け，馴染みのない用語には簡単な解説文を加えて，専門英語辞典として充実させました。そして，過去10年間の2級海技士および1級海技士の英語試験問題で使用されている用語がほぼ100%掲載されていることが確認できており，海技士英語試験対策には必携の辞典と言えます。

　このように，専門英語の授業や海技士試験における学習用辞典として，あるいは海事に関係する専門家はもちろん広く一般に用いられて便利な辞典として自信を持ってお勧めできます。しかし，不備や誤りがあれば，使用者各位のご叱正を得て他日改めたいと存じます。

　終わりに，本辞典の刊行にあたり編集委員諸氏の努力はもとより海文堂出版の熱意に対して厚くお礼を申し上げます。

　　令和7年6月吉日

商船高専機関英語研究会代表　中　島　邦　廣

凡　例

1．見出し語の配列はアルファベット順とした。

2．品詞名は略号を用いて示した。
　　名：名詞　　代：代名詞　　関代：関係代名詞　　形：形容詞
　　副：副詞　　動：動詞　　助：助動詞　　前：前置詞
　　接：接続詞　　接頭：接頭語

3．その他の略号
　　関：関連語　　反：反意語　　同：同義語　　記：元素記号
　　《米》：主に米国用法　《英》：主に英国用法

4．括弧の用法
　　（　）：括弧の中の語と前の語が置き換えられるとき
　　　　a great（large）amount of
　　　　「a great amount of」または「a large amount of」の意
　　［　］：括弧の中を省いてもよいとき
　　　　機関［軸］出力
　　　　「機関出力」または「機関軸出力」の意
　　【　】：語句の解説文
　　〈　〉：関連分野を示す
　　　　圧力面〈プロペラ〉
　　　　「プロペラ」分野の意

5．長音は「ー」で示したが，人名以外で用語が3字以上の場合，最後の長音は原則として省略した。

A-a

A アンペア【電流の強さの単位】
a 形一つの，〜というもの，〜につき
  a great (large) amount of 〜 形多量の〜
  a great (large) number of 〜 形多数の〜
  a number of 〜 いくらかの〜，多数の〜
  関 the number of 〜：〜の数
  a small amount of 〜 形少量の〜
  a small number of 〜 形少数の〜
a contact (connection) a接点
A_1 transformation A_1変態【鋼の組織形態の変化の一つ】
AA battery 単三電池
AB (American Bureau) of Shipping アメリカ船級協会
abaft 副船尾の方に，後方に
abatement 減退，軽減，減少，除去
abbreviate 動省略する，短縮する
abbreviation 略語
ABC (Automatic Boiler Control system) 自動ボイラ制御装置【ボイラの空燃比と給水量を制御する装置】
abeam 副真横に
aberration 収差
ability 能力，才能
ablation 消耗，削摩，焼摩
ablator アブレータ【熱遮断材の一種】
able 形有能な 関 be able to 〜：〜することができる
abnormal 形異常な，変則な
  abnormal combustion 異常燃焼
  abnormal condition 異常状態
aboard 副船内に，船上に，舷側に
abound 動満ちている，富んでいる
about 〜 前〜について，〜の周りに，およそ，約 副まわりに，およそ，約〜 関 be about to 〜：まさに〜しようとしている
above 〜 前〜の上に(で) 形上述の〜 副上に(の)，前述の，上記の，上位にある，それより上の 名上記のこと 反 below
  above all とりわけ
  above mentioned 形上述の
abrasion 摩耗，研磨
  abrasion failure 摩耗破壊
  abrasion hardness 摩耗硬さ
  abrasion limit 摩耗限度
  abrasion loss 摩耗減量，摩耗量
  abrasion proof 耐摩耗
  abrasion resistance 耐摩耗性
  abrasion resistant 摩耗抵抗
  abrasion resistant steel 耐摩耗鋼
  abrasion test 摩耗試験
  関 abrasion tester：摩耗試験機
abrasive 形すり減らす，剥離の 名研削材，研磨材，砥石
  abrasive cloth 布やすり，研磨布
  abrasive finishing 研磨
  abrasive friction 研磨摩擦
  abrasive machining 砥粒加工
  abrasive paper 研磨紙，紙やすり
  abrasive substance 研磨材
  abrasive wear アブレシブ摩耗，摩滅損耗，摩損
  abrasive wheel 砥石車
abrasives 研磨材
abrupt 形急な
  abrupt contraction pipe 急縮小管
  abrupt expansion pipe 急拡大管
abruptly 副急に
abscissa 横座標
absence 不在，欠乏
absolute 形絶対の，純粋な
  absolute alcohol 無水アルコール
  absolute error 絶対誤差
  absolute humidity 絶対湿度
  absolute manometer 絶対圧力計
  absolute maximum 絶対最大［値］
  absolute minimum 絶対最小［値］
  absolute number 無名数
  absolute pressure 絶対圧力

absorb

absolute size　実寸
absolute specific gravity　絶対比重, 真比重
absolute speed　絶対速度, 対地速度
absolute speed indicator　対地速度計
absolute system of units　絶対単位系
absolute temperature　絶対温度
absolute unit　絶対単位
absolute vacuum　絶対真空
absolute value　絶対値
absolute velocity　絶対速度
absolute viscosity　絶対粘度
absolute work　絶対仕事
absolute zero　絶対零度
absorb　動吸収する
absorbed energy　吸収エネルギ
absorbent　吸収剤, 吸着材
absorber　吸収装置, 吸収材
absorption　吸収, 吸着
　absorption band　吸収帯
　absorption dynamometer　吸収動力計
　absorption phenomenon　吸着現象
　absorption rate　吸収率
　absorption refrigerating machine　吸収冷凍機
　absorption spectrum　吸収スペクトル
absorptive power　吸収力, 吸収能
absorptivity　吸収能, 吸収率
abstract　名アブストラクト, 要旨　形抽象的な　動取り出す, くみ上げる, 要約する, 分離する
abundance　多量, 大量, 多数
　関 in abundance：大量に, 豊富に
　abundance of 〜　多数の, 豊富な
AC (Alternating Current)　交流
　関 DC：直流
　AC arc welding　交流アーク溶接
　AC bridge　交流ブリッジ
　AC circuit　交流回路
　AC commutator motor　交流整流子モータ
　AC compound motor　交流整流子電動機
　AC electromotive force　交流起電力
　AC exciter　交流励磁機
　AC generator　交流発電機
　AC machine　交流機
　AC motor　交流電動機
　AC power supply　交流電源
　AC series motor　交流直巻電動機
　AC servomotor　交流サーボモータ
　AC shunt motor　交流分巻電動機
　AC source　交流電源
　AC-to-DC converter　交流・直流変換器
　AC variable speed motor　AC可変速モータ, ACサーボモータ
　AC voltage　交流電圧
　AC welder　交流溶接機
ACB (Air Circuit Braker)　気中遮断器
ACC (Automatic Combustion Control system)　燃焼制御装置【ボイラの空燃比を制御する装置】
accelerate　動加速する
accelerating　加速, 増速
　accelerating characteristics　加速特性
　accelerating device　加速器
　accelerating force　加速力
acceleration　加速, 加速度
　acceleration of gravity　重力の加速度【記号：g】
　acceleration pump　加速ポンプ
accelerative　形加速的な
accelerator　アクセル, 加速装置, 加速器
accelerometer　加速度計
accept　動受け入れる, 認める
acceptability limit　許容限度
acceptable　形容認できる, 受け入れられる, 許容できる, 満足な
　acceptable concentration　許容濃度
acceptor　アクセプタ【半導体に正孔を増加する目的で加える不純物】, 受容体
access　名アクセス, 出入り, 接近, 通路, 呼出し　動（コンピュータ）にアクセスする, 検索する, に近づく, を利用する, 接近する
　access code　アクセスコード
　access door　点検窓, 検査ふた

accessible 形入手できる，近づきやすい，利用できる，理解できる，影響されやすい

accessory アクセサリ，装備品，付属品

accident 事故，災害，故障
関 by accident：偶然に
accident frequency rate 災害率，事故発生率
accident severity rate 災害強度率

accidental 形偶然の，不慮の
accidental error 偶然誤差

accommodate 動適応する，順応する，合わせる

accommodation 便宜，調節，適応，居住設備
accommodation space 居住区

accompanying drawing 付図

accomplish 動達成する，完成する

accord 動一致する，調和する

accordance 一致，調和
関 in accordance with：～に従って，～と一致して

according to ～ 前～に従って，～によれば

accordingly 副それに応じて，状況に応じて，それゆえに，したがって

account 名説明，報告 動説明する，報告する 関 on account of ～：～のため，～の理由で／take A into account：Aを考慮する

accumulate 動蓄積する，ためる，積もる

accumulation 蓄積
accumulation test 蓄気試験

accumulator アキュムレータ，蓄圧器，蓄電池，～溜【エネルギ蓄積装置の一種】同 surge drum

accuracy 精度，正確さ

accurate 形正確な，精密な
accurate measuring instrument 精密測定装置

accurately 副正確に，精密に

accustom 動慣れる，慣らす，順応する・させる，馴染ませる

acescence 弱酸性

acescent 形弱酸性の，酸味の

acetaldehyde アセトアルデヒド【可燃性の無色の液体】

acetate アセテート，酢酸塩

acetic 形酢の，すっぱい
acetic acid 酢酸

acetone アセトン【無色・揮発性の可燃性液体】

acetylene アセチレン
acetylene cutting アセチレン切断
acetylene gas generator アセチレンガス発生器
acetylene welding アセチレン溶接

achieve 動達成する

acicular 形針状の
acicular cast iron アシキュラ鋳鉄，針状鋳鉄【高級鋳鉄の一種】
acicular structure 針状組織

acid 名酸，酸性 形酸[性]の
acid attack 酸腐食
acid cleaning 酸洗い
acid content 酸分
acid corrosion 酸腐食
acid degree 酸度
acid neutralizer 酸中和剤
acid number 酸価，酸率
acid pickling 酸洗い
acid proof 名耐酸，耐酸性 形耐酸性の
acid proof alloy 耐酸合金
acid radical 酸基
acid rain 酸性雨
acid resisting alloy 耐酸合金
acid resisting paint 耐酸ペイント
acid salt 酸性塩
acid smut アシッドスマット【硫黄分を含んだすす】
acid steel 酸性鋼
acid treatment 酸処理
acid value 酸価

acidic 形酸性の
acidic group 酸性基

acidification 酸性化

acidity 酸性，酸性度，酸度
acidity test 酸価試験

acidizing 酸処理

acknowledge 動認める，承認する
acme thread アクメねじ
acoustic 形音響の，聴覚の，防音の
 acoustic coupler 音響カプラ，音響結合装置
 Acoustic Doppler Velocimeter ADV, 超音波流速計
 Acoustic Emission AE, アコースティックエミッション，超音波放射【非破壊検査の一つ】
 acoustic inspection 音響検査【ハンマを用いた欠陥検査】
 acoustic velocity 音速
 acoustic wave 音波
acquire 動得る，樹得する，入手する
across 〜 前〜を横切って，〜の向かいに，〜と交差して 副幅で，直径で，横切って，交差して
 across from 〜 前〜の反対側に
acrylic 形アクリルの
 acrylic resin アクリル樹脂
act 動働く，作用する，実行する，行動する 名行為，作用，動作，法
 act on (upon) 〜 動〜に作用する
acting 作用
 acting face 圧力面
 acting time 動作時間
action アクション，作用，運動，動作，働き
 action coil 作動コイル
 action cycle 動作サイクル
 action rate 微分動作
activate 動活性化する，作動させる，浄化する，放射化する，起動する・させる
activated 形活性化した・された，活性〜
 activated alumina 活性アルミナ
 activated hydrogen 活性化水素
activation 活性化
 activation energy 活性化エネルギ
active 形活性な
 active coil 有効巻線
 active current 有効電流
 active gas 活性ガス
 active group 活性基
 active material 活性物質
 active oxygen 活性酸素
 active power 有効電力
activity アクティビティ，使用，活性，活動度，放射能，運動，機能
actual 形実際[上]の，事実上の
 actual capacity 実容量，実能力
 actual efficiency 実際効率
 actual head 実揚程
  関 total head：全揚程
 actual horsepower 実馬力
 actual pitch 実測ピッチ【推進器のピッチ】
 actual pump head 実揚程
 actual size (dimension) 実寸法
 actual stress 真応力，実応力
 actual throat 実際のど厚
 actual thrust 有効スラスト
 actual value 実効値
 actual velocity 絶対速度
 actual weight 実重量
actually 副実際に，事実，本当に，実は
actuate 動動かす，作動させる
actuating signal 動作信号
actuation point 作動点
actuator アクチュエータ，作動器【流体のもつエネルギを機械的仕事に変換する装置】
acute 形鋭い，鋭角の，重大な
acute angle 鋭角
AD (Analog-to-Digital) converter AD変換器，アナログディジタル変換器
Adamson's joint アダムソン継手【伸縮継手の一種】
adapt 動適応する・させる，適合する・させる，順応する・させる，改造する
adaptability 適応性，順応性
adaptation 順応，適合
adapter アダプタ，加減装置，中継ぎ部品
adaptive 形順応できる，適応性のある
 adaptive control 適応制御

adjusting

adaptive control constraint　拘束適応制御
adaptive control optimization　最適化適応制御
adaptor　⇒ adapter
add　動加える
added mass　付加質量
addendum　アデンダム，歯先，歯先のたけ【ピッチ径より外方にでている歯の部分の高さ】，付録
　addendum circle　歯先円【歯車の外径】
adder　加算器
addition　加算，加法，添加（物），付加，足し算　(関) in addition to：～に加えて，～のほかに
　addition agent　添加剤
　addition product　付加物
　addition theorem　加法定理
additional　形追加の，付加された，付加的な
　additional function　補助機能
　additional strength　補強
　additional strengthening　補強構造
additive　名添加剤，添加物　形付加的な，追加の，加法の
　additive polarity　加極性
address　アドレス，番地，あて名，住所
adequate　形適切な，妥当な，適当な，十分な，適した
adequately　副十分に，適切に
adhere　動接着する，付着する
adherent　形粘着する，付着性の，接着性の
adhesion　粘着，接着，付着
　adhesion brake　粘着ブレーキ
　adhesion coefficient　粘着係数
adhesive　形粘着性の，付着する　名接着剤
　adhesive bonding　接着，接着結合
　adhesive joint　接着継手
　adhesive material (agent, bond)　接着剤
　adhesive property　接着性
　adhesive strength　接着力
　adhesive tape　接着テープ
　adhesive wear　凝着摩耗
adhesiveness　接着性，付着性
adiabatic　形断熱的な　名断熱曲線
　adiabatic change　断熱変化
　adiabatic compression　断熱圧縮
　adiabatic condition　断熱条件
　adiabatic constant　断熱係数，断熱指数
　adiabatic cooling　断熱冷却
　adiabatic efficiency　断熱効率
　adiabatic engine　断熱機関
　adiabatic expansion　断熱膨張
　adiabatic exponent　断熱指数
　adiabatic heat drop　断熱熱落差
　adiabatic index　断熱指数
　adiabatic process　断熱過程
　adiabatic saturation temperature　断熱飽和温度
　adiabatic temperature efficiency　断熱温度効率
　adiabatic thermal efficiency　断熱熱効率
adiabaticity　断熱性
adjacent　形隣接した，面する
adjust　動調節する，調整する・される，順応する，整える
adjustable　形調整可能な
　adjustable bearing　調整軸受
　adjustable guide vane　調整案内羽根
　adjustable resistance　加減抵抗
　adjustable speed governor　オールスピード調速機
　adjustable speed motor　加減速度電動機
　adjustable thrust block　調整スラスト軸受
adjuster　アジャスタ，調整器
adjusting　調整
　adjusting block　調整軸受
　adjusting device　調整装置
　adjusting nut　調整ナット
　adjusting ring　調整環
　adjusting screw　調整ねじ
　adjusting spring　調整ばね
　adjusting valve　調整弁

adjustment 調整, 修正
administration 管理, 政府
admiralty アドミラルティ
 admiralty brass　アドミラルティ黄銅【黄銅合金の一種】
 admiralty coefficient　アドミラルティ係数【計画速力に対する馬力を求める式に用いる係数】
 admiralty constant　アドミラルティ係数
 admiralty test　アドミラルティ試験
admissibility 許容性
admissible 形 許容すべき
 admissible load　許容荷重, 許容負荷
admission 給気, 進入, 注入
 admission passage　給気道, 進入路
 admission port　吸入口, 進入口
 admission stroke　吸入行程
 admission valve　給入弁, 進入弁
admit 動 入れる, 認める, 通す, 導く, 許可する, 通じる
admittance 入場, アドミタンス【交流回路における電流の流れやすさを表す量】
 関 No admittance：立入禁止
admixture 混合, 混合材, 混合物, 添加剤, 混合剤, きょう雑物
adopt 動 採用する, 導入する
adsorb 動 吸着する
adsorbent 吸着剤
adsorption 吸着
 adsorption compound　吸着化合物
ADV (Acoustic Doppler Velocimeter) 超音波流速計
advance 動 前進する, 進める, 早める, 出す, 促進する
名 進歩, 発達, 前進, 進み
 関 in advance：事前に, あらかじめ／in advance of：〜に先立って, 〜の前に
 advance coefficient (constant)　前進係数
advanced 形 [前に]進んだ, 進歩した, 先進の
 advanced control　先回り制御, 先行制御
 advanced ignition　早め点火
 advanced speed　前進速度
 advanced starting valve　操縦元弁
 advanced technology　先進技術, 先端技術
 advanced timing　時期進み
advantage 長所, 好都合, 便利, メリット, 利点, 有利
advection 移流
adverse 形 反対の, 逆の, 不利な
adversely 副 逆に, 不利に
advise 動 アドバイスする
adze 手斧
AE (Acoustic Emission) アコースティックエミッション, 超音波放射【非破壊検査の一つ】
AED (Automated External Defibrillator) 自動体外式除細動器【保安用具】
aeration エアレーション, 曝気, 通気, 空気混入
aerator エアレータ, 通気装置
aerial 形 空中の, 気体の, 空気の, 大気の 名 アンテナ, 空中線
 aerial discharge　気中放電
 aerial pollution　大気汚染
aero- 接頭「空気, 空, 航空機」の意
aerodynamics 空気力学
aerofoil エーロフォイル, 翼, 翼形
 aerofoil profile　翼形断面
 aerofoil section　エーロフォイル断面
 aerofoil tip　翼端
aerometer エーロメータ, 気体密度計, 比重計, 気体計, 空気計
aeroplane 航空機
aerosol エアゾール, エーロゾル, 煙霧質
affect 動 影響する, 作用する
affected 形 影響を受けた, 変質した
 affected zone　変質部
affinity 親和力, 親和性
aft 副 船尾に 形 船尾の
 aft engine　船尾機関
after 〜 接 〜したあとに(で)
前 〜のあとに(で)

形後の，船尾の，後部の
after all 結局
after-burning あと燃え
after-condenser 後部復水器
after-contraction 再加熱収縮，後収縮
after-cooler アフタクーラ，後部冷却器，給気冷却器
after-fire アフタファイア
after-peak tank 船尾タンク
afterward[**s**] 副あとで，以後
again 副再び，さらに，もう一度
against 前〜に対して，〜に対抗して，〜に逆らって
age hardening 時効硬化
aged deterioration 経年劣化，経年変化
ageing 同 aging
agent 〜剤，薬品，代理人，因子
agglomeration 凝集，かたまり
aggregation 集合，凝集
　関 heat of aggregation：凝集熱
aging 時効，経年変化【時間の経過に伴って性質が変化する現象】
aging crack 時効割れ
agitation 攪拌
agitator かき混ぜ機，攪拌機
agree 動同意する，一致する
agreement 一致，協定，合意，同意
　関 in agreement with 〜：〜と一致する
aground 副座礁して 形座礁した
　関 run aground：座礁する
Ah（**Ampere hour**） アンペア時
ahead 名前進 副前方に，前に
ahead cam 前進カム
ahead exhaust cam 前進排気カム
ahead maneuvering valve 前進操縦弁
ahead power 前進力
ahead stage 前進段
ahead turbine 前進タービン
　反 astern turbine：後進タービン
AI（**Artificial Intelligence**） 人工知能
aid 動援助する，助ける，力添えする 名助力，救援，援助，助手

aim 名目的，目標，意図 動意図する，目指す，目的とする，志す
air エア，空気，大気
air accumulator 空気アキュムレータ，空気溜，空気蓄積装置
air acetylene welding 空気アセチレン溶接
air adjuster 空気調節器
air atomizing burner 空気噴霧バーナ
air bearing 空気軸受
air blast atomizer 空気噴射噴霧器
air blast circuit breaker 空気遮断器
air bleed valve 抽気弁
air bleeder 空気抜き
air blow エアブロー，通気
air blower 送風機
air borne dust 浮遊粉じん
air borne particles 浮遊粒子状物質
air brake 空気ブレーキ，空気制動機
air break switch 気中開閉器，気中遮断器
air bubble 気泡，空気泡
air cell 空気室，空気電池
air cell type combustion chamber 空気室式燃焼室
air chamber エアチャンバ，空気室
air chamber type combustion chamber 空気室式燃焼室
air change 換気
air channel 風胴，風路
air charge 充気
Air Circuit Breaker ACB，気中遮断器
air circulation system 空気循環式
air cleaner 空気清浄器，空気ろ過器
air cock 空気コック，空気弁
air compressor 空気圧縮機
air conditioner エアコン，空気調和機，空気調和装置
air conditioning 空気調和
air conditioning apparatus 空気調和装置
air cooled 形空冷の，空冷式
air cooled engine 空冷機関

air cooled type　空冷式
air cooler　エアクーラ，空気冷却器
air cooling　空冷，空気冷却
air core　空心
air core coil　空心コイル
air damper　空気ダンパ，空気制振器
air damping　空気制動
air density　空気密度
air duct　空気ダクト
air ejector　空気エゼクタ，空気抽出器【流体噴射にとって復水器などから，空気などのガスを除去する装置】
air escape valve　空気逃し弁
air excess ratio　空気過剰率
air extractor　空気抜き
air filter　空気フィルタ，空気ろ過器
air foil　エーロフォイル，翼，翼形
air foil fan　翼形ファン【遠心送風機の一種】
air fuel mixture　混合気
air fuel ratio　空燃比
air gap　エアギャップ，空隙
air gauge　気圧計
air hammer　空気ハンマ
air hardening steel　自硬鋼，空気焼入れ鋼
air heater　空気加熱器
air horn　エアホーン，汽笛
air injection　空気噴射，空気注入
air injection diesel engine　空気噴射式ディーゼル機関
air inlet valve　空気入口弁
air intake　空気取入れ口
air inter cooler　空気中間冷却器
air jet oil burner　空気噴射式バーナ
air leakage test　空気漏れ試験
air lift pump　空気揚水ポンプ，気泡ポンプ
air lock　エアロック
air manifold　空気マニホールド
air micrometer　空気マイクロメータ【比較測長器の一種】
air motor　エアモータ
air packing　空気パッキン
air pipe　空気管

air plane　飛行機　Ⓔaeroplane《英》
air pollution　大気汚染
air preheater　空気予熱器【煙道ガスの廃熱を利用して燃焼用空気を加熱する装置】
air pressure　空気圧
air proof　㊝気密の
air pump　空気ポンプ
air quenching　空気焼入れ
air ratio　空気比，空気率
air register　エアレジスタ，整風器
air reservoir　空気溜め
air resistance　空気抵抗
air sending valve　空気送出弁
air separator　空気分離器
air slide　風格子
air space　気室，空隙，空気層
air space cable　空気絶縁ケーブル
air speedometer　風速計
air standard cycle　空気標準サイクル
air starting cam　空気始動カム
air starting valve　空気始動弁
air storage tank　空気ストレージタンク，圧縮空気タンク
air supplied respirator　送気マスク
air suspension　空気ばね
air temperature　気温
air test　気密試験
air thermometer　空気温度計
air throttle valve　空気絞り弁
air tight　㊝気密の
air tight joint　気密継手
air tight ring　気密リング
air tight test　気密試験
air tightness　気密性
air trunk　風路，空気トランク
air turbulence　乱気流，空気の乱れ
air valve　空気弁
air vent　エアベント，空気抜き
air vessel　空気室
air volume displacement　排気量
air washer　空気洗浄装置
air whistle　汽笛
airless injection burner　油圧噴霧バーナ
airless injection diesel engine　無気

噴射式ディーゼル機関
Airy point エアリ点
alarm 名アラーム，警報，警報器 動警報を発する
 alarm bell　警報ベル
 alarm circuit　警報回路
 alarm device　警報装置
 alarm indicator　警報計
 alarm lamp　警報ランプ
 alarm panel　警報盤
 alarm signal　警報信号
 alarm system　警報装置
 alarm valve　警報弁，用心弁
 alarm whistle　非常汽笛，警笛
alclad アルクラッド【耐食性を向上させたアルミニウム系合板】
alcohol アルコール
 alcohol thermometer　アルコール温度計
aldrey アルドレイ【アルミニウム合金の一種】
alert 形油断のない，用心深い 名警報，警告，警戒 動警告する，警報を出す，注意を喚起する
alertness 警戒
algebra 代数，代数学
align 動心出しする，軸心調整をする，一直線に並べる，整列させる，調整する
aligned arrangement 基盤配列
alignment アライメント，心出し，調整，心合わせ，一列，整列，位置合わせ，一直線【2台以上の回転機械の回転中心線を一致させること】
 alignment key　位置決めキー
 alignment test　精度検査
aliphatic 形脂肪性の
 aliphatic hydrocarbon　脂肪族炭化水素
alitieren アリティエレン【アルミ[ニウム]皮膜を浸透させる金属浸透法】
alkali アルカリ　関acid：酸
 alkali catalyst　アルカリ触媒
 alkali chloride matrix　アルカリ金属塩化物
 alkali cleaning　アルカリ洗浄
 alkali corrosion　アルカリ腐食
 alkali dry cell　アルカリ乾電池
 alkali embrittlement　アルカリ脆性
 alkali extraction　アルカリ処理
 alkali manganese dry cell　アルカリマンガン乾電池
 alkali metal　アルカリ金属
 alkali meter　アルカリメータ，炭酸定量器，アルカリ比重計
 alkali proof　耐アルカリ，耐アルカリ性
alkaline 形アルカリ性の
 alkaline accumulator　アルカリ蓄電池
 alkaline additive　アルカリ添加物
 alkaline battery (cell)　アルカリ電池
 alkali[ne] earth　アルカリ土類
 alkali[ne] earth metal　アルカリ土類金属
 alkaline paper　中性紙
 alkaline storage battery　アルカリ蓄電池
alkalinity アルカリ性，アルカリ度 関P(M) alkalinity：P(M)アルカリ度
alkalinization アルカリ化
all 形全ての，全体の，全部の 副まったく，すっかり
 関at all times：いつも，常に／no A at all：まったくAがない
 all round reversing gear　全周逆転装置
 all over　副一面に
 all the time　副いつも，常に
Allen screw 六角穴付きボルト
alleviate 動軽減する
alley 通路，狭い通り
 関shaft alley：軸路
 alley way　通路
alligator clip わに口クリップ
allot 動分配する，割り当てる
allotter 分配器
allow 動許す，可能にする，させておく，認める，考慮する
 関be allowed to ～：～が許される

allow A to ～　動 A が～するのを許す
allowable　形許容の，許容できる
　allowable bending stress　許容曲げ応力
　allowable concentration　許容濃度
　allowable current　許容電流
　allowable error　許容差，許容誤差
　allowable limit　許容限度
　allowable maximum load　許容最大荷重
　allowable power　許容電力
　allowable pressure　許容圧力
　allowable shearing stress　許容せん断応力
　allowable stress　許容応力
　allowable temperature　許容温度
　allowable tensile stress　許容引張応力
　allowable torsional stress　許容ねじり応力
　allowable value　許容量，許容値
　allowable variation　許容偏差
　allowable voltage　許容電圧
allowance　許容，許可，許容差，許容値，許容量，許容度，公差，余裕
　allowance for contraction　縮みしろ
　allowance time　余裕時間
allowed time　許容時間，標準時間
alloy　名合金　動合金を作る
　alloy addition　金属添加物
　alloy cast iron　合金鋳鉄
　alloy element　合金成分
　alloy for bearing　軸受用合金
　alloy steel　特殊鋼，合金鋼
alloying element　合金元素
almost　副ほとんど，ほぼ，たいてい
alnico magnet　アルニコ磁石
along ～　前～に沿って，～を通って
　along with ～　前～と一緒に，～に加えて
alongside　前～の横側に，～と並んで　副横付けに
alpha iron　α(アルファ)鉄【強磁性の鉄の同素体】
alpha ray　α線【放射線の一種】

already　副すでに
also　副また，～もまた，同様に
alter　動変える，変更する，改造する
alterable memory　可変メモリ
alteration　修正，変更，変化，論理和，交互，交替，循環，変質，変更
alternate　形交互の，代わりの
　名交互，交替
　動交互にする，交替する
　alternate (alternative) fuel　代替燃料
　alternate (alternative) load　交番荷重
　alternate long and short dash line　一点鎖線
　alternate method　代替方法
　alternate stress　交番応力
alternately　副交互に
alternating　形交流の，交互の
　関 three phase alternating current：三相交流
　Alternating Current　AC，交流
　関 Direct Current：DC，直流
　alternating current arc welding　交流アーク溶接
　alternating current circuit　交流回路
　alternating [current] electromotive force　交流起電力，交番起電力
　alternating current exciter　交流励磁機
　alternating current generator　交流発電機
　Alternating Current motor　ACモータ，交流電動機
　alternating current power　交流電力
　alternating current power supply　交流電源
　alternating current series motor　交流直巻電動機
　alternating current shunt motor　交流分巻電動機
　alternating field　交番磁界
　alternating flux　交番磁束
　alternating load　交番荷重，両振荷重
　alternating pressure　交番圧力
　alternating stress　交番応力，繰返し応力
　alternating voltage　交流電圧，交

番電圧
alternating voltage generator　交流電圧発電機

alternation　交代, 交替, 交番, 交互, 入替え

alternative　形 交互の, 代わりの, 一方の, 二者択一の
名 代替, 二者択一
alternative current　交流
alternative energy　代替エネルギ

alternatively　副 その代わり

alternator　交流発電機, 同期発電機

although ～　接 [～である]けれども, ～にもかかわらず

alumel-chromel　アルメル-クロメル【熱電対で用いる金属】

alumina　アルミナ【水酸化アルミニウムを加熱して得られる粉末】
alumina oxide　アルミナ酸化物
alumina silica　アルミナシリカ

aluminium　同 aluminum

aluminizing　アルミめっき

aluminum　アルミニウム, アルミ
同 aluminium《英》
aluminum alloy　アルミ[ニウム]合金
aluminum base grease　アルミニウム基グリース
aluminum brass　アルミ[ニウム]黄銅
aluminum bronze　アルミ[ニウム]青銅
aluminum calorizing　アルミカロライジング【アルミニウムを浸透させた部品】
aluminum foil　アルミ箔
aluminum oxide　アルミナ, 酸化アルミニウム
aluminum piston　アルミ製ピストン

alumite　アルマイト【アルミニウムの表面に酸化アルミニウムを被覆したもの】

always　副 いつも, 常に, 必ず

AM (Ante Meridiem)　午前
関 PM (Post Meridiem)：午後

amalgam　合成物, 混合物, アマルガム【水銀と他の金属との合金】

amateur　アマチュア, 素人

amber　琥珀(こはく)

ambient　形 周囲の, 周辺の, 環境の
ambient [air] temperature　大気(周囲)温度, 室温
ambient condition　環境条件, 周囲条件
ambient pressure　大気圧, 周囲圧力

ambiguity　あいまいさ, 多義性

amendment　修正, 訂正, 改正
amendment plan　訂正図

amenity　アメニティ, 快適性
amenity space　快適空間

American Bureau of shipping　AB, アメリカ船級協会

American Petroleum Institute hydrometer　API比重計

American Society of Mechanical Engineers　ASME, アメリカ機械学会

American Standard Code for Information Interchange　ASCII, アスキーコード【情報交換用アメリカ標準コード】

ammeter　電流計

ammonia　アンモニア
ammonia absorption refrigerating machine　アンモニア吸収冷凍機
ammonia anhydrite　無水アンモニア
ammonia corrosion　アンモニア腐食
ammonia water (solution)　アンモニア水

among ～　前 ～の間に(で), ～の中に

amorphous　名 アモルファス, 非結晶, 非晶質
形 無定形の, 非晶質の, 非結晶の
amorphous alloy　アモルファス合金, 非晶質合金
amorphous carbon　無定形炭素
amorphous magnetic material　アモルファス磁性材料
amorphous metal　アモルファス金属, 非晶質金属

amortisseur winding　制動巻線

amount　名 量, 総量　動 総計～に

なる 関 a small (large) amount of ～：小(多)量の～／the amoun of：～の量／the tiniest amount of：ごく少量の
amount of carbon　炭素量
amount of combustion gas　燃焼ガス量
amount of deflection　たわみ量
amount of energy　エネルギ量
amount of evaporation　蒸発量
amount of heat　熱量
amount of fuel　燃料量
amount of information　情報量
amount of substance　物質量
amount of theoretical combustion air　理論空気量
ampere　アンペア【電流の強さの実用単位：A】
ampere capacity　アンペア容量
ampere hour　アンペア時【電気量の単位：Ah】
ampere hour meter　積算電流計
ampere meter　電流計
ampere turn　アンペア回数
Ampere's right handed screw rule　アンペアの右ねじの法則
ample　形 十分な，広大な，豊富な
amplification　増幅，拡大
amplification degree　増幅度
amplification factor　増幅率
amplifier　アンプ，増幅器，拡大鏡
amplifier action　増幅作用
amplifier circuit　増幅回路
amplify　動 拡大する，増幅する
amplitude　振幅，許容度，広さ，偏角
amplitude modulator　振幅変調器
amplitude ratio　振幅比
Amsler's universal material testing machine　アムスラー万能材料試験機
analog　アナログ，相似形，相似体，類似[体]
analog circuit　アナログ回路
analog control　アナログ制御
analog derivative　アナログ微分
analog display　アナログ表示
analog instrument　アナログ計器
analog output　アナログ出力
analog servo system　アナログサーボ系
analog signal　アナログ信号
analog switch　アナログスイッチ
Analog [to] Digital converter　AD変換器，アナログディジタル変換器
analogous　形 類似の，相似の
analogous element　同属元素
analogue　同 analog
analogy　アナロジ，類似，相似[則]
analysis　分析，分解，解析
analysis pitch　実測ピッチ【推進器のピッチ】
analytical　形 解析的な
analyze　動 分析する，解析する
analyzer　アナライザ，分析計，分析器，解析器
anchor　名 アンカ，錨　動 固定する，投錨する，停泊する
anchor bolt　アンカボルト，基礎ボルト
anchorage　錨地，固定
and　接 そして，および，～と(や)～
AND circuit　AND回路，論理積回路　関 OR circuit：OR回路
AND gate　ANDゲート
and/or　接 および(あるいは)または
and so on　～など
android　アンドロイド，人造人間，人間型ロボット
anemometer　風力計，風速計
anergy　アネルギ，無効エネルギ
aneroid barometer　アネロイド気圧計，空ごう気圧計
angle　アングル，角，角度，山形鋼，視点，見方，すみ
angle bar　山形鋼，L形鋼
angle gauge　角度ゲージ，角度計
angle gear　アングル歯車
angle index　角度目盛
angle joint　アングル継手
angle meter　角度計，傾斜計

angle of action　作用角
angle of advance　前進角，進み角
angle of approach　進入角
angle of attack　迎え角
angle of bend　屈曲角，曲げ角度
angle of contact (wrap)　接触角
angle of deflection　たわみ角
angle of delay　遅れ角
angle of eccentricity　離心角
angle of encounter　出会い角
angle of friction　摩擦角
angle of heel　横傾斜角
angle of incidence　入射角
angle of lag　遅れ角
angle of lead　進み角
angle of obliquity　圧力角
angle of rake　傾斜角
angle of reflection　反射角
angle of refraction　屈折角
angle of relief　逃げ角
angle of repose　[安]息角
angle of roll　横揺れ角
angle of skew back　スキュ角
angle of stagger　食違い角
angle of thread　ねじ山角
angle of torsion (twist)　ねじれ角
angle of trim　トリム角
angle piece　角度ゲージ
angle section　山形断面
angle steel　山形鋼，L形鋼
angle valve　アングル弁【止め弁の一種で流体の流れが直角に変わる弁】
angled piston pump　斜軸式ピストンポンプ
angular　形角の
angular acceleration　角加速度
angular advance　前進角
angular ball bearing　段付き玉軸受
angular bevel gear　斜交かさ歯車
angular contact ball bearing　アンギュラ玉軸受，アンギュラコンタクト玉軸受【転がり軸受の一種】
angular displacement　角変位
angular frequency　角周波数
angular impulse　角力積

angular lead　前進角
angular moment　回転モーメント
angular momentum　角運動量
angular momentum equation　角運動量方程式　同 moment of momentum equation
angular motion　角運動，回転運動
angular velocity　角速度
angularity　直角度
aniline　アニリン【無色で有害な油状液体】
aniline number　アニリン価
aniline point　アニリン点【アニリンとガソリンのような物質が完全に混合するための最低温度】
animal oil　動物油
animation　アニメーション，動画
anion　アニオン，陰イオン
anisometric drawing　不等角図，不等角画法
anisotropic magnet　異方性磁石
anisotropy　異方性
anneal　名焼なまし　動焼なます
annealed copper wire　軟銅線
annealing　焼なまし【鋼の軟化を目的とした熱処理】
annealing furnace　焼なまし炉
annex　名付属書　動添付する
announce　動発表する，知らせる
annual　形年一回の，年次の，例年の，毎年の
annual maintenance schedule　年間整備計画
annual survey　年次検査
annular　形環状の
annular flow　環状流
annular valve　蛇の目弁，輪形弁
annunciator　アナンシエータ，警報器，信号表示器
anode　アノード，陽極
anode current　陽極電流
anodic　形陽極の
anodic coating　陽極被膜，陽極酸化
anodic protection　アノード防食
anodization　陽極酸化
anodizing　陽極酸化，陽極処理

anomalous 形異常な，変則的な
anomaly 異常，偏差
another 形他の，異なった，別の，同様の，もう一つの
　関 one another：互いに，お互い
　another way　別の方法
anoxia 酸欠
answer 名答え，解答　動答える
answering lamp 応答ランプ
answering time 応答時間
Ante Meridiem A.M., 午前
　関 P.M.：午後
antechamber 予燃室，副室
anti- 接頭「反対」「排斥」「敵対」「対抗」の意
　anti-catalyst　負触媒
　anti-clockwise　反時計回り，反時計方向，左回り
　anti-corrosion　防食，耐食
　anti-corrosion alloy　耐食合金
　anti-corrosive　形さび止めの　名防食剤
　anti-corrosive composition　さび止め剤
　anti-corrosive paint　防錆塗料，さび止めペイント
　anti-corrosive treatment　さび止め処理
　anti-detonation fuel　アンチノック燃料
　anti-detonator　アンチノック剤
　anti-disaster facility　防災設備
　anti-foaming　形あわ立ち防止の
　anti-foaming agent　消泡剤
　anti-foaming tendency　消泡性
　anti-fouling　防汚ペイント
　anti-freeze　不凍液
　anti-freezing admixture　凍結防止剤
　anti-freezing solution　不凍液
　anti-friction　減摩，耐摩耗，減摩剤
　anti-friction bearing　ころがり軸受
　anti-friction composition　減摩剤
　anti-friction metal　減摩メタル
　anti-hunting　乱調防止
　anti-jamming　妨害対策
　anti-knock　形アンチノック性の　名アンチノック剤，耐爆剤
　anti-knock agent (compound)　アンチノック剤
　anti-knock dope　耐ノック剤
　anti-knock fuel　アンチノック燃料
　anti-knock property (quality)　アンチノック性，耐ノック性
　anti-oxidant　酸化防止剤
　anti-phase　逆位相
　anti-polishing ring　カーボンスクレーパリング
　anti-priming pipe　沸水防止管，水け止め管
　anti-rolling apparatus　揺れ止め装置
　anti-scales　清缶剤
　anti-seize compound　焼付き防止剤
　anti-septic　名防腐剤　形防腐作用のある
　anti-static agent　帯電防止剤
　anti-wear　耐摩耗性
anticipate 動予想する，予期する，未然に防ぐ
antimony アンチモン　記 Sb
any 形何らかの，いくらかの，どの(どんな)〜でも［訳す必要がない場合が多い］
　関 at any time：いつでも／in any case：どんな場合でも，とにかく
　any of　いずれか，どれでも
anyone 代だれか，だれでも
apart 副ばらばらに，離れて，一方　形離れた，無関係の，別個の
　apart from 〜　前〜は別として，〜を除いて，〜から離れて
aperture 穴，すきま
apex 頂点，先端
API (American Petroleum Institute) gravity API 度
API hydrometer API 比重計
apical angle 頂角，尖角
apparatus 装置，器具，道具
　関 experimental apparatus：実験装置
apparent 形明白な，見かけの
　apparent capacity　皮相容量
　apparent pitch　見かけピッチ
　apparent power　皮相電力

apparent slip　見かけスリップ
apparent specific gravity　見かけ比重
apparent viscosity　見かけ粘度
apparently　副明らかに，見かけは
appeal　名アピール　動訴える
appear　動現れる，見える
appearance　外観，外形，出現
appendage　付加物
appendix　付録，補足資料
appliance　電気器具，家電製品，装置，器具，設備
applicable　形適応(応用)できる
application　応用，適用，利用，用途，塗布，[熱・力などを]加えること
application of heat　加熱[処理]
application of load　負荷投入
application software　応用ソフト
applied mechanics　応用力学
apply　動適用する・される，応用する，塗る，加える，当てる
apply ～ to…　動～を…にあてがう
appraise　動評価する，査定する
appreciate　動感謝する，察する，認識する，評価する，理解する
apprehend　動理解する
apprentice　見習い，実習生
approach　動近づく，接近する，取組む　名アプローチ，取組み，入口，進入，接近，手引き，入門
approach angle　進入角
　同 angle of approach
approach speed　進入速度
approach vector　接近ベクトル
appropriate　形適当な，適切な
　関 if appropriate：適切な場合
approval　承認
approve　動承認する，許可する
approved　形承認された，認可された
approved drawing　承認図
approximate　形近似の，おおよその，近接の　動～に近似する，おおよそ～になる，近づける
approximate calculation　近似計算
approximate formula　近似公式
approximate method　近似，近似法
approximate value　近似値

approximately　副おおよそ
approximation　近似，近似値
Arabic figure　アラビア数字
Arago's disc　アラゴの円板
arbitrary　形任意の
arbor　アーバ，軸，回転軸
arc　アーク，電弧，円弧，弧，弓形
arc cutting　アーク切断
arc discharge　アーク放電
arc hardening　放電硬化法
arc length　アーク長
arc of action　作用弧
arc voltage　アーク電圧
arc welding　アーク溶接
Archimedes' principle　アルキメデスの原理
area　面積，域，範囲，～部門
area coefficient　面積係数
area flow meter　面積流量計
area of heat transfer surface　伝熱面積
area of midship section　中央横断面積
area ratio　面積比
areal velocity　面積速度
areometer　浮きばかり，比重計
argon　アルゴン　記 Ar
argon arc welding　アルゴンアーク溶接
argument　偏角
arise　動生じる，起こる
arithmetic　演算
arithmetic circuit　演算回路
arithmetic mean　算術平均
arithmetic unit　演算器，演算装置
arm　アーム，腕，腕のようなもの
armature　電機子
armature arm　電機子アーム
armature characteristic curve　電機子特性曲線
armature coil　電機子コイル
armature core　電機子鉄心
armature current　電機子電流
armature leakage reactance　電機子漏れリアクタンス
armature ohmic loss　電機子銅損
armature reaction　電機子反作用

armor

armature reaction reactance 電機子反作用リアクタンス
armature resistance 電機子抵抗
armature voltage control 電圧制御
armature winding 電機子巻線

armor 動防具をつける 名外装
armored 形外装した
armored cable 外装ケーブル【強度を増したケーブル】
armored hose 外装ホース【強度を増したホース】

aromatic 形芳香の, 芳香族の
aromatic compound 芳香族化合物
aromatic hydrocarbon 芳香族炭化水素
aromatic series 芳香族

aromaticity 芳香性, 芳香族性
around 〜 前〜の周りに, 〜のあたりに(で), およそ, 約 副周りに(を), およそ
arrange 動配置する, 配列する, 準備する, 整える, 取り決める
arrangement 配置, 配列, 整理, 設備, 装置, 準備, 用意
arrangement drawing (plan) 配置図
array 配列
arrest point 停[止]点
arrester 防止器, 避雷針
arrow head 矢印
article 品物, 品目
関 heavy article：重量物
articulated pipe 関節管
artificial 形人工の, 人工的な
artificial draft 人工通風
artificial illumination 人工照明
Artificial Intelligence AI, 人工知能
artificial petroleum 人造石油
artificial ventilation 人工通気
as 〜 接〜(する)ように, 〜する時, 〜しながら, 〜するにつれて, 〜なので, 〜する限りでは, 〜であるのと同じように, 〜ではあるが 副同じくらい 前〜として, 〜のように, 〜なので, 〜のときに
関 not as A as B：BほどAではない
as 〜 as… …と同じほど〜

as 〜 as possible 副出来るだけ〜
as a result 副結果として
as a rule 副概して, 一般に
as far as 〜 接〜する限りでは 前〜まで(も), 〜するところまでは
as far as practicable できるだけ
as follows 次のとおり
as for 〜 前〜に関しては, 〜について言えば
as if 〜 接まるで〜であるかのように
as long as 〜 前〜の間 接〜である限りは, 〜する間は, 〜なので
as more than 〜以上
as much as possible できる限り, 極力
as soon as 〜 接〜するとすぐに
as soon as possible できるだけ早く
as to 〜 前〜に関して(は), 〜について
…as well 副その上…も, …もまた
…as well as 〜 〜同様…も
asbestos アスベスト, 石綿
ascertain 動確かめる
ASCII (**American Standard Code for Information Interchange**) アスキーコード【情報交換用アメリカ標準コード】
ash 灰, 灰分
ash content 灰分
aside from 〜 前〜はさておき, 〜は別として
as-made drawing 正式図, 最終図
ASME (**American Society of Mechanical Engineers**) アメリカ機械学会
aspect ratio アスペクト比, 縦横比
asperity 面の凹凸, 荒さ, でこぼこ
asphalt アスファルト
asphaltene アスファルテン
aspirate 動吸い込む
aspiration 吸気, 吸引
aspirator アスピレータ, 吸引器
assay 動分析する, 検定する 名分析, 検定

assaying 分析試験
assemble 動組み立てる
assembling アセンブリング【2つ以上の部品を組み立てること】
  assembling drawing　組立図
  assembling jig　組立ジグ
  assembling machine　組立機械
assembly 組立[品], 集合体
  assembly drawing　組立図
  assembly line operation　流れ作業
  assembly operation　組立作業
assess 動査定する, 評価する
assessment アセスメント, 査定, 評価
assign 動指定する, 任命する, 配属する, 選任する, 割り振る
assist 動助ける, 援助する, 補助する
assistant cylinder 緩衝シリンダ
associate 動関係させる, 結合する・させる, 組織する, 提携する, 連合させる
associated equipment 関連機器
association 協会
assume 動仮定する, ～だとみなす, 想定する, 憶測する
assuming [that] ～ 接～と仮定して, ～とすれば
assurance 保証
assure 動確信する, 保証する
astatic 形無定位の, 不安定の
astern 副船尾に, 後方へ
  astern cam　後進カム
  astern maneuvering valve　後進操縦弁
  astern nozzle　後進ノズル
  astern power　後進力
  astern stage　後進段
  astern turbine　後進タービン
  astern valve　後進弁
ASTM (American Society for Testing and Materials) アメリカ材料試験協会
asymmetry 非対称
asynchronous generator 非同期発電機
athwart 副横切って

前横切って, に反して
athwartships 副船体を横切って
atmosphere 大気, 気圧, 雰囲気
atmospheric 形大気の, 空気の
  atmospheric air　大気
  atmospheric condenser　大気圧コンデンサ, 大気圧復水器
  atmospheric contamination　大気汚染
  atmospheric corrosion resistant steel　耐候性鋼
  atmospheric diffusion　大気拡散
  atmospheric distillation　常圧蒸留
  atmospheric pollutant　大気汚染物質
  atmospheric pollution　大気汚染
  atmospheric pressure　大気圧
  atmospheric pump　吸上げポンプ
atom 原子
  atom physics　原子物理学
atomic 形原子の, 原子力の
  atomic energy　原子エネルギ
  atomic nucleus　原子核
  atomic number　原子番号
  atomic power　原子力
  atomic structure　原子構造
  atomic symbol　原子記号
  atomic weight　原子量
atomization 霧化, 噴霧, 微粒化
atomize 動霧化する, 噴霧する, 原子にする, 細分化する
atomized firing 噴霧燃焼
atomizer 噴霧器
atomizing pressure 噴霧圧
attach 動取り付ける
attachment 付属品, 取付, 付着
  attachment plug　差込みプラグ
attack 浸食, 腐食, 作用
  attack angle　迎え角
  同 angle of attack
attain 動得る, 達成する
attemperator 過熱低減器
attempt 動企てる, 試みる
attention 注意, 手当, 配慮, 気配り
attenuation 減衰, 希釈
  attenuation coefficient　減衰係数, 減衰定数
attenuator 減衰器

attitude 姿勢，態勢，態度
  attitude angle 姿勢角，偏心角
  attitude control 姿勢制御
attract 動引き付ける
attraction 引力，吸引力
  関 repulsion：斥力
attractive 形魅力的な，引力のある
  attractive force 引力
attribute 属性
attrition 摩擦，摩滅，減少，縮小
audible 形可聴の
  audible and visual alarm 音と光による警報
audio 形音声の，可聴周波の 名オーディオ，音声
  audio frequency 可聴周波数
  Audio Visual 名 AV，オーディオビジュアル【音響と映像の組合せ】形視聴覚の
audiometer 聴力計
augment 動増加(大)する
augmented resistance 増加抵抗
augmenter 増大因子
augularity 傾斜度
aus-forming オースフォーミング【焼入れ熱処理の一種】
austemper オーステンパ【熱処理の一種】
austenite オーステナイト【鋼の組織名の一つ】
authorized pressure 認可圧力
auto- 接頭「自動推進の」の意
autoalarm 自動警報機
autocollimator 真直度，オートコリメータ【直角度，平行度を測定する計器】
autoconverter 直流変圧器
autogenous ignition 自然発火
autogenous welding ガス溶接
autographic recorder 自動記録器
automatic 形オートマチックの，自動の，自動的，機械的な，無人の
  Automatic Boiler Control system ABC，自動ボイラ制御装置【ボイラの空燃比と給水量を制御する装置】
  automatic brake valve 自動ブレーキ弁
  automatic change 自動切換え
  Automatic Combustion Control system ACC，自動燃焼制御装置【ボイラの空燃比を制御する装置】
  automatic control 自動制御
  automatic control system 自動制御系
  automatic controller 自動制御装置，自動調節計
  automatic expansion valve 自動膨張弁
  automatic feed water regulator 自動給水調節器
  automatic gear 自動装置
  automatic interlock device 自動連動装置
  automatic nonreturn valve 自動逆止め弁
  automatic operation オートメーション
  automatic pilot 自動操舵装置
  automatic processing 自動処理
  automatic regulation valve 自動調整弁
  automatic regulator 自動調整器
  automatic scavenging valve 自動掃気弁
  automatic signaling apparatus 自動信号装置
  automatic starter 自動始動器
  automatic switch 自動スイッチ
  automatic synchronizer 自動同期装置
  automatic transformer 単巻変圧器
  automatic transmission 自動変速機
  automatic trip 自動引きはずし
  automatic valve 自動弁
  automatic voltage control 自動電圧調整
  Automatic Voltage Regulator AVR，自動電圧調整器
automatically 副自動的に，機械的に
automation オートメーション，自動(操作)化

autopilot オートパイロット，自動操舵装置，自動舵取り装置
autoradiography オートラジオグラフ
autospinning オートスピニング
autosteerer 自動舵取り装置
autotransformer 単巻変圧器
aux.（auxiliary） 形補助の，予備の 名補助，補助機関，補機
　auxiliary blower　補助ブロワ
　auxiliary boiler　補助ボイラ
　auxiliary circulating pump　補助循環ポンプ
　auxiliary condenser　補助復水器
　auxiliary contact　補助接点
　auxiliary device　補助装置
　auxiliary diesel engine　補助ディーゼル機関
　auxiliary engine　補助機関【主機関以外の機関の総称】
　auxiliary equipment　付属装置
　auxiliary exhaust　補機排気
　auxiliary feed check valve　補助給水弁
　auxiliary feed pump　補助給水ポンプ
　auxiliary generator　補助発電機
　auxiliary machine（machinery）　補機
　auxiliary turbine　補助タービン
　auxiliary unit　補助単位
　auxiliary valve　補助弁
AV（Audio Visual） 名オーディオビジュアル【音響と映像の組合せ】形視聴覚の
avail 動役に立つ
availability 有効性，稼働率，可用性，役に立つこと
　availability destruction　エクセルギ損失
available 形有効な，役に立つ，利用できる，手に入る
　available energy　有効エネルギ
　available head　有効落差，有効水頭
　available heat　有効熱量
　available heating surface　有効加熱面
　available hydrogen　有効水素
　available NPSH　有効吸込みヘッド

average 名平均 形平均の 動平均する
　average coefficient of thermal expansion　平均熱膨張係数
　average error　平均誤差
　average heat transfer coefficient　平均熱伝達率
　average load　平均負荷
　average speed（velocity）　平均速度
　average value　平均値
Avogadro's number アボガドロ数
avoid 動避ける
　avoid 〜 ing 動〜するのを避ける
AVR（Automatic Voltage Regurator） 自動電圧調整器
aware 形気づいて 関 be aware of：認識している
away 副離れて，取り去って 形離れて
　関 wearing away of metal：金属の摩耗
awkward to 〜 形〜しにくい
ax[e] 斧
axes 「ax」または「axis」の複数形
axial 形軸の，軸方向の
　axial bearing　アキシャル軸受
　axial blade clearance　軸方向すきま
　axial blower　軸流送風機
　axial clearance　軸向きすきま
　axial compressor　軸流圧縮機
　axial deflection　軸方向のたわみ
　axial flow　軸流
　axial flow compressor　軸流圧縮機
　axial [flow] fan　軸流ファン
　axial flow impulse turbine　軸流衝動タービン
　axial [flow] pump　軸流ポンプ
　axial flow reaction turbine　軸流反動タービン
　axial flow supercharger　軸流型過給機
　axial flow turbine　軸流タービン
　axial load　軸方向荷重
　axial motor　アキシャルモータ
　axial piston pump　アキシャルピストンポンプ

axial pitch　軸方向ピッチ
axial plunger pump　アキシャルプランジャポンプ
axial stress　軸応力，軸方向応力
axial thrust　軸方[向]推力
axial velocity ratio　軸流速度比
axially　副軸方向に
axiom　原理，原則，公理
axis　軸，軸線，中心線
　関 x-axis：x軸／neutral axis：中立軸
axis of abscissa[s]　横軸，横座標軸
axis of coordinater　座標軸
axis of elasticity　弾性軸
axis of ordinate　縦軸
axis of oscillation　動揺軸
axis of symmetry　対称軸
axle　アクスル，軸，車軸
axle ratio　減速比
azeotropic mixture　共沸混合物
azimuth　方位，方位角
azimuthal　形方位[角]の
azimuthal angle　方位角
azimuthal error　方位角誤差

B-b

b contact b接点
Babbitt metal バビットメタル【錫（すず）を主成分とする軸受用合金】
Babcock and Wilcox boiler バブコックボイラ【水管ボイラの一種】
back 名背面, 背 形背後の, 後方の, 逆の 副後方へ
  back-board 背板
  back-cavitation 背面キャビテーション
  back-cone 背円すい
  back-current 逆電流
  back-electromotive force 逆起電力
  back-elevation 背面図
  back-end plate 後部鏡板〈ボイラ〉
  back-face 背面
  back-fire バックファイア, 逆火
  back-flow 逆流
  back-gear 後部歯車
  back-hand welding 後進溶接【溶接が溶加材の前を進む溶接法】
  back-header 後部管寄せ
  back-lash バックラッシ, がた, あそび【歯車の歯面間の遊び】
  back-metal 裏金
  back-plate 背板〈ボイラ〉
  back-pressure 背圧
  back-pressure turbine 背圧タービン
  back-pressure valve 背圧弁
  back-resistance 逆抵抗
  back-scatter 後方散乱
  back-to-back duplex bearing 背面組合せ軸受
  back-tube plate 後部管板
  back-up ring バックアップリング
  back up protection バックアップ保護, 予備保護
  back-washing 逆流洗浄
  back-water 戻り水, 逆流
  back-water valve 背水弁
backing バッキング, 裏打ち, 裏当て, 背面, 反転, 後進, 支持体
  backing power 後進力
  backing ring 裏当て金, 裏当て輪
backward 形後方の, 逆の, 逆行する
  backward current 逆電流
  backward curved vane 後向き羽根
  backward direction 逆方向
  backward gear 後進装置
bacteria バクテリア, 細菌
bad 形悪い, 不良, 粗悪な, 劣った, 厳しい, 不適当な, 不完全な, 不十分な
  bad conductor 不良導体
badly 副ひどく, とても
baffle バッフル, 隔壁
  baffle board (plate) 整流板, そらせ板, じゃま板, 衝突板
bag filter バグフィルタ【ろ過集じん装置の一種】
bahn metal バーンメタル【軸受合金の一種】
bainite ベイナイト【鋼の焼入れ組織の一種】
bake 動焼く
bakelite ベークライト【熱硬化性合成樹脂の一種】
baking ベーキング, 焼き付け, 焼成, 加熱乾燥
balance 名バランス, 平衡, 調和, つり合い, 天秤, はかり, 残り 動つり合う, 調和させる
  balance cylinder つり合いシリンダ
  balance disc バランスディスク, つり合い盤
  balance dynamometer はかり動力計
  balance lever つり合いてこ
  balance weight つり合いおもり
balanced 形つり合いのとれた, 平衡の
  balanced draft 平衡通風
  balanced load つり合い荷重
  balanced reaction 可逆反応
  balanced valve つり合い弁, 両座弁
  balanced voltage 平衡電圧
balancer 平衡装置, 均圧装置

balancing つり合い
　balancing hole　つり合い穴
　balancing piston　バランスピストン
　balancing test　つり合い試験
ball　ボール，玉，球
　ball and roller bearing　ころがり軸受
　ball bearing　ボールベアリング，玉軸受
　ball joint　玉継手
　ball race　ボールレース【ボールベアリングのころやボールをはさむ輪】
　ball [screw] thread　ボールねじ【鋼球で摩擦を小さくしたねじ】
　ball thrust bearing　スラスト玉軸受
　ball valve　ボール弁，玉弁
ballast　バラスト，底荷，安定器
　ballast coil　安定コイル
　ballast pump　バラストポンプ【船体の喫水や傾斜を調節するポンプ】
　ballast resistor　安定抵抗器
　ballast tank　バラストタンク
band　图バンド，帯，周波数帯域　動縛る
　band brake　帯ブレーキ
　band pass filter　帯域通過フィルタ
　band plate　帯板
　band reject filter　帯域除去フィルタ
　band steel　帯鋼
　band width　帯域幅
banded corrosion　層状腐食
banded structure　縞状組織
bang-zero-bang control　バンゼロバン制御　同 bang-bang-off control
bank　バンク［規則的に配置された装置などの群，列］
　bank angle　バンク角，傾斜角
banking fire　埋め火
bar　棒，バール【圧力の cgs 単位】，かん抜き
　bar gauge　棒ゲージ
　bar graph (chart)　棒グラフ
　bar iron　棒鉄，鉄棒
　bar magnet　棒磁石
　bar steel　棒鋼，条鋼
　bar thermometer　棒状温度計

barb bolt　鬼ボルト
Barcol hardness tester　バーコル硬さ計
bare　形覆いがない，裸の
　bare tube wall　裸水管壁
barge　バージ，はしけ
barium　バリウム　記 Ba
　barium carbonate　炭酸バリウム
　barium chloride　塩化バリウム
　barium sulfate　硫酸バリウム
barometer　気圧計
barometric　形気圧の
　barometric condenser　大気復水器
　barometric pressure　気圧，大気圧
barrage　ダム，せき
barrel　バレル，円筒，胴，樽，管【1 バレル＝約 159 リットル】
　barrel cam　筒形カム
　barrel finishing　バレル仕上げ
　barrel fitting　バレル継手
　barrel type pump　バレル形ポンプ【高圧多段ポンプの一種】
barricade tape　隔離ロープ
barrier　バリヤ，障壁
barring gear　ターニング装置，クランク変位装置
base　图ベース，塩基，底，基礎，土台，つけ根，基地　動基づく
　base alloy　ベース合金
　base circle　基礎円
　base current　ベース電流
　base line　基線
　base load　ベース負荷
　base metal　卑金属，母材
　base oil　基油
　base pitch　法線ピッチ
　base pitch deviation　法線ピッチ誤差
　base plate　基板
　base quantity　基本量
　base station　基点，準点，基地局
　base unit　基本単位
based　形ベース（基礎）にした
basic　图基礎，原理，基本　形基礎の，基(根)本的な，初歩の，塩基の，塩基性の
　関 on the basic of：～に基づいて

basic block　基準ブロック
basic cycle　基本サイクル
basic design　基本設計
basic dimension　基準寸法
basic oil　塩基性油
basic operation　基本操作
basic profile　基準山形〈ねじ〉
basic rack　基準ラック
basic size　基本寸法
basic steel　塩基性鋼
basic strength　塩基強度
basically　形基本的に，要するに
basicity　塩基度
basin　ドック，洗面器，係船地
 basin trial　係留運転
basis　基礎，基準，基底，土台，根本，原則，根拠，主成分
 関 on the basis of：〜に基づいて，〜の観点から
basket　バスケット，籠
bastard cut file　荒目やすり
batc　バッチ【バッチとは「一束」の意】
 batch control　バッチ制御
 batch processing　バッチ処理，一括処理
bath-tub curve　バスタブ曲線，故障率曲線【縦軸に故障率，横軸に時間をとった寿命特性曲線】
bathe　動浸す
batten　バッテン【定規の一種】
 batten plate　帯板，目板
battery　バッテリ，蓄電池
Baume's hydrometer　ボーメ比重計
Bauschinger's effect　バウシンガー効果【金属材料を塑性変形させた時に生じる現象】
bauxite　ボーキサイト【アルミニウムの原鉱】
bayonet base　差込み口金
bayonet joint　差込み継手
bayonet tube exchanger　差込み管式熱交換器
BDC (Bottom Dead Center)　下死点
 関 TDC：上死点
beach mark　ビーチマーク

bead　ビード【溶接の際にできる波形の模様】
beaker　ビーカ
beam　ビーム，梁（はり），けた
 関 cantilever：片持ばり／simple beam：単純ばり／fixed beam：固定ばり／continuous beam：連続ばり
 beam compasses　ビームコンパス【製図器の一種】
bear　動もつ，提示する，耐える，運ぶ
bearer　支えるもの
 bearer bar　支え棒
bearing　ベアリング，軸受，方位，支え，支点
 bearing alloy　軸受合金
 bearing bar　支持棒
 bearing blue　ベアリングブルー【けがき用の塗料】
 bearing bore　軸受内径面
 bearing brass　軸受メタル
 bearing bush　軸受ブッシュ
 bearing cap　軸受押え
 bearing capacity　支持力，支圧強度
 bearing clearance　軸受すきま
 bearing factor　面圧係数
 bearing housing　軸受箱
 bearing load　軸受荷重
 bearing lubrication　軸受潤滑
 bearing material　軸受材料
 bearing metal　軸受メタル
 bearing modulus　軸受定数
 bearing number　呼び番号
 bearing pressure　軸受圧力，支え圧
 bearing shim　ベアリングシム
 bearing sleeve　軸受スリーブ
 bearing stand　軸受台
 bearing steel　軸受鋼
 bearing strength　支圧強度，面圧強さ
 bearing stress　ベアリング応力，支持応力，側圧応力
 bearing surface　軸受面，軸受表面
 bearing temperature　軸受温度
 bearing wear　軸受摩耗
beat　名うなり　動打つ，たたく

because ~ 接なぜならば~だから，~だから，~なので
　because of ~ 前~のために
becket ベケット，索（つな）止め
become 動~になる，よく似合う
bed ベッド，台，床，層
　bed in 動はめ込む，固定する
　bed plate ベッドプレート，台板
before ~ 前~の前に
　接~する前に 副以前に
　before and after 前後
beforehand 副あらかじめ
　形前もっての
begin 動始まる・める 関 to begin with : そもそも，まず第一に
beginning 名始め，開始，起源
　形最初の，初歩の，基礎の
behave 動振る舞う
behavior 挙動，態度，動作，行動，行為，習性，性状
belay 動巻きつける
belaying cleat クリート，耳形索止め
belaying pin ビレーピン，索止め栓
belief 確信，意見，信用，信頼
believable 形信じられる，信用できる
believe 動信じる，~だと考える
bell ベル，鐘，鐘状のもの
　bell and spigot joint 差込み継手
　bell chuck ベルチャック
　bell metal ベルメタル，鐘銅【青銅の一種】
　bell mouth ラッパ口，ベルマウス
Belleville spring 皿ばね
bellofram ベローフラム【圧力を変換するプロセス用変換器の一種】
bellows ベロー，ベローズ，ふいご
　bellows coupling ベローズ継手
　bellows expansion joint ベローズ形伸縮管継手
　bellows manometer ベローズ圧力計
　bellows type pressure gauge ベローズ形圧力計
belong 動所属する
below ~ 前~より下に，未満で，副下に，以下で 反 below

belt ベルト，帯
　belt conveyer ベルトコンベア
　belt drive ベルト駆動，ベルト伝動
　belt joint ベルト継手
　belt pulley ベルト車
　belt tension ベルト張力
　belt training roller 調心ローラ【Ｖベルトなどの片寄りを調整するローラ】
　belt transmission ベルト伝導
bench 作業台
　bench mark 水準点，基準，ベンチマーク
bend 名ベンド，曲がり，わん曲，曲管 動曲がる，曲げる
bender ベンダ，曲げ機械
bending 曲げ
　bending action 曲げ作用
　bending angle 曲げ角度
　bending failure 曲げ破壊
　bending load 曲げ荷重
　bending moment 曲げモーメント
　bending moment diagram 曲げモーメント図
　bending resistant moment 曲げ抵抗モーメント
　bending roll 曲げロール
　bending strength 曲げ強さ
　bending stress 曲げ応力
　bending test 曲げ試験
　　関 bending tester : 曲げ試験機
　bending torsion 曲げねじり
　bending vibration 曲げ振動
　bending work 曲げ加工
benefit 名利益，長所，恩恵，手当 動利する，利益を得る
Benson boiler ベンソンボイラ【貫流ボイラの一種】
bent 動「bend」の過去・過去分詞形 形曲がった
　bent axis type axial piston pump 斜軸式ピストンポンプ
　bent pipe 曲がり管
bentness 曲がり
bentone grease ベントングリース
benzene C_6H_6, ベンゼン

biomass

benzene series hydrocarbon　ベンゼン系炭化水素
benzine　ベンジン【燃料，溶剤，しみ抜きなどに用いる】
Bernoulli's equation　ベルヌーイの式
Bernoulli's theorem　ベルヌーイの定理　同 Bernoulli's law
beryllium　ベリリウム　記 Be
beside　～　前～のそばに，～と比べると，～を外れて
besides　～　前～のほかに，～に加えて，～を除いて，～以外に　副その上
best　形「good」「well」の最上級，最良の，最大の　関 at best：[いくら]よくても，せいぜい
beta ray　ベータ線【放射線の一種】
better　形「good」の比較級，優れている，～より良い，大半の
　better than　より良い
between　～　前～の間に
　between A and B　AとBの間に
bevel　名ベベル，傾斜，斜角，斜面　動傾斜を付ける，面取りする
　bevel gear　ベベルギア，かさ歯車
　bevel protractor　角度定規
　bevel square　角度定規，斜め定規
　bevel wheel　かさ歯車，斜め歯車
beveling　面取り，開先
　beveling frame　ベベル測定器
beyond　～　前～を越えて，～の向こうに(へ)，～を過ぎて
B-H curve　B-H曲線，磁束－磁化力曲線
BHP (Brake HorsePower)　制動馬力，ブレーキ馬力，実馬力
bi-　接頭「2」「2倍」「2重」の意
biannual　形年2回の，半年ごとの
bias　バイアス，偏り
　bias circuit　バイアス回路
　bias error　偏り誤差
　bias voltage　バイアス電圧
biaxial　形二軸の
bib cock　水栓，蛇口
bifurcation point　分岐点
big (**bigger/biggest**)　形大きい，激しい
　big end　[連接棒の]ビッグエンド，大端[部]
　　関 small end：小端[部]
bilateral　バイラテラル【サーボ機構の一種】
bilge　ビルジ【船底にたまる汚水／船底の湾曲部】
　bilge injection valve　ビルジ吸込み弁
　bilge pump　ビルジポンプ
　bilge suction pipe　ビルジ吸入管
　bilge water　ビルジ水
　bilge well　ビルジウェル，ビルジ溜り
billet　ビレット【圧延や鍛造用の素材】
billion　10億　形10億の，無数の
bimetal　バイメタル【熱膨張率が異なる2種類の金属板を張り合わせスイッチなどに用いる】
　bimetal bearing　バイメタル軸受
　bimetal regulator　バイメタル調整器
　bimetal thermometer　バイメタル温度計【温度変化によるバイメタルの変位を利用した温度計】
binary　形2つの，2成分の，2進[数]の，2元の
　名2進[数]，2進法，2値
　binary alloy (metal)　二元合金，2成分合金
　binary compound　二元化合物
　binary logic element　二値論理素子
　binary number system　2進法
　binary refrigerating system　二元冷凍方式
　binary signal　2値信号
bind electron　拘束電子
binder　接着剤，結合剤，固着剤
binding　バインディング
　binding material　接着剤，結合材
binomial distribution　二項分布
binomial expression　二項式
binomial theorem　二項定理
biological　生物学の
biomass　バイオマス【生物群からエネルギを取り出して利用すること】

biomechanics バイオメカニクス【生物の運動を研究する学問】
bionics バイオニクス，生体工学
biopolymer 生体高分子【生体を構成する高分子の総称】
Biot number ビオ数
biotechnology バイオテクノロジ，生物工学，生命工学
biotribology バイオトライボロジ【生物の関節などの摩擦や潤滑を扱う学問】
bipolar 形双極の 名バイポーラ【二つの極性のこと】
bird's-eye view 鳥瞰(かん)図
bismuth ビスマス 記Bi
bit ビット【「binary digit」の略で2進数字のこと】，小片
関 a bit of ～：少量の／the slightest bit of：ほんのわずかの／even a little bit of：ほんの少しの
bit brace クランクボール
black 形黒い 名黒
black body 黒体
black body radiation 黒体放射
black body surface 黒体面
black body temperature 黒体温度
black box ブラックボックス【「中身のわからない箱」の意】
black heart malleable cast iron 黒心可鍛鋳鉄
black lead 黒鉛
black sheet 薄鋼板
blackness 黒さ
blackout ブラックアウト，停電
blacksmith welding 鍛接
blade 羽根，翼，刃先
blade angle 羽根取付け角
blade arrangement 翼配列
blade back 羽根後進面，羽根の背
blade clearance 翼列すきま
blade efficiency 羽根効率
blade face 羽根前進面，羽根の腹
blade following edge 羽根後縁
blade height 羽根高さ
blade inlet angle 羽根入口角
blade lattice 翼列
blade leading edge 羽根前縁
blade length 羽根長さ
blade loss 羽根損失
blade outlet angle 羽根出口角
blade pressure side 羽根前進面
blade profile 翼形
blade ring 羽根輪
blade root 翼付根，羽根根元
blade section 羽根断面，翼断面
blade stopper 羽根止め金
blade suction side 羽根後進面
blade thickness 翼厚，羽根厚さ
blade thickness ratio 翼厚比
blade tip 翼端，羽根先
blade tip clearance 羽根先すきま
blade velocity coefficient 翼速度係数
blading 羽根，羽根植付，翼配列
blame 動非難する，責任にする
blank 形空白の，空の 名空白
blank cap めくら蓋
blank flange めくらフランジ
blanking 空白化，消去，打抜き，打抜き加工
blast 名突風，送風，爆発 動爆破する
blast air 噴射空気
blast fuel injection 燃料空気噴射
blast pipe 送風管
blast pressure 送風圧
blasting ブラスト【金属表面などの清浄法の一つ】
bleaching 脱色
bled 「bleed」の過去，過去分詞
bleed 名ブリード，抽気 動抜く，取り出す，注ぐ
bleed air 抽気
bleed off circuit ブリードオフ回路【油圧回路における速度制御回路の一種】
bleed off pipe 給水管
bleeder feed water heater 抽気給水加熱器
bleeder heater 抽気加熱器
bleeder plug 空気抜きプラグ
bleeder turbine 抽気タービン

bleeder valve　吹出し弁
bleeding　抽気
　bleeding cycle　抽気サイクル
　bleeding pipe　気泡管
　bleeding turbine　抽気タービン
blend　名ブレンド，混合，混合物　動混合する，混ぜる
blended fuel oil　混合燃料油
blending　ブレンディング，混合
　blending naphtha　混合用ナフサ
blew　動「blow」の過去形
blind　名盲（めくら）板　形行止まりの，出口のない
　blind joint　めくら継手
　blind patch　めくら板
blister　ブリスタ，膨出，気泡
　blister copper　粗銅，荒銅
　blister steel　ブリスタ鋼，浸炭鋼
block　名ブロック，角材，台木，滑車　動閉鎖する，塞ぐ
　block brake　ブロックブレーキ，まくらブレーキ【摩擦ブレーキの一種】
　block coefficient [of fineness]　方形係数
　block diagram　ブロック線図
　block gauge　ブロックゲージ【長さ測定の標準となるゲージ】
blocked resistance　制止抵抗
blocking　ブロッキング，閉塞，粘着【はく離しにくくなる現象】
bloom　ブルーム
blooming　分塊圧延【インゴットを圧延して成型すること】
blow　動吹く，送風する，爆発する，破裂する，[ヒューズが]とぶ　名ブロー，送風，一撃，ひと吹き，打撃，破裂
　blow-back　ブローバック，逆流，ガス漏れ
　blow-by　ブローバイ，吹き抜け【シリンダ内のガスがクランク室に吹き抜けること】
　blow-by gas　ブローバイガス
　blow-cock　排気コック
　blow-down　ブローダウン，吹出し，排出，抽気
　blow-down pressure of safety valve　安全弁吹下がり圧力
　blow-down turbine　ブローダウンタービン
　blow-down turbocharging system　動圧過給
　blow-hole　ブローホール，ガス孔，気孔，巣【鋳物の中の空洞】
　blow-lamp　ブローランプ，トーチランプ，小型発炎装置
　blow-off　ブローオフ，吹消し，吹消え，吹出し【ボイラ水を排出すること】
　blow-off pipe　吹出し管
　blow-off valve　吹出し弁
　blow out　動ヒューズが飛ぶ
　blow-out　ブローアウト，噴出，破裂，溶断，吹消え，吹出し，吹止め
　blow-pipe　トーチ，ブローパイプ，吹管
　blow-torch　トーチランプ
blower　ブロワ，送風機
blown　形（ヒューズ）がとんだ，破壊された，膨れた　動「blow」の過去分詞
blue　形青い　名青
　blue annealing　青熱焼鈍
　blue brittleness　青熱脆性，青熱脆（もろ）さ【200〜300℃の温度範囲で，鋼の延性，靭性が低下する現象】
　blue heat　青熱
　blue light　青色光
　blue shortness　青熱脆性，青熱脆（もろ）さ
blu[e]ing　ブルーイング，青焼法【鋼の外観・耐食性を向上させるための熱処理】
blueprint　青写真，青図
blushing　かぶり【塗料の白化現象】
board　名ボード，盤，板，基盤　動乗り込む，乗船させる　関 on board：（船などに）に乗って
　on-board computer　搭載コン

ピュータ
on-board operation　船上作業
bob　下げ振り【糸で吊り下げた円錐のおもり】
Bode diagram　ボード線図
body　名物体，胴体，本体，〜体
  body centered cubic lattice　体心立方格子
  body centered cubic structure　体心立方構造
  body plan　正面線図
  body tilting control　振り子制御
boil　動沸騰する
boiled oil　ボイル油【乾性油の一種】
boiler　ボイラ
  boiler air heater　ボイラ空気加熱器
  boiler blow-off pipe　ボイラ水吹出し管
  boiler blow-off system　ボイラブロー装置【ボイラ水の濃縮を防ぐためボイラ水の一部を排出する装置】
  boiler bracket　ボイラ支え
  boiler casing　ボイラケーシング
  boiler circulation pump　ボイラ循環ポンプ
  boiler cleaning　ボイラ洗浄
  boiler compound　ボイラ清浄剤
  boiler draft　ボイラドラフト(通風)
  boiler drum　ボイラドラム，ボイラ胴
  boiler efficiency　ボイラ効率
  boiler exhaust gas　ボイラ排ガス
  boiler feed pump　ボイラ給水ポンプ
  boiler feed water　ボイラ給水
  boiler feed water regulation　ボイラ給水調節
  boiler fittings　ボイラ取付物【圧力計，水面計，安全弁など】
  boiler heat balance　ボイラ熱平衡，ボイラ熱勘定
  boiler horsepower　ボイラ馬力
  boiler hydrostatic test　ボイラ水圧試験
  boiler incrustation　ボイラスケール
  boiler lagging　ボイララギング
  boiler main stop valve　ボイラ主蒸気止め弁
  boiler mountings　ボイラ付属品
  boiler plate　ボイラ板
  boiler pressure　ボイラ圧[力]
  boiler room　ボイラ室
  boiler setting　ボイラ据付け
  boiler shell　ボイラ胴
  boiler soda boiling　ボイラソーダ煮
  boiler steel　ボイラ用鋼材
  boiler superheater　ボイラ過熱器
  boiler test　ボイラ試験
  boiler tube　ボイラ管，ボイラ水管
  boiler wall　ボイラ壁
  boiler water　ボイラ水
  boiler water circulation　ボイラ水循環
  boiler water level　ボイラ水位
boiling　沸騰
  boiling curve　沸騰曲線
  boiling flow　沸騰流
  boiling heat transfer　沸騰熱伝達
  boiling limitation　沸騰限界
  boiling point　沸点，沸騰点
  boiling transition　沸騰遷移
bold　形太い
bollard　ボラード，係船柱
bolometer　ボロメータ【放射温度計の一種】
bolster spring　まくらばね【楕円形のばね】
bolt　名ボルト　動ボルトで締める
  bolt hole　ボルト穴
  bolt flange　ボルト締めフランジ
  bolt head　ボルト頭
  bolt heater　ボルトヒータ【ボルトを加熱膨張させる装置】
  bolt joint　ボルト継手
  bolt with reduced shank　伸びボルト
Boltzmann's constant　ボルツマン定数
bomb　ボンベ
  関 oxygen bomb：酸素ボンベ
  bomb calorimeter　ボンベ熱量計【固体の燃焼熱を測る熱量計】

bombardment　衝撃
bond　結合，結合剤，目地材
　bond energy　結合エネルギ
　bond joint　接着継手
　Bond number　ボンド数
　bond stress　付着応力
bonderizing　リン酸被膜処理
bonding　結合，接合
　bonding material (agent)　結合材，接着剤
　bonding power　結合力
　bonding strength　接合強度
bonnet　ボンネット，エンジンフード
book　名ブック，図書，本　動予約する，記入する
boom　ブーム【柱状の構造物】
boost　名増加，上昇　動強化する，上昇させる，増加させる
　boost pressure of inlet air　給気圧
　boost temperature of inlet air　給気温度
booster　ブースタ，昇圧器
　booster amplifier　ブースタ増幅器
　booster circuit　ブースタ回路
　booster pump　ブースタポンプ【加圧用ポンプ】
boral　ボーラル【遮蔽材の一種】
borate　硼(ほう)酸
borax　硼(ほう)砂
bore　名ボア，穴，内径，口径　動穴を開ける
　bore diameter　ボア径，内径，穴径
　bore gauge　ボアゲージ
　bore scope inspection　内視鏡検査
boride　硼(ほう)化物
boring　ボーリング，中ぐり
　boring machine　中ぐり盤
born off　目こぼれ
borne　動「bear」の過去分詞
boron　ボロン，硼(ほう)素　記B
　boron steel　ボロン鋼
boss　ボス【回転体で軸がはまる部分】
　boss ratio　ボス比
both　形両方の　代両方，双方
　both A and B　接AもBも，ABいずれも
bottle　ボトル，ボンベ，容器，びん
bottom　名ボトム，底[部]，船底，根本(底)　形底の，最下の
　bottom blowoff　ボイラ底吹出し
　bottom center　底部中心
　Bottom Dead Center　BDC，下死点　関TDC：上死点
　bottom end　ボトムエンド，大端部　同big end
　bottom land　歯底面
　bottom of thread　ねじの谷底
　bottom paint　船底ペイント
　bottom plug　船底プラグ
　bottom rake　後逃げ角
　bottom ring　ボトムリング
　bottom tank　下部タンク
　bottom view　下面図
bottoming tap　仕上げタップ
bound energy　束縛エネルギ
boundary　境界
　boundary condition　境界条件
　boundary dimension　主要寸法
　Boundary Element Method　BEM，境界要素法
　boundary film　境界膜
　boundary friction　境界摩擦
　boundary layer　境界層
　boundary layer separation　境界層はく離
　boundary layer thickness　境界層厚さ
　boundary layer transition　境界層遷移
　boundary lubrication　境界潤滑，不完全潤滑
　boundary plank　縁板
　boundary surface　界面，境界面
　boundary value　境界値
Bourdon tube　ブルドン管
　Bourdon tube pressure gauge　ブルドン管圧力計
bow　船首，弓状，円弧状　動しなる，たわむ，曲がる
　bow shock　弓形衝撃波
　bow thruster　バウスラスタ，船首横押しプロペラ

bowing 歪

bowl ボール, 鉢, 碗
  関 separation bowl：清浄機の回転体
  bowl nut　袋ナット

box 箱
  box beam (girder)　箱形げた【箱形をした断面を持つ梁（はり）】
  box coupling　スリーブ継手
  box nut　袋ナット
  box spanner　ボックススパナ【箱形のスパナ】
  box type piston　箱形ピストン【中空のピストン】

Boyle-Charles' law　ボイル・シャルルの法則

brace 支柱, 突っ張り, 中括弧
  動 補強（支持, 固定）する, きつく締める

bracing 控え
  bracing tube　控え管

bracket ブラケット, 腕木, 括弧
  bracket bearing　ブラケット軸受

braided packing　ブレードパッキン【ひも状にしたパッキン】

brake 名 ブレーキ, 制動装置
  動 制動する
  brake band　ブレーキ帯
  brake cylinder　ブレーキ筒
  brake disk　ブレーキ板
  brake drum　ブレーキ胴
  brake dynamometer　ブレーキ動力計
  brake efficiency　ブレーキ効率
  brake gear　ブレーキ装置
  Brake HorsePower　BHP, ブレーキ馬力, 軸馬力, 軸出力, 制動馬力, 正味馬力
  brake leverage　ブレーキ倍率
  brake lining　ブレーキライニング【ブレーキバンド・ディスクの表面に貼るもの】
  brake load　ブレーキロード, 正味出力
  brake magnet　制動磁石
  brake mean effective pressure　正味平均有効圧力, ブレーキ平均有効圧力
  brake motor　ブレーキモータ
  brake nozzle　制動ノズル
  brake output　ブレーキ出力
  brake percentage　ブレーキ率
  brake ring　ブレーキリング
  brake shoe　ブレーキシュー, ブレーキ片, 制輪子
  brake thermal efficiency　正味熱効率, ブレーキ熱効率

braking 制動
  braking ratio　制動率
  braking time　制動時間
  braking torque　ブレーキトルク, 制動トルク
  braking work　制動仕事

branch ブランチ, 枝, 枝管, 分岐
  動 分ける
  branch box　分岐箱
  branch circuit　分岐回路
  branch pipe　枝管, 分岐管
  branch[ing] point　分岐点

brass ブラス, 黄銅, 真鍮, 軸受金
  brass solder　黄銅ろう【硬ろうの一種】

Braun tube　ブラウン管

Brayton cycle　ブレイトンサイクル【ガスタービンの定圧燃焼サイクル】

braze 名 ろう付け結合, 真鍮の継手　動 ろう付けする
  braze welding　ブレイズ溶接, ろう接

brazing ろう付け【接合部の地金を溶かさず, 低融点の合金を溶かして金属を接合】

breadth 幅
  breadth depth ratio　幅深さ比
  breadth extreme　最大幅
  breadth maximum　全幅

break 名 ブレーク, 破壊（断）, 切断, 亀裂, 中断, 開路, 休み時間　動 破断する, 遮断する, こわす, 折る, 引き外す, 開く, 切る
  break-away　はく離, 分離
  break-away torque　始動トルク
  break contact　ブレーク接点, b接点, 開接点

break down 動故障する，破壊する
break-down ブレークダウン，故障，破壊，分解，崩壊，降伏，絶縁破壊，損傷，破損
break-down maintenance 事後保守，故障後保全
break-down potential 破壊電圧
break-down test 破壊試験
break-down voltage 絶縁破壊電圧，降伏電圧，放電開始電圧
break frequency 折点周波数
break-in 慣らし運転
break joint 食違い継手
break line 破断線
break off 動もぎ取る，ちぎり取る
break out 動（突然）起こる
break-out friction 始動摩擦
break point 降伏点
break potential 破壊電圧
break up 動粉々にする，分割する，解体する
break-up 破壊，崩壊
breakable 形破損の恐れのある
breakage 破損，切断，断線，折損
　breakage model 破壊モデル
breaker ブレーカ，遮断器
breaking 破壊，切断，破断
　breaking current 遮断電流
　breaking [down] test 破壊試験
　breaking elongation 破断伸び
　breaking energy 破壊エネルギ
　breaking joint 食違い継手
　breaking load 破壊荷重，破断荷重
　breaking of emulsion エマルジョンの破壊
　breaking of wire 断線
　breaking point 破壊点
　breaking radiation 制動放射
　breaking strain 破壊ひずみ
　breaking strength 破壊強さ，破壊力
　breaking stress 破壊応力
　breaking test 破壊試験
breakover voltage ブレークオーバ電圧
breathe 動呼吸する
breather ブリーザ，息つき
　breather pipe 息抜き管
　breather valve ブリーザ弁，呼吸弁
breathing 呼吸，ガス抜き
　breathing apparatus 呼吸装置
breeding 増殖
brick れんが
bridge ブリッジ，船橋
　bridge circuit ブリッジ回路
　bridge console 船橋制御盤
　bridge gauge ブリッジゲージ，橋形ゲージ
　bridge method ブリッジ法
　bridge rectifier ブリッジ整流器
　bridge type relay ブリッジ型継電器
bridging ブリッジング【ホッパの出口で排出が悪くなる現象】
bright 形輝く，鮮やかな
　bright finish 光輝仕上げ
brighten 動輝かせる，明るくする・なる，みがく
brightness 輝度，明るさ
brim 縁，へり
動溢れそうになる
関 fill ～ to the brim：～で溢れんばかりになる
brine ブライン，塩水【不凍液の一種】
　brine cooler ブライン冷却器
　brine freezing ブライン凍結
　brine pump ブラインポンプ
　brine system ブライン式
Brinell hardness ブリネル硬さ
　Brinell hardness test ブリネル硬さ試験 関 Brinell hardness tester：ブリネル硬さ試験機
brinelling ブリネリング【浸炭層が薄い場合に変形する現象】
bring 動もっていく，もってくる，導く，もたらす
　bring about 動～をもたらす
　bring into 動～に至らせる
brittle 形脆（もろ）い
動脆くなる，劣化する
　brittle coating 脆（ぜい）性塗料【塗装によってひずみの方向や大きさを測る】

brittleness

brittle ductile transition 脆性－延性遷移
brittle failure (fracture, rupture) 脆性破壊
brittle material 脆性材料
brittle [ness] temperature 脆化温度 ⇨ brittle point
brittleness 脆(もろ)さ，脆性
broach ブローチ，穴ぐり器，穴あけ錐
broad 形広い
broadly 副大まかに，幅広く
broke 動「break」の過去形
broken 動「break」の過去分詞 形故障した，損傷している，半端の
broken line 破線，折れ線
bromine 臭素 記Br
bronze ブロンズ，青銅
brought 動「bring」の過去・過去分詞
Brownian motion ブラウン運動
brush ブラシ，刷子【発電機などで回転子と固定子を接続する接触子の役割を持った導体】
brush away 動払いのける
brush holder ブラシ保持器
brushless DC motor ブラシレスDCモータ
bubble 気泡
bubble nuclei 気泡核
bubbling point 泡立ち点
bubbly flow 気泡流
buck stay バックステー【炉壁のための支持構造物】
bucket バケット，バケツ，おけ
bucket pump バケットポンプ
bucket valve バケット弁
buckling 座屈【長柱が縦方向の荷重で曲がる現象】
buckling length 座屈長さ
buckling load 座屈荷重
buckling stress 座屈応力
buff バフ【表面仕上げ材】
buffer 名バッファ，緩衝器 動緩和する，衝撃を和らげる
buffer action 緩衝作用

buffer amplifier 緩衝増幅器
buffer gas バッファガス，緩衝気体
buffer layer バッファ層，緩衝層
buffer resister バッファレジスタ
buffer solution 緩衝液
buffer spring 緩衝ばね
buffering 名緩衝[作用] 形緩衝の
buffing バフみがき，バフ加工，バフ研磨，バフ仕上げ
buffing compound バフ研磨剤
bug バグ【プログラムの不良箇所】，故障原因
build 動建てる，組み立てる，起こす，作る，製造する
build up 動蓄積する，集める，強まる
builder 建造者，製造者
building ビル，建物，建造，建造物
building yard 造船所
build-up 名増強，強化 動築く，蓄積する，仕上げる，増強する
build-up time 立ち上がり時間
build-up welding 肉盛溶接
built 形組立の 動「build」の過去・過去分詞
built-on 成り立つ，設置される
built-in 形組み込まれた，はめ込みの，内蔵された，一体型，固有の
built-in amplifier アンプ内蔵
built-in beam 固定梁(ばり)
built-in compression ratio 固有圧縮比
built-in edge 固定端
built-in stabilizer 自動安定装置
built-in stress 残留内部応力
built-in type 作り付け形式
built-up 形組立ての 名増大，蓄積，集合，付着
built-up beam 組立梁(ばり)
built-up construction 組立構造
built-up crank 組立クランク
built-up crank shaft 組立クランク軸
built-up nozzle 組立ノズル
built-up piston 組立ピストン
built-up propeller 組立プロペラ

built-up type　組立式
bulb　バルブ, 球, 球状部, 電球
　bulb plate　球板
bulbous bow　球状船首
bulge　名バルジ, 膨(ふく)れ
　動膨らむ
bulk　容積, 大量, 大半
　bulk boiling　バルク沸騰
　bulk-head　隔壁
　bulk-head door　隔壁戸
　bulk-head valve　障壁弁
　bulk modulus [of elasticity]　体積弾性係数　関 modulus of elasticity of volume
　bulk strain　体積ひずみ
bull gear　ブルギア【小歯車群の中の大きい主動歯車】
bull ring　パッキン保持器
bulletin board　掲示板
bump　バンプ, 衝突【衝撃の一種】
bumped head　さら形鏡板
bumper　バンパ, 緩衝器
bundle　バンドル
bunker　名バンカ【船の燃料をいう】
　動燃料を積み込む
　bunker capacity　燃料庫容積
　bunker oil　バンカ油【容器に入れず, ばら積みの油】
bunkering　燃料搭載作業, 補油
buoy　ブイ, 浮標
buoyancy　浮力
　関 center of buoyancy：浮心
　buoyancy apparatus　救命浮器
burn　動燃える, 焼ける, 燃焼する
　名焼損, 焼成, 燃焼
　burn damage　焼損
　burn out　動燃え尽きる
burnable　名可燃物　形可燃性の
burned　形燃えた, 焼けた, 焦(こ)げた
　burned gas　燃焼ガス
　burned-out　形燃え尽きた, 機能を果たさなくなった
burner　バーナ, 燃焼器【ガスバーナ, 油バーナなど】
　burner characteristics　バーナ特性
　burner master valve　バーナ主弁
　burner tip　バーナチップ, バーナ口金
burning　バーニング, 燃焼, 焼成, 焼損, 焼付き
　burning characteristics　燃焼特性
　burning oil　燃料油, 燃えている油
　burning point　発火点, 燃焼点
　burning process　燃焼過程
　burning pump　噴燃ポンプ
　burning quality　燃焼性
　burning rate　燃焼率, 燃焼速度
　burning velocity　燃焼速度
burnish　動磨く, 研ぐ　名つや出し
burnisher　研磨器
burnishing　バニシング【表面を平滑にする加工法】, 光沢仕上げ
　burnishing surface　つや出し面
burnout　バーンアウト, 焼損
　burnout heat flux　バーンアウト熱流束, 臨界熱流束
　burnout point　バーンアウト点
burnt　動「burn」の過去・過去分詞
　形焼いた, 焦(こ)げた, 燃焼した
　burnt gas　燃焼ガス
burnup　バーンアップ,燃焼,燃焼度
　burnup fraction　燃焼率
burr　名バリ, まくれ, かえり, 砥石, きり　動ぎざぎざに傷つける
burring　バーリング加工【穴広げ加工のこと】
burst　名バースト, 破裂, 爆発
　動割れる, 破裂する
　burst pressure　破裂圧, 破壊圧力
bursting disk　破裂板
bursting force　破裂力
bursting pressure　破裂圧
bursting speed　崩壊速度
bursting strength　破裂強度
bursting stress　破裂応力
bus　バス, 母線
　bus bar　母線
　bus line (bar)　母線
bush[ing]　ブッシュ, はめ輪【穴の内面に挿入する薄肉の円筒】
but　接～だが…, ～しかし…

前～以外，～を除いて
　関 not A but B：A でなく B
　but for ～　前 もし～がなかったら
butane　ブタン【無色で可燃性の気体炭化水素で燃料や合成ゴムの原料となる】
butt　名 突合せ　動 接合する
　butt joint　突合せ継手
　butt seam welding　バットシーム溶接【突合せ抵抗溶接の一種】
　butt strap　目板
　butt weld (welding)　突合せ溶接，突合せ鍛接
butterfly nut　蝶（ちょう）ナット
butterfly valve　バタフライ弁，蝶（ちょう）形弁
buttering　バタリング【溶接によって表面を処理する手法の一つ】
Butterworth　バタワース【油タンクを高温高圧の海水で掃除すること】
　Butterworth pump　バタワースポンプ
button　ボタン
　button head rivet　丸リベット
　button switch　ボタンスイッチ
buttress thread　のこ歯ねじ

butyl rubber　ブチルゴム【ガス不透過性合成ゴム】
buzzer　ブザー
　buzzer oscillator　ブザー発信器
by ～　前 ～によって，～のそばに（で），～単位で，まで，だけ
　関 A by B：A × B
　by hand　副 手動で
　by means of ～　前 ～によって，～を用いて
　by the head　副 船首に
　by the stern　副 船尾に
　by way of ～　前 ～を通って，～経由で，～として，～するために，～するつもりで
by-pass　名 バイパス，回避，側路，側管，分路，迂回路
　動 迂回する，バイパスする
　by-pass damper　バイパスダンパ
　by-pass diode　バイパスダイオード
　by-pass governing　バイパス調速法
　by-pass purifying　側流清浄
　by-pass valve　バイパス弁
by-product　副産物
byte　バイト【コンピュータ処理の単位】

C-c

C battery 単二電池
C clamp えび万力, しゃこ万力, Cクランプ
C [fuel] oil C重油
CA alloy CA合金【耐食性のばね材】
CA thermocouple CA熱電対
cab キャブ, 管制室, 運転室
cabinet 戸棚, 保管庫, キャビネット
  cabinet panel 分電盤
  cabinet projection drawing キャビネット図
cable ケーブル, 電線, 太索, 1/10海里
  cable band 電線帯金
  cable clench ケーブルクレンチ
  cable clip 電線帯金
  cable duct ケーブルダクト
  cable grand 電線貫通金物
  cable hanger 電線吊下げ支持金物
  cable run 電路
  cable saddle 電線支持金物
  cable tray 電線みち板【電線敷設のための取付け板】
  cable way 電路
cabtyre cable キャブタイアケーブル【ゴムで被覆された電線】
CAD (Computer Aided Design) キャド, コンピュータ援用設計
cadet 訓練生, 実習生
cadmium カドミウム ㊝ Cd
cage ケージ, 保持器, かご, 箱
  cage motor かご形誘導電動機
  cage pulley かご形プーリ【滑車の一種】
  cage rotor かご形回転子
  cage-type induction motor かご形誘導電動機
CAI (Computer Asisted Instruction) コンピュータを用いた教育システム
cake ケーク【「かたまり」の意】 ⑩固まる, 塊になる
caking coal 粘結炭
caking power 粘結力

calcined lime 生石灰, 焼石灰
calcium カルシウム ㊝ Ca
  calcium carbide 炭化カルシウム, カーバイド
  calcium carbonate 炭酸カルシウム
  calcium chloride 塩化カルシウム
  calcium hydroxide 水酸化カルシウム, 消石灰
  calcium oxide 酸化カルシウム, 生石灰
  calcium phosphate 燐酸カルシウム
  calcium silicate 珪酸カルシウム
  calcium soap base grease カルシウム石けん基グリース
  calcium sulfate 硫酸カルシウム
calculate ⑩計算する
Calculated Carbon Aromaticity Index CCAI, 計算芳香族炭素指数
calculating machine 計算機
calculation 計算
calculator 計算機
calculus 計算法, 微積法
caliber キャリバ【口径や穴径のこと】
calibrate ⑩較正する
calibrated hole in starter valve 始動空気[弁]ブリード穴
calibrating constant 計数定数
calibrating lever 検定用てこ
calibration 較正, 校正, 目盛定め
  calibration curve 較正曲線
calibrator 較正器
caliper キャリパ, パス【寸法, 厚みなどを計測する測定器】
  caliper gauge はさみ尺
calking コーキン, かしめ
  calking piece 植え金
call ⑧コール, 合図, 呼子, 呼び出し音 ⑩呼ぶ, 呼び出す
  call bell 呼鈴
  call for ⑩必要とする, 求める
  call sign (letter) 呼出し符号
  call signal 呼出し信号
calling device 呼出し装置

calling lamp 呼出しランプ
calm 形穏やかな，静かな，風のない 名無風状態，凪(なぎ)
caloric 形熱の 名熱素
calorie カロリー，熱量【熱量の単位：cal】
calorific 形熱の，発熱の，熱カロリーの
　calorific capacity 熱容量，熱負荷
　calorific intensity 発熱度
　calorific power (value) 発熱量
calorification 熱発生，発熱
calorifier カロリファイア，温水器
calorimeter 熱量計
calorimetry 熱分析，熱量測定法
calorizing カロライジング【鋼の表面にアルミニウムを拡散させる処理】
CAM(**Computer Aided Manufacturing**) キャム，計算機援用製造
cam カム【偏心器の一種】
　cam case カム囲い
　cam clearance カムすきま
　cam closed nut カム締付ナット
　cam diagram カム線図
　cam follower カム従動節，カム従動子
　cam lever カムレバー，カムてこ
　cam mechanism カム装置，カム機構
　cam movement rack カム移動用歯板
　cam ring カムリング
　cam roller カムローラ，カムころ
　cam roller guide カムローラ案内
　cam shaft カム軸
　cam shaft controller カム軸制御器
camber キャンバ，反(そ)り
　camber angle 反り角
　camber line 反り曲線
　camber ratio 厚さ幅比
camcorder キャムコーダ
camplasto meter カムプラストメータ【変形抵抗を測定する圧縮試験機】
camshaft カム軸
　camshaft gear カム軸歯車
　camshaft lubricating oil pump カム軸潤滑油ポンプ
　camshaft pump カム軸付き噴射ポンプ
can 缶 助〜できる，〜する可能性がある
　can filter 筒形濾器
　can type chamber 筒形燃焼室
cancel 名キャンセル，打切り，取消し 動[取り]消す，削除する
candela カンデラ【光度の単位：cd】
canned motor pump キャンドモータポンプ【水中ポンプの一種】
canonical 形標準的な
canopy 天蓋(てんがい)，ひさし
cant 傾斜，斜面，隅
　cant beam 斜め梁(はり)
　cant frame 斜肋骨
cantilever 片持[梁(はり)]
　cantilever crane 片持クレーン
　cantilever spring 片持ばね
　cantilever tank カンチレバータンク，トップサイドタンク
canvas キャンバス，帆布
　canvas hose 布ホース
cap キャップ，押え
　cap bolt キャップボルト
　cap nut 袋ナット
　cap screw 押えねじ
capability 能力，特性，性能，機能，可能性
capable 形有能な，可能で
capacitance キャパシタンス，静電容量
　capacitance type level gauge 静電容量液面計
capacitive circuit 容量性回路
capacitive load 容量負荷
capacitive reactance 容量リアクタンス
capacitor キャパシタ，コンデンサ，蓄電器
　capacitor motor コンデンサモータ
　capacitor start induction motor コンデンサ始動誘導電動機
capacity 能力，容量，容積

capacity ground　容量接地
capacity loss　電力損
capacity plan　容量図，容積図
cape chisel　えぼしたがね【キー溝などを削るたがね】
capillarity　毛管現象，毛細管現象
capillary　形毛状の，毛細管の　名キャピラリ，細管，毛細管
capillary action　毛管現象，毛管作用
capillary chemistry　界面化学
capillary condensation　毛管凝縮
capillary electric phenomenon　界面電気現象
capillary pressure　毛管圧
capillary tube　毛管，毛細管
capillary viscometer　毛管粘度計
capital letter　頭文字
capped bearing　密封軸受
capsize　動転覆する
capstan　キャプスタン【縦軸ウインチ】
capsular manometer　空ごう式圧力計
capture　捕獲
car　自動車，乗用車
car carrier　自動車運搬船
carat　カラット【ダイヤモンドなどの目方の単位；1カラット＝0.205g】
carbide　カーバイド，炭化物【水と反応してアセチレンを発生する炭化物】
carbide annealing　炭化物なまし
carbide tip　超硬チップ
carbide tool　超硬工具
carbon　カーボン，炭素　記C
carbon arc cutting　炭素アーク切断
carbon arc welding　炭素アーク溶接
carbon brush　カーボンブラシ【直流機の整流作用に用いる】
carbon content　炭素含有量
carbon deposit　カーボンデポジット，炭素堆積物【燃焼生成物の一種】
carbon dioxide　CO_2，炭酸ガス，二酸化炭素
carbon dioxide cylinder　炭酸ガスボンベ
carbon dioxide fire extinguisher　炭酸ガス消火器
carbon dioxide machine　炭酸ガス冷凍機
carbon electrode　炭素電極，炭素電極棒
carbon fiber　炭素繊維
Carbon Fiber Reinforced Plastic　CFRP，炭素繊維強化プラスチック
carbon monoxide　CO，一酸化炭素
carbon oxide　酸化炭素，炭素酸化物
carbon packing　炭素パッキン
carbon residue　残留炭素［分］
carbon ring　カーボンリング
carbon steel　炭素鋼
carbon steel for machine structure use　機械構造用炭素鋼
carbon steel pipe for high pressure service　高圧配管用炭素鋼鋼管
carbon steel pipe for ordinary piping　配管用炭素鋼鋼管
carbon steel pipe for pressure service　圧力配管用炭素鋼鋼管
carbon steel tube for hydraulic line service　油圧配管用炭素鋼鋼管
carbon tetrachloride　四塩化炭素【消火剤などに使用】
carbon tool steel　炭素工具鋼
carbon trumpet　カーボンフラワ，ノズル花咲き
carbonic acid gas　炭酸ガス
carbonitrided steel　浸炭窒化鋼
carbonitriding　浸炭窒化法【表面硬化法の一つ】
carbonization　炭化，乾留
carbonization gas　乾留ガス
carbonize　動炭化する
carbonizing　浸炭，炭化，拡散浸透めっき，拡散めっき法
carborundum　カーボランダム【研磨剤，耐火材】
carborundum oil stone　油砥石
carborundum paste　バルブコンパウンド
carburetion　気化

carburetor キャブレタ, 気化器, 揮発器
  carburetor body 気化器本体
  carburetor engine 気化器機関, 火花点火機関
carburization 浸炭, 炭化
carburized case depth 浸炭硬化層深さ
carburizer 浸炭剤
carburizing 浸炭, はだ焼き【鋼の表面層の炭素量を高める表面硬化法】
  carburizing agent 浸炭剤
  carburizing and quenching 浸炭焼入れ
  carburizing compound 浸炭剤
  carburizing steel 浸炭鋼
card カード
cardinal number 基数
care 图注意 動注意する
  care for 動〜に関心を持つ, 〜の手入れをする
careen 傾船手入れ
careenage 傾船修理, 傾船修理費
careening 傾船
careful 形注意深い, 慎重な
carefully 副慎重に, 注意深く
careless 形不注意な
  careless miss ケアレスミス【不注意によるやりそこない】
cargo 貨物, 積荷
  cargo boil-off system 貨物ボイルオフ装置〈LNG船〉
  cargo control console 荷役制御盤
  cargo [handling] gear 荷役装置
  cargo hatch カーゴ・ハッチ
  cargo hold 貨物倉
  cargo lamp (light) カーゴランプ, 荷役灯
  cargo oil 貨物油
  cargo oil collecting pump 貨物油集合ポンプ
  cargo oil heating pipe 貨物油加熱管
  cargo oil pump 貨物油ポンプ
  cargo oil stripping pump 貨物油浚えポンプ
  cargo oil tank 貨物油タンク
  cargo oil transferring system 貨物油移送装置
  cargo refrigerator 貨物冷凍機
  cargo winch 荷役ウインチ, 揚貨機
Carnot cycle カルノーサイクル【熱サイクルの一つ】
carpenter 大工
carriage 往復台
carrier キャリア, 回し金, 輸送船
carriet 回し金
carry 動運ぶ, 保持する, 通過する, 通す, 支える, 伝える, 処理する
  carry on 動続ける
  carry out 動行う, 実行する
carryover キャリオーバ【汽水共発現象のこと】
  carryover loss 持逃げ損失, 排出損【エネルギ損失の一つ】
cartridge カートリッジ【本体に容易に着脱できる交換用の部品】
  cartridge fuse 筒型フューズ
cascade カスケード【「滝」の意】, 翼列, 縦続
  cascade blade 翼列
  cascade combination system カスケード方式
  cascade connection カスケード結合, 縦続接続
  cascade control カスケード制御【主調節器の出力信号によって2次調節器の目標値を変更する制御方式】
  cascade effect カスケード効果, 翼列影響
  cascade pump カスケードポンプ【回転ポンプの一種】
  cascade tank カスケードタンク【給水ろ過器の一種】
case 容器, 場合, 事例
  関 in any case：どんな場合でも, とにかく／in case：万が一〜の場合は, 念のため／in [the] case of：〜の場合には／in extreme cases：極端な場合は／in other case：別のケースでは／in some cases：場

合によっては，時として／in such a case：そんな場合は(に)
case hardening　はだ焼き
case hardening steel　はだ焼用鋼
casing　ケーシング，外箱，車室，容器，枠
cassette　カセット【小箱の意】
cast　名鋳型，鋳物　動鋳造する
　cast crank shaft　鋳造クランク軸
　cast-in nozzle　鋳込みノズル
　cast iron　鋳鉄
　cast iron section boiler　鋳鉄ボイラ
　cast iron surface plate　鋳鉄製定盤
　cast steel　鋳鋼
　cast structure　鋳造組織
castability　可鋳性
castable refractory　キャスタブル耐火物
castellated nut　菊ナット
caster　キャスタ，脚輪，足車【方向自在の車輪】
casting　鋳物，鋳造，鋳込
　casting defect　鋳造欠陥
　casting fin　鋳ばり
　casting metal　鋳込みメタル
　casting strain　鋳造ひずみ
　casting stress　鋳造応力
　casting surface　鋳はだ
　casting temperature　鋳込温度
castings　鋳物，鋳込み，鋳造　関 mold：鋳型
castle nut　溝付きナット，菊ナット
castor oil　ひまし油
casualty　海難，災害，死傷[者]
cat walk　常設歩路
catalog　カタログ
catalysis　触媒作用，接触反応
catalyst　触媒
catalytic　形触媒作用の，促進する
　catalytic action　触媒作用
　catalytic combustion　触媒燃焼，接触燃焼
　catalytic converter　触媒コンバータ
　catalytic cracked gasoline　接触分解ガソリン
　catalytic cracking　接触分解
　catalytic cracking unit　接触分解装置
　catalytic denitrification　触媒脱硝法
　catalytic ignition　触媒点火
　catalytic oxidation　接触酸化
　catalytic polymerization　接触重合
　catalytic reaction　触媒反応
　catalytic reforming　接触改質
catalyze　動触媒作用を及ぼす
catalyzer　触媒
catapult　カタパルト，射出機
catastrophic failure　破局故障，壊損
catch　動捕まえる，[エンジンが]かかる　名捕えること
　catch water　気水分離器，集水路
categorize　動分類する
category　カテゴリ，範囲，範疇，分類，類別
catenary angle　カテナリ角【電線のたわみで生じる角度】
caterpillar　キャタピラ，無限軌道
cathode　カソード，陰極
cathodic protection　カソード式防食，陰極防食，電気防食
cation　カチオン，陽イオン
　cation exchange　カチオン交換
　cation exchange resin　陽イオン交換樹脂
CATV (CAble Tele Vision)　ケーブルテレビ
caulking　コーキング，かしめ
　caulking　かしめたがね
causality　因果律
cause　名原因，理由　動原因として〜となる，[引き]起こす
　cause and effect　原因と結果
　cause less　動起こしにくい，少なくする
　cause 〜 to…　動〜に…させる
caustic　形苛性の，腐食性の　名苛性アルカリ
　caustic action　腐食作用
　caustic alkali　苛性アルカリ
　caustic attack　アルカリ腐食
　caustic brittleness　アルカリ脆性，苛性脆化

caustic corrosion　アルカリ腐食
caustic cracking　アルカリ割れ
caustic embrittlement　苛性脆化，アルカリ脆性
caustic lime　生石灰，酸化カルシウム
caustic potash　苛性カリ，水酸化カリウム
caustic resistance　耐アルカリ性
caustic soda　苛性ソーダ，水酸化ナトリウム
caustic wash　苛性洗浄［液］
causticity　腐食性，苛性度，苛性化率
caution　名注意，用心，警告　動注意する，警告する
caution plate　注意板
caution tape　トラテープ，標識テープ
cavitation　キャビテーション，空洞現象
cavitation coefficient　キャビテーション係数
cavitation control　キャビテーション制御
cavitation damage　キャビテーション損傷，キャビテーション侵食，壊食
cavitation noise　キャビテーション騒音
cavity　キャビティ，空洞，空孔，巣
cavity radiator　空洞放射体
cavity resonator　空洞共振器
cavity shape　キャビティ形状
CD (Compact Disk)　コンパクトディスク
cease　動止める，無効となる
ceiling　天井，内張
ceiling crane　天井クレーン
ceiling light　天井灯
cell　電池
cell corrosion　電池腐食
cell motor　セルモータ，電池式電動機
cell structure　セル組織
cell voltage　電池電圧，槽電圧，浴電圧

cellophane tape　セロハンテープ
cellular phone　携帯電話
celluloid　セルロイド【プラスチックの一種】
cellulose　セルロース【セロハン，セルロイド等の原料】
Celsius degree (temperature)　摂氏温度
Celsius thermometer　摂氏温度計，C温度計
cement　セメント，接着剤
cementation　浸炭，拡散浸透めっき，拡散めっき法
cementation process　浸炭法
cemented carbide alloy　超硬合金
cemented carbide tool　超硬工具
cemented joint　張合せ継手
cemented steel　浸炭鋼
cementing　接着
cementite　セメンタイト【炭化鉄 Fe_3C の組織名】
center　名センタ，中心　動集中する
center distance　中心距離
center division　中心線仕切り
center gauge　センタゲージ【中心の位置を調べるゲージ】
center line　中心線
center of buoyancy　浮心【浮力の中心】
center of curvature　曲率中心
center of figure　図心
center of floatation　浮面心
center of gravity　重心
center of inertia　慣性の中心
center of mass system　重心系
center point　中心点
center punch　センタポンチ
center rest　振れ止め
center shaft　中央軸，中心軸
center square　心出し定規
centering　センタリング，心出し
centering device　中心合わせ装置
centerless grinding　心なし研削
centi-　センチ【「百分の一」の意】
centigrade　百分度，百分目盛
centigrade scale　摂氏目盛，百分

目盛
centigrade temperature　摂氏温度
centigrade thermometer　摂氏温度計
Centimeter Gram Second unit　CGS単位系【基本単位】
centistokes　センチストークス【動粘度の単位：cSt】
central　形中心の，中央の，中心的，主要な
　central air conditioning system　中央空気調和システム
　central control room　中央制御室
　central cooling system　集中冷却システム
　central fresh water cooling system　セントラル清水冷却システム
　central lateral plan　中心線縦断面
　central lubrication　集中注油
　central operating console　中央操作盤
　central valve　中央弁
centralized　形集中の
　centralized control system　集中制御システム
　centralized lubrication system　集中潤滑方式
　centralized monitoring　集中監視
centre　「center」に同じ
centrifugal　形遠心の
　centrifugal action　遠心(力)作用
　centrifugal advance　遠心進角装置
　centrifugal air compressor　遠心空気圧縮機
　centrifugal blower　遠心送風機
　centrifugal brake　遠心ブレーキ
　centrifugal casting machine　遠心鋳造機
　centrifugal clutch　遠心クラッチ
　centrifugal compressor　遠心圧縮機
　centrifugal control　遠心進角装置
　centrifugal dust collector　遠心力集塵装置
　centrifugal fan　遠心送風機
　centrifugal force　遠心力
　centrifugal force tachometer　遠心力式回転速度計
　centrifugal governor　遠心調速機
　centrifugal lubricator　遠心注油機
　centrifugal machine　遠心分離機
　centrifugal pump　遠心式ポンプ，渦巻きポンプ
　centrifugal purifier　遠心清浄機
　centrifugal relay　遠心継電器
　centrifugal separator　遠心分離機
　centrifugal stress　遠心応力
　centrifugal supercharger　遠心過給機
　centrifugal tachometer　遠心力式回転計
　centrifugal timer　遠心進角装置
centrifuge　動遠心分離する　名遠心分離機
centripetal　形求心性の，求力性の
　centripetal acceleration　向心加速度
　centripetal force　向心力，求心力
centroid　セントロイド，重心，質量中心，図心
centuple　形百倍の
ceramic　形セラミックの，陶磁器の　名セラミック，陶磁器
　ceramic coating　セラミック被覆
　ceramic engine　セラミックエンジン
　ceramic tool　セラミック工具
ceramics　セラミックス，陶磁器
cerium　セリウム　記Ce
cermet　サーメット【セラミックスと金属を混合したもの】
　cermet tool　サーメット工具
certain　形確かな，確信して，いくらかの，きっと〜する
　関 a certain amount of：ある程度の／make certain：よく確かめる
certainly　副確かに，確実に
certainty factor　確信度
certificate　証明書，証憑
　certificate of competency　海技免状
　certificate of ship's survey　船舶検査証書
certification　認証，資格，証明書
certified　形証明(保証)された
certify　動認定する，証明する
certified safe type apparatus　防爆保護機器

cesium セシウム ㊘ Cs
cetane セタン
  cetane number (value) セタン価【着火性を示す数値】
CFC (**Chloro Fluoro Carbon**) クロロフルオロカーボン, フロン
CFR (**Cooperative Fuel Reserch engine**) CFR 機関
CFRP (**Carbon Fiber Reinforced Plastics**) 炭素繊維強化プラスチック
CG (**Computer Graphics**) コンピュータグラフィクス
CGS (**Centimeter Gram Second unit**) CGS 単位系【基本単位】
chafe ㊐すり減らす, 摩擦する
chafing チェーフィング, 擦損
  chafing fatigue 摩擦疲労
  chafing gear 擦れ止め
  chafing lip 擦らせ板
  chafing mat 擦れ止めマット
chain チェーン, 鎖
  chain block チェーンブロック
  chain cable compressor (controller) 制鎖器
  chain coupling 鎖継手
  chain drive 鎖駆動
  chain hoist チェーンホイスト
  chain intermittent fillet welds 並列断続隅肉溶接
  chain line 鎖線
  chain of link 連鎖
  chain reaction 連鎖反応
  chain sprocket 鎖歯車【鎖駆動に用いる歯車】
  chain stopper チェーンストッパ (錨鎖止め)
  chain wheel 鎖車 (錨鎖)
chalk チョーク, 白墨
chamber チャンバ, 室, 空間
chamfer ㊇面取り ㊐面取りをする
chamfered edge 面取り縁
chamfering 面取り
chamotte brick シャモットれんが【酸性耐火れんがの一種】
chance failure 偶発故障
change ㊇変化, 変更, 取り替え ㊐変える, 変化する, 交換する, 交替する
  change at constant pressure 定圧変化
  change gear 変速装置, 変速ギア, 換え歯車
  change-over 切替え, 転換
  change-over contact C 接点, 切換え接点
  change-over switch 切換スイッチ
  change-over valve 切換弁
  change point 思案点
  change speed motor 多段速度電動機
  change valve 切換弁
changeable ㊋交換可能な, 変わりやすい
channel チャンネル, 通路, 水路, 流路, 溝, 開きょ, 通信路, 経路 ㊐導く, 通す
  ㊘ open channel：開水路／closed channel：閉水路, 暗きょ
  channel section みぞ形断面
  channel steel みぞ形鋼
channeling チャネリング, 偏流
chaplet けれん, 中子押え
chapter 章
character キャラクタ, 特性, 性質, 特色, 記号, 字, 文字
  character of classification 船級符号
characteristic ㊇特性, 特色, 特徴, 性質, 性能 ㊋特有の, 独特の
  characteristic curve 特性曲線
  characteristic equation 状態式, 特性方程式
  characteristic length 代表長さ
  characteristic value 固有値, 特性値
characterization 特性
characterize ㊐特徴づける, 特徴を述べる
charge ㊇チャージ, 給気, 充電, 電荷, 帯電, 供給 ㊐詰める, 充電する, 満たす
  ㊘ in charge of～：～を担当している, ～の責任を持っている／opposite charge：反対の電荷
  charge air temperature 給気温度
  charge cooling 給気冷却

chill

Charge Coupled Device　CCD，電荷結合素子
charge cut-off voltage　充電終止電圧
charger　充電器
charging　充てん，給気
　charging and discharging switchboard　充放電用配電盤
　charging current armature　充電コイル
　charging device　充電装置
　charging efficiency　充てん効率
　charging equipment　充電装置
　charging port　給気孔
　charging pressure　充てん圧力
　charging rate　充電率
　charging valve　給気弁
Charles' law　シャルルの法則
Charpy impact test　シャルピー衝撃試験
chart　図，海図，図表，線図，表
chassis　シャーシ，車台
chatter　びびり
　chatter vibration　びびり振動
chattering　チャタリング，びびり振動
cheap　形安価な，ちゃちな
cheaper　「cheap」の比較級
check　名検査，点検，停止，妨害，割れ　動チェックする，阻止する，調べる，検査する
　関visual check：肉眼検査
　check bolt　止めボルト
　check gauge　チェックゲージ，検定ゲージ
　check nut　止めナット
　check pin　止めピン
　check plate　止め板
　check ring　制限リング
　check test　確認試験
　check valve　チェック弁，逆止め弁【逆流を防ぐ弁】
checking devices of polarity　極性検知装置
chemical　形化学の，化学的な　名化学製品

chemical action　化学作用
chemical analysis　化学分析
chemical attraction　親和力
chemical bond　化学結合
chemical cleaning　化学洗浄
chemical composition　化学組成
chemical compound　化学物質，化合物
chemical deposition　化学蒸着
chemical energy　化学エネルギ
chemical engineering　化学工業
chemical equation　化学反応式
chemical equilibrium　化学平衡
chemical equivalent　化学当量
chemical fire extinguisher　化学消火器
chemical formula　化学式
chemical industry　化学工業
chemical plating　化学めっき
chemical potential　化学ポテンシャル
chemical property　化学的性質
chemical reaction　化学反応
chemical reactor　化学反応器
chemical resistance　耐薬品性
chemical tanker　ケミカルタンカ
chemical treating　化学処理
chemical tube　ケミカルチューブ
Chemical Vapor Depositions　CVD，化学蒸着法
chemically correct mixture ratio　理論混合比，理論空燃比
chemistry　化学
chest　木箱，箱
chief　形最高の，主要な　名チーフ，長
　chief engineer　機関長
　chief officer　一等航海士
chiefly　副主に，ほとんどが，まず第一に
chill　名冷硬，チル【鋳鉄が急冷によって硬くなる現象】
　動冷える，冷やす
　chill casting　チル鋳造
　chill hold　冷蔵倉庫
　chill mold　チル鋳型【チル鋳物

chilled

を作るための鋳型】
 chill ring　裏当て輪
chilled　形冷却された，冷蔵の，冷硬された
 chilled cargo　冷蔵貨物
 chilled cast iron　チルド鋳鉄，急冷鋳鉄
 chilled water system　冷水系
chiller　冷却(冷蔵)装置
chilling　形冷える
 chilling machine　冷却機
 chilling treatment　低温処理
chimney　煙突
Chinese character　漢字
chinsing iron　コーキンたがね
chip　チップ，素子，切りくず
chipping　チッピング, 剥離(はくり), はつり, さび落とし
 chipping goggle　防塵めがね
 chipping hammer　チッピングハンマ
chisel　たがね，のみ
 chisel steel　たがね鋼
chit drill　一文字ぎり
chloride　塩化物，塩素イオン
chlorination　塩素処理
chlorine　塩素　記Cl
 chlorine disinfection　塩素消毒
 chlorine ion　塩素イオン
 chlorine water　塩素水
chloro-ethylene　塩化ビニル
Chloro-Fluoro-Carbon　CFC，クロロフルオロカーボン，フロン
chock　チョック，くさび，架, 導索器
 chock liner　調整ライナ
choice　選択
choke　名チョーク，絞り 動チョークする, ふさぐ, 詰まらす
 choke coil　チョークコイル，塞流コイル
 choke coupling　チョーク結合
 choke line　閉塞線〈ガスタービン〉
 choke relief valve　チョークリリーフ弁
 choke system　始動系統
 choke tube　絞り管, ベンチュリ管
 choke valve　チョーク弁

choking　チョーク, 閉塞
 choking coil　チョークコイル，塞流コイル
 choking flow　チョーク流量
choose　動選ぶ, 選択する
chopper　チョッパ【直流電流を交流にする装置】
 chopper control　チョッパ制御
chord　弦, 翼弦
 chord length　翼弦長さ
 chord of blade　翼弦
chordal pitch　弦ピッチ
chordal thickness　弦歯厚
chosen　動「choose」の過去分詞 (choose-chose-chosen)
chroloy nine　クロロイ9【クロムステンレス鋼の一種】
chromansil　クロマンシル【Cr-Mn-Si合金鋼】
chromate treatment　クロメート処理【防錆被膜処理の一種】
chromatograph　クロマトグラフ【クロマトグラフィ法で分離を行うための装置】
chromatography　クロマトグラフィ　関 gas chromatography：ガスクロマトグラフィ
chrome　クロム　記Cr
 chrome molybdenum steel　クロムモリブデン鋼
 chrome plated cylinder liner　クロムめっきライナ
 chrome plated piston ring　クロムめっきピストンリング
 chrome plating　クロムめっき
 chrome steel　クロム鋼
chromel　クロメル【Ni-Cr合金】
 Chromel-Alumel thermometer　クロメル - アルメル温度計
chromite　クロマイト，クロム鉄鉱
chromium　同 chrome
chronograph　クロノグラフ【時間を図形的に記録する装置】
chronometer　クロノメータ【船で使用する基準時計】
chronometric tachometer　時計式回

転計
chuck チャック, つかみ
chugging チャギング, 断続燃焼
churning 撹拌
circle 名サークル, 円, 円周, 環, 循環, 輪, 圏
動旋回する, 回る, 取り囲む
関 pitch circle diameter：ピッチ円径
circle diagram 円線図
circle graph 円グラフ
circuit 回路　関 short circuit：短絡
circuit analyzer 回路計, 回路分析器
circuit breaker ［サーキット］ブレーカ, 回路遮断器
circuit changing switch 切替スイッチ
circuit controller 回路制御器
circuit diagram 回路図
circuit tester 回路計
circuitry サーキットリ, 回路, 回路構成, 回路図
circular 形円形の, 環状の, 循環の
circular angle of obliquity 周傾斜角
circular arc cam 円弧カム
circular back 円弧形背面
circular cone 円すい
circular cutting 曲線切断
circular cylinder 円柱, 円筒
circular disc cam 円板カム
circular disc coil 円板コイル
circular frequency 角周波数, 円振動数
circular gear 円弧歯車
circular measure 弧度法
circular motion 円運動
circular pitch 円周ピッチ
circular polarization 円偏波
circular section 円弧形断面
circular thickness 円弧歯厚
circular tooth thickness 歯厚
circular tube 円管
circularity 真円度
circulate 動循環する
circulating 形循環する
circulating lubrication 循環注油
circulating oiling 循環給油

circulating pump 循環ポンプ
同 circulation pump
circulating water 循環水
circulation 循環
circulation ratio 循環比
circumference 円周, 周辺, 周囲
circumference work 周辺仕事
circumferential 形円周の, 周辺の
circumferential efficiency 周辺効率
circumferential joint 周継手
circumferential pitch 周ピッチ
circumferential speed 周速度
circumferential strain 周ひずみ, 周方向ひずみ
circumferential stress 周応力, 周方向応力
circumferential velocity 周速度
circumferential work 周辺仕事
circumferentially 副円周に, 周囲に
circumscribed circle 外接円
circumstance 状況, 事情, 環境
cistern システン, タンク, 水槽
cite 動引用する
clack valve 逆止め弁, ちょう形弁, 羽打弁
clad クラッド, 被覆, 燃料被覆
clad steel 合せ鋼板【普通鋼にクロム鋼などを合わせた鋼板】
cladding 合せ板法, クラッド法, 被覆加工
claim 名クレーム, 要求, 賠償請求, 苦情　動要求する, 請求する
claimed 形主張した
clamp 名クランプ, 締金, 留め金, つかみ　動留める, 固定する
clamp coupling クランプ継手
clamp meter クランプメータ
clamp screw 締付けねじ
clamping クランプ, クランピング
clamping bolt 締付けボルト
clamping coupling クランプ継手
clarified oil 浄化油, 清浄油
clarifier クラリファイヤ, 清澄機【油中の固形分・不純物を除く装置】
clarify 動澄ませる, 浄化する, 不純物を除去する, 明らかにする

clarity 清澄性

Clark engine クラーク機関【2 サイクル内燃機関の別名】

clash 名衝突 動衝突する，対立する

clasp 留め金

class クラス，等級，種類，船級
  class A insulation　A 種絶縁
  class of finish　仕上げ程度
  class of insulation　絶縁の種類
  class survey　船級検査，入級検査

classification タイプ，区分，分類，分級，船級
  classification certificate　船級証書
  classification society　船級協会

classifier 分級機

classify 動分類する，等級に分ける

Clausius cycle クラウジウスサイクル

claw つめ
  claw clutch　かみ合いクラッチ
  claw coupling　かみ合い継手

clay 粘土
  clay tile　陶磁器タイル

clean 形きれいな，清浄な
  副きれいに，全く
  動掃除する，きれいにする・なる
  clean ballast　クリーンバラスト
  clean ballast pump　クリーンバラストポンプ
  clean bearing　クリーン軸受
  clean energy　クリーンエネルギ【大気汚染の少ない動力源をいう】
  clean off　動(汚れを)落とす
  clean oil　清澄油
  clean petroleum products　白物石油製品
  clean room　クリーンルーム
  clean tanker　白物石油製品タンカ，クリーンタンカ
  clean water　清浄水

cleaning 洗浄，掃除
  cleaning action　清浄作用
  cleaning agent　洗浄剤
  cleaning gear　掃除器具
  cleaning material　洗浄剤

cleanliness 清浄度

clear 形澄んだ，離れた，透明な，空の，明瞭な，はっきりした，妨げるものがない 副はっきりと，離れて 動クリアする，きれいにする，清掃する
  clear lamp　透明電球
  clear view screen　旋回窓

clearance クリアランス，すきま，間隙
  clearance adjuster　すきま調整器
  clearance angle　逃げ角
  clearance fit　動きばめ，すきまばめ
  clearance for insulation　絶縁間隙
  clearance for packing　パッキンしろ
  clearance gauge　すきまゲージ
  clearance hole　ばか穴
  clearance of bearing　軸受すきま
  clearance ratio　すきま比
  clearance volume　すきま容積

cleat クリート，綱止め

cleavage 裂け目，割れ目，へき開
  cleavage fracture　へき開破壊
  cleavage plane　へき開面

clench 留め具 動締める

clerical personnel 事務職員

Cleveland tester クリーブランド試験器

clevis U 字形金具

click 名クリック，つめ，カチッという音 動クリックする

climate 気候

climb 動登る，よじ登る

clinker クリンカ

clinometer クリノメータ，傾斜計

clip 名クリップ，留め金具 動留める

clipper joint クリップ継手【ベルト継手の一種】

clock 時計
  clock valve　クロックバルブ【一方向だけに流す蝶形弁】

clockwise 形／副時計回りの(に)

clog 動詰まらせる

close 動閉じる 形接近した，緻密な，閉じた，綿密な，密集した 副接近して，すぐ近くに，ぴった

りと，密集して
close-coiled helical spring 密巻コイルばね
close examination 綿密な検査
close fit 締まりばめ
close grained 形緻密な
close off 動止める，閉鎖する
close-packed hexagonal lattice ちゅう密六方格子
closed 形閉じた，密閉式，循環式の
　closed circuit 閉回路
　closed circuit voltage 閉路電圧
　closed crank chamber type クランク室密閉式
　closed cycle 密閉サイクル
　closed feed water system 密閉給水装置
　closed gap 合口すきま
　closed impeller クローズ羽根，クローズ羽根車
　closed joint めくら継手
　closed loop 閉ループ
　closed loop control 閉ループ制御
　closed loop system 閉ループ系
　closed loop transfer function 閉ループ伝達関数
　closed nozzle 閉止弁，閉止噴射弁，密閉ノズル
　closed system 閉じた系
　closed type 密閉型
　closed type bearing 密閉型軸受
　closed type fuel valve 密閉型燃料弁
　closed type thrust bearing 密閉型スラスト軸受
closely 副正確に，ぴったりと，厳密に，密接に
closing クロージング，クローズ，閉鎖，終了，締切り
　closing apparatus (appliance) 閉鎖装置
　closing pressure 圧縮圧
cloth クロス，布
　cloth scissors ラシャ鋏(はさみ)
cloudy 形濁った，曇った
clove hitch クラブヒッチ，巻き結び
cluster クラスタ【同種類の集団】

clutch 名クラッチ，連動器，駆動　動クラッチを入れる
　clutch case クラッチ室
　clutch coupling クラッチ継手，かみ合い継手
　clutch lining クラッチライニング
　clutch magnet クラッチ電磁石
　clutch plate クラッチ板
　clutch shaft クラッチ軸
clutching test クラッチかん脱試験
CM (**Condition Monitoring**) 状態監視
CMOS (**Complementary Metal Oxide Semiconductor**) 相補型金属酸化膜半導体
CNC (**Computerized Numerical Control**) 計算機数値制御
CNT (**Carbon Nano Tube**) カーボンナノチューブ
CO (**carbon monoxide**) 一酸化炭素
CO_2 (**carbon dioxide**) 二酸化炭素，炭酸ガス
　CO_2 meter CO_2メータ【CO_2量の測定器】
coagulant 凝固剤
coagulation 凝固，凝結，凝集
coal 石炭
　coal tar コールタール
Coalite コーライト【コークスの一種】
coaming コーミング，縁材
coarse 形きめの粗い，不正確な
　coarse both cutter 荒刃フライス
　coarse grain 粗粒
　coarse paper 目の粗いペーパ
　coarse saw 歯数の荒いのこぎり
　coarse screw thread 並目ねじ
coarseness (きめの)粗さ
coast 海岸
　coast guard 沿岸警備隊
coat 動塗る，覆う，かぶせる　名塗装
coated electrode 被覆アーク溶接棒
coating コーティング，被覆，被覆材，めっき，塗装【表面を被膜で覆うこと】
　coating material 塗料

coaxial 形同軸の
  coaxial cable　同軸ケーブル
coaxing effect　コーキング効果【材料が硬化する現象】
cobalt　コバルト　記Co
Cochran boiler　コクランボイラ
cock　コック，活栓，蛇口
cocked hat　誤差三角形
cockpit　コックピット，操縦室
cocoanut oil　椰子油
co-current flow　並流
code　コード，符号，記号，規格，規則
  code book　コードブック，信号書
  code letters　信号付字
coefficient　係数，率
  coefficient of capacity　容量係数
  coefficient of contraction　収縮係数
  同 contraction coefficient
  coefficient of cubical expansion　体膨張係数，体積膨張率
  coefficient of discharge　流量係数
  同 flow coefficient
  coefficient of elasticity　弾性係数，弾性率
  coefficient of expansion　膨張係数
  coefficient of fluctuation　変動率
  coefficient of friction　摩擦係数
  coefficient of heat convection　熱対流係数
  coefficient of heat transfer　熱伝達係数，熱伝達率
  coefficient of heat transmission　熱貫流率，熱貫流係数
  coefficient of induction　誘導係数
  coefficient of inertia　慣性係数
  coefficient of kinematic friction　動摩擦係数
  coefficient of kinematic viscosity　動粘性係数
  coefficient of linear contraction　線収縮率
  coefficient of linear expansion　線膨張係数
  coefficient of mutual induction　相互誘導係数
  coefficient of overall heat transmission　熱貫流率
  Coefficient Of Performance　COP，成績係数，動作係数
  coefficient of Pitot tube　ピトー管係数
  coefficient of restitution　反発係数
  coefficient of self-induction　自己誘導係数
  coefficient of speed fluctuation　回転不整率
  coefficient of static friction　静止摩擦係数
  coefficient of thermal expansion　熱膨張率，熱膨張係数
  coefficient of use　利用率
  coefficient of utilization　照明率，利用率
  coefficient of variation　変化係数
  coefficient of velocity　速度係数
  coefficient of viscosity　粘性係数
  coefficient of volumetric expansion　体積膨張係数
coercive force　保磁力
cofferdam　コファダム，空所
cog　はめ歯
  cog belt　はめ歯ベルト
  cog wheel　はめ歯歯車
  cog wheel pump　歯車ポンプ
cogeneration system　コージェネレーションシステム，熱電併給システム
cogging mill　分塊圧延機
coherence　コヒーレンス【光の干渉性の度合いを示す】
  coherence function　コヒーレンス関数
coherent state　コヒーレント状態
cohesion　凝集[力]，粘着[力]
cohesive strength　結合力
coil　名コイル，ソレノイド，巻線，うず巻き，線輪
  動ぐるぐる巻く，輪状にする
  coil tube evaporator　コイル蒸発器
  coil type glow plug　コイル形グロープラグ
coiled pipe cooler　コイル冷却器
coil[ed] spring　コイルばね

coincide 動一致する
coining コイニング，圧印加工【仕上げ法の一種】
coir mat シュロマット
coke コークス
  coke oven gas　コークス炉ガス
coking coal 原料炭，コークス用炭，瀝青炭
cold 形冷たい，低温の
  cold air blast system　冷風式
  cold bending　常温曲げ，冷間曲げ
  cold bending test　冷間曲げ試験
  cold brittleness　低温脆(もろ)さ，低温脆性
  cold crack　低温割れ，冷間割れ
  cold drawn pipe　冷間引抜管
  cold extrusion　圧出，冷間押出し
  cold flow　コールドフロー【クリープ現象】
  cold forging steel　冷間鍛造鋼
  cold insulation material　保冷剤
  cold junction　冷接点
  cold machine　低温切削
  cold point　寒点
  cold proof　耐寒性
  cold rolling　冷間圧延
  cold rolling mill　冷間圧延機
  cold setting　常温硬化，低温硬化，冷間硬化
  cold shortness　低温脆(もろ)さ，低温脆性
  cold spring　コールドスプリング【熱膨張を考慮した配管の取付け法】
  cold start feed water pump　始動用給水ポンプ
  cold start fuel oil burning pump　始動用噴燃ポンプ
  cold start fuel oil heater　始動用燃料加熱器
  cold starting　常温起動
  cold storage　冷蔵
  cold welding　冷間圧接
  cold working　冷間加工
    関 hot working：熱間加工
  cold working hardening　冷間加工硬化
collapse 名コラップス，圧壊，圧潰，くぼみ，へこみ，崩壊
　動つぶれる，崩壊する
  collapse load　崩壊荷重
  collapse stress　崩壊応力
collar カラー，鍔(つば)，継ぎ輪
  collar bearing　つば軸受
  collar head screw　つば付頭ねじ
  collar journal　つばジャーナル
  collar oiling　つば給油
  collar plate　カラープレート【スロット部分などをふさぐための小形板材】
  collar thrust bearing　カラースラスト軸受
collect 動集まる・める，回収する，ためる
collecting pipe 集合管
collecting tank 集合タンク
collection 収集，回収，堆積，集塵
collector コレクタ，採取装置，集電体，トランジスタの電極
  collector pipe　集気管
  collector ring　集電環
collide 動衝突する・させる
collimater コリメータ【光学器具】
collision 衝突
  collision avoidance radar system　レーダ衝突予防装置
  collision mat　防水マット
  collision warning system　衝突警報システム
colloid コロイド【微粒子が液体などの中で分散している状態】
  colloid solution　コロイド溶液
colloidal dispersion コロイド分散
colloidal electrolyte コロイド電解質
colloidal fuel コロイド燃料
colloidal purifier コロイダル油清浄機
color 名色　動着色する，塗る
  color change　変色
  color chart　色表
  color check　カラーチェック

color code　カラーコード, 色分け
color comparison tube　比色管
color conditioning　色彩管理
color emissivity　色放射率
color filter　色フィルタ
color former　発色剤
color number　色価
color paint　色[彩]ペイント
color pyrometer　色調高温計
color rendering properties　演色性
color scale　色合表
color temperature　色温度
colorimeter　比色計, 色度計
colorimetry　比色, 測色, 色彩測定
column　コラム, 柱, 長柱, 支柱, 列, 欄
column of mercury　水銀柱
column of water　水柱
columnar structure　柱状組織
combatting oil spills　油防除
combination　組合せ, 連結, 結合, 化合
　㋕ in combination with：〜と組み合わさって
combination control　結合制御
combination cycle　複合サイクル, 混合サイクル
combination gauge　組合せゲージ
combination ring　組合せリング
combination turbine　混式タービン
combine　⑩結合する, 組合す
combined　㋻結合した, 組合せた, 複合の, 化合した
combined blade　根付羽根
combined carbon　化合炭素, 結合炭素
combined convection　複合対流
combined cycle　複合サイクル, サバテサイクル
combined cycle engine　複合サイクル機関
combined cycle power generation　複合サイクル発電
combined efficiency　総合効率
combined feed water heater　総合熱交換器
combined flow　混流
combined impulse turbine　組合せ衝動タービン
combined lathe　万能工作機
combined load　複合荷重
combined oil　混成油
combined processing　複合加工法
combined resistance　合成抵抗
combined stress　組合せ応力
combined system　組合せ式
combined turbine　混式タービン
combined type air cleaner　組合せ式空気清浄機
combusted　㋻燃焼された
combustibility　燃焼性
combustible　㋻可燃性の
combustible component　可燃成分
combustible gas　可燃性ガス
combustible gas detector　可燃性ガス検知器
combustible gas mixture　可燃混合気
combustible liquid　可燃性液体
combustible loss　未燃焼損失
combustible material　可燃性材料
combustible solids　可燃性固体
combustible substance　可燃性物質
combustion　燃焼
combustion aids　助燃剤
combustion air　燃焼[用]空気
combustion at constant volume　定容燃焼, 等積燃焼
combustion chamber (space)　燃焼室
combustion chamber temperature　燃焼室温度
combustion chamber volume　燃焼室容積
combustion condition　燃焼状態
combustion control　燃焼制御
combustion diagnostics　燃焼計測
combustion efficiency　燃焼効率
combustion engine　燃焼機関
combustion engineering　燃焼工学
combustion equipment　燃焼装置
combustion gas　燃焼ガス
combustion heat　燃焼熱
combustion improver　助燃剤

compared

combustion knock　燃焼ノック
combustion limit　燃焼限界
combustion performance　燃焼性能
combustion period　燃焼期間
combustion pressure　燃焼圧力
combution process　燃焼過程
combustion products　燃焼生成物
combustion properties　燃焼性, 燃焼特性
combustion quality　燃焼性
combustion rate　燃焼率, 燃焼速度, 燃焼負荷率
　⦿ rate of combustion
combustion reaction　燃焼反応
combustion recorder　炭酸ガス記録計
combustion speed　燃焼速度
combustion stroke　燃焼行程
combustion temperature　燃焼温度
combustion test　燃焼試験
combustion theory　燃焼理論
combustion turbine　燃焼タービン
combustor　燃焼器, 燃焼装置
come about　動起こる, 生じる
come from　動～による, 起こる, 生じる
come in contact with　動接触する
come into ～　動～になり始める
come out of　動出てくる, 回復する
come to rest　動停止する, 静止する
come up to ～　動～まで達する
comfort　快適さ
　comfort chart　快感線図, 快適線図
comfortable　形快適な, 心地よい
command　名コマンド, 目標値, 命令　動命じる, 指揮する
commence　動開始する
commercial　形商業[上]の, 市販の
　commercial bearing　コマーシャル軸受【精度の低い軸受】
　commercial speed　巡航速度
　commercial test　商用試験
commissioning　初運転, 試運転
common　形共通の, 普通の, 一般的な
　⦿ most common　最も一般的な, 最も頻度の高い
　common bilge piping　共通ビルジ管系
　common logarithm　常用対数
　common rail injection system　コモンレール燃料噴射システム, 共通配管燃料噴射システム
　common sense　常識
　common slide valve　普通型すべり弁
　common to A and B　AとB共通の
　common to all　全てに共通した
commonly　副ふつう, 一般に, 通例
communal aerial system for receiver　受信空中線共用装置
communicate　動伝える, 伝達する, 知らせる
communication　コミュニケーション, 通信, 伝達, 連絡, 情報
　communication pipe　連絡管
　communication valve　連絡弁
commutate　動整流する, 直流にする
commutating capacitor　整流コンデンサ, 転流コンデンサ
commutating field　整流磁界
commutating pole winding　補極巻線
commutating reactor　整流リアクタ, 転流リアクタ
commutation　整流, 転換, 交換
　commutation curve　整流曲線
commutator　整流子, 整流器
　commutator motor　整流子モータ
compact　形コンパクトな, 小型の, こぢんまりした　動圧縮する　名コンパクト, 成形体, 緻密
compaction　圧縮, 圧密
companion　天窓, 昇降口
　companion way　昇降口
comparable　形比較できる, 類似の, 匹敵する, 同等の
comparative measurement　比較測定
comparatively　副比較的, かなり
comparator　コンパレータ, 比較器
compare　動比較する, たとえる
　compare A to B　AをBにたとえる
　compare A with B　AをBと比較する
compared with (to) ～　前～と比較

すると，〜と比べて
comparison 比較
　comparison means 比較部
compartment 収納スペース，区画，仕切り，防水隔室
compass コンパス，羅針盤【方位計器の一種】
compasses コンパス，両脚器
compatibility 互換性，両立性，適合性
compatible 形共用の，適合性のある，互換性のある
　名コンパチブル，互換機
　compatible machine 互換機
compel 動強いて〜させる
compensate 動補正(償)する
compensating 補整，補正，補償
　compensating arm 補整腕
　compensating device 補正装置，補償装置
　compensating lead wire 補償導線
　compensating mechanism 復原機構【動作の安定を得る装置】
　compensating plate 補強板
　compensating winding 補償巻線
compensation 補償，修正，補強
　compensation circuit 補償回路
　compensation ring 補強リング
　compensation temperature 補償温度
　compensation theorem 補償の定理
compensator コンペンセータ，補償器，補正器，補償装置
compete 動競争する，匹敵する
compiler コンパイラ【翻訳プログラムの一種】
complement 名余角，余集合，補足，定員　動補足する
complementary angle 余角
Complementary Metal Oxide Semiconductor CMOS，相補型金属酸化膜半導体
complete 形完全な，十分な
　動終える，完了する，仕上げる
　complete annealing 完全焼なまし
　complete combustion 完全燃焼
　complete combustion limit 完全燃焼限界
　complete cycle 完全サイクル
　complete equation 完全方程式
　complete joint penetration 完全溶け込み
　complete lubrication 完全潤滑
　complete survey 完成検査
　complete thread 完全ねじ部
　complete unit 完成品一式
completely 副完全に
completion 仕上げ，完成，終了
　completion drawing 完成図
complex 形複合の，複雑な
　名複合体，合成物，錯体
　complex impedance 複素インピーダンス
　complex number 複素数
　complex potential 複素ポテンシャル
　complex response 複素応答
　complex variable method 複素変数法
　complex velocity 複素速度
compliance コンプライアンス【ばね定数の逆数】，遵守
　熟 in compliance with：〜に従って，〜を遵守して
　compliance control コンプライアンス制御
complicate 形複雑な　動複雑にする
comply 動応じる，遵守する
component 名コンポーネント，成分，構成，構成部品，構成要素，分力，部品　形構成している
　component distillation 共沸蒸留
　component of force 分力
　component of velocity 分速度，速度成分
　component part 構成部品
　component stress 分応力
compose 動構成する，成り立たせる
composite 形コンポジット，合成の，混合の，複合の，組み合わせた
　名合成，複合物，複合材料，多層
　composite boiler コンポジットボイラ【排気ガスボイラ＋油焚き

ボイラ】
composite chart 合成図
composite circuit 複合回路
composite cycle 合成サイクル
composite cylinder 多層円筒
composite flange 組合せフランジ
composite material 複合材料
composite piston 合成ピストン,組立ピストン
composite plane wall 多層平板
composite plating 多層めっき,複合めっき
composite sphere 多層球殻
composite steel 複合鋼,合せ鋼
composite tooth form コンポジット歯形
composition 構成,組成,成分,配合,合成,溶剤 (関) antifriction composition：減摩剤
composition of force 力の合成
composition rubber 合成ゴム
compound 形合成の,複合の,化合した 名化合物,混合物 動調合する
compound air compressor 複式空気圧縮機
compound angle 合成角
compound DC generator 複巻直流発電機
compound dynamo 複巻直流発電機
compound engine 複合機関,二段膨張機関
compound gauge 連成計
compound generator 複巻発電機
compound impulse turbine 連成衝動タービン
compound lubricating oil 混成潤滑油
compound motor 複巻電動機
compound pressure gauge 連成計
compound reaction turbine 連成反動タービン
compound screw 両ねじ
compound stress 組合せ応力
compound turbine 複式タービン
compound winding 複巻コイル
compounded oil 配合油

comprehend 動理解する,含む
compress 動圧縮する,縮む・める
compressed 形圧縮した,加圧した
compressed air 圧縮空気
compressed air starting 空気始動
compressed gas 圧縮ガス
compressed liquid 圧縮液
compressed oxygen 圧縮酸素
compressed self ignition 圧縮自己着火
compressed water 圧縮水
compressibility 圧縮,圧縮率,圧縮性
compressible 形圧縮性の
compressible flow 圧縮性流れ
compressible fluid 圧縮性流体
compression 圧縮
compression efficiency 圧縮効率
compression gas 圧縮ガス
compression gasket 圧縮ガスケット
compression ignition 圧縮点火
compression ignition [oil] engine 圧縮点火機関,ディーゼル機関
compression member 圧縮材
compression molding 圧縮成形【プラスチックの成形方法】
compression pressure 圧縮圧力
compression ratio 圧縮比
compression refrigerating machine 圧縮式冷凍機
compression ring 圧縮リング【ピストンリングの一種】
compression space すきま容積,圧縮室
compression space volume 有効すきま容積
compression spring 圧縮ばね
compression stress 圧縮応力
compression stroke 圧縮行程
compression temperature 圧縮温度
compression test 圧縮試験
compression volume すきま容積
compression work 圧縮仕事
compressional axis 圧縮軸
compressional wave 圧縮波,疎密波
compressive 形圧縮の,圧縮力のある

compressive fluid　圧縮性流体
compressive force　圧縮力
compressive load　圧縮荷重
compressive strain　圧縮ひずみ
compressive strength　圧縮強さ
compressive stress　圧縮応力
compressive wave　圧縮波
compressometer　縮み計
compressor　圧縮機
compressor oil　圧縮機油
compressor rotor　圧縮機ロータ
comprise　動～から成る
関 be comprised in：～に含まれる／be comprised of：～から成る
compromise　妥協, 折衷
compute　動計算する
Computed Tomography　CT, コンピュータ断層撮影
computer　コンピュータ, 計算機, 電子計算機
Computer Aided Design　CAD, キャド, コンピュータ援用設計
computer control　コンピュータ制御
Computer Graphics　CG, 計算機図形処理
computing control　コンピュータ制御, 計算制御
computing logger　コンピューティングロガー【演算機能を持つデータロガー】
concave　形凹面の, くぼんだ 名凹面
concave cam　凹面カム
concave lens　凹レンズ
concave mirror　凹面鏡
concentrate　動集中する, 濃縮する
concentrated load　集中荷重
関 uniform load：等分布荷重
concentrated winding　集中巻
concentration　濃度, 濃縮, 集中
関 stress concentration：応力集中
concentration cell　濃淡電池
concentration factor　集中係数
concentration meter　濃度計
concentration of dissolved oxygen　溶存酸素濃度

concentration pump　濃縮ポンプ
concentrator　濃縮器, 集配線装置
concentric　形同心の
concentric circle　同心円
concentricity　同軸度, 同心性
concept　コンセプト, 概念
concern　動関係がある, 影響する
conchoidal fracture　貝殻状破面
conclude　動終える, 完成させる
conclusion　結末, 結論, 帰結
concrete　名コンクリート, 凝固物 形凝固した, 固まった 動固める・まる, 凝固する・させる
concurrent heating　補熱〈溶接〉
condensate　復水, 凝縮液, 凝縮水
condensate pump　復水ポンプ
condensate temperature　復水温度
condensation　凝縮, 凝結, 復水, 濃縮, 結露
condensation cooling　凝縮冷却
condensation heat transfer　凝縮熱伝達
condensation nucleus　凝結核
condensation polymer　縮重合体
condensation polymerization　縮重合
condensation rate　復水率
condense　動凝縮する, 液化する
condenser　コンデンサ, 凝縮器, 復水器, 蓄電器, 集光レンズ
condenser motor　コンデンサモータ【単相誘導電動機の一種】
condenser relief valve　復水器安全弁
condenser tube brush　チューブブラシ【復水器管掃除用具】
condensing engine　復水器付き機関
condensing plant　復水装置, 凝縮装置
condensing pressure　凝縮圧力, 復水圧力
condensing turbine　復水タービン【タービンの排気を大気に放出しないで, 復水させるタービン】
condensing unit　凝縮器を持つ冷凍ユニット
condensive load　進相負荷【進み電流負荷】
condition　状態, 条件

(関) design condition：設計条件
condition curve　状態曲線〈蒸気タービン〉
condition diagnosis technique system　自動診断システム
conditioning　調節(整), 貯蔵, 保管
conditioning tank　調整タンク
conduct　動案内する, 実行する, 導く　名案内, 行為, 行動, 指揮
conduct of trials　試験の実施
conductance　コンダクタンス
conducting surface　伝熱表面
conduction　伝導, 導通
conduction of heat　熱伝導
conductive　形伝導性の
conductive part　導電部
conductivity　電気伝導度, 伝導性, 伝導度, 電気伝導率, 導電率, 誘電率
conductometer　伝導度測定器
conductor　導体, 伝導体, 電線, 導管
conduit　管路
conduit tube　導管
cone　コーン, 円すい, 炎心
(関) right circular cone：直円すい
cone angle　円すい角
cone brake　円すいブレーキ
cone clutch　円すいクラッチ
cone end　円すい継手
cone friction clutch　円すいクラッチ
cone gauge　コーンゲージ
cone journal　円すい継手
cone key　円すいキー
cone pulley　円すいベルト車
cone screw　テーパねじ
confidence limits of availability　安全使用限界
configuration　配置, 構造, 構成
confine　動限定する, 閉じ込める
confirm　動確認する
conform　動順応する・させる, 適合する・させる
conformability　順応性, 正角性
conformal　形等角の, 共形の, 正角の
conformal mapping　等角写像
conformity　一致性, 適合
confuse　動混同する
congruence　合同, 適合
conical　形円すいの
conical cam　円すいカム
conical coupling　円すい継手
conical friction clutch　円すいクラッチ
conical pendulum governor　振り子調速機
conical plug　円すいプラグ
conical roller　円すいころ
conical spring　円すいコイルばね
conical valve　円すい弁
conically seated valve　円すい座弁
conjugate　名共役　形接合した, 対になった, 共役の　動活用させる, 共役させる
conjunction　結合, 接続
(関) in conjunction with：～とともに, ～と連動して, ～と同時に
connect　動接続する, 関連する
connecting　形接続の
connecting pipe　接続管, 連絡管
connecting rod　コンロッド, 連接棒
connecting rod big-end　連接棒大端
connecting rod body　連接棒本体
connecting rod bolt　連接棒ボルト
connecting rod cap　連接棒キャップ
connecting rod large-end　連接棒大端
connecting rod shank　連接棒本体
connecting rod small-end　連接棒小端
connecting rod small-end bush　連接棒小端軸受
connecting rod small-end cap　連接棒小端キャップ
connecting sleeve　接続スリーブ
connection　接続, 結合, 結線
connection box　接続箱
connection diagram　結線図, 接続図
connector　コネクタ, 接続器, 結合子, 接合具
conscious　形意識して
consciously　副意識して, 故意に
consent　名同意, 承諾

動同意する，一致する
consequence 成果,結果,結論,帰結
consequent 結果 形結果として生じる，それに伴う，必然的な
consequential 形結果的，間接的，派生的，重要(大)な
consequently 副従って，その結果
conservation 保存
  conservation law 保存則
  conservation of energy エネルギ保存
  conservation of momentum 運動量保存
conserve 動節約する，保存(護)する
consider 動みなす，検討する，考える，考慮する
considerable 形かなりの，相当な，過大な，考慮される
considerably 副かなり
consist of ~ 動~からなる
consistency 軟度，ちょう度
consistent 形一貫した，変わらない
console コンソール【運転制御盤の一種】
consolidate 動統合する，強化する
consolidation 統合，強化，集約
conspire 動重なる，同時に発生する
constant 形一定の，不変の
 名コンスタント，定数，係数
 関 gas constant：ガス定数
  constant acceleration 等加速度
  constant amplitude 一定振幅
  constant current circuit 定電流回路
  constant enthalpy change 等エンタルピ変化
  constant error 定誤差
  constant flow 一定流量
  constant loss 一定損失
  constant of integration 積分定数
  constant of proportionality 比例定数
  constant pitch 一定ピッチ
  constant pitch propeller 一定ピッチプロペラ，固定ピッチプロペラ
  constant pressure 等圧，定圧
  constant pressure change 等圧変化，定圧変化
  constant pressure combustion 等圧燃焼，定圧燃焼
  constant pressure gas turbine 定圧ガスタービン
  constant pressure line 等圧線
  constant pressure turbocharging system 静圧過給[方式]
  constant rate 一定速度，一定割合
  constant rate pump 定量ポンプ
  constant rating 連続定格
  constant setpoint control 定値制御
  constant speed 定速，一定速度
  constant speed governor 定速調速機
  constant speed motor 定速度電動機
  constant temperature cycle 等温サイクル，定温サイクル
  constant temperature line 等温線
  constant temperature oven 恒温槽
  constant value control 定値制御
  constant velocity 等速，等速度
  constant voltage 定電圧
  constant voltage circuit 定電圧回路
  constant volume and constant pressure cycle 定容定圧サイクル
  constant volume change 定容変化，等積変化
  constant volume combustion 定容燃焼
  constant volume cycle 定容サイクル
  constant volume line 定容線
Constantan コンスタンタン【銅とニッケルの合金で熱電対温度計などに用いる】
constantly 副絶えず,いつも,常に,しばしば
constituent 成分
 関 sulfur constituent：硫黄成分
constitute 動構成する
constitution 組成，構造(成)
constitutional diagram 状態図
constitutional formula 構造式
constitutive equation 構成方程式
constrained 形限定の，強制的な
  constrained motion 拘束運動，束縛運動，限定運動

constraint 拘束，束縛，強制，圧迫
construct 名作成物
　動構成する，組み立てる
construction 構造，作図，建造
　construction plan　構造図
constructor 造船技士
consult 動調べる，参考にする
consulting engineer 技術士
consumable stores 消耗品
consume 動消費する
consumer 消費者
consumption 消費，消費量
　関 fuel consumption：燃料消費量
contact 名接触，接点
　動接触する，接する
　関 in contact with：〜と接触して
　contact angle　接触角
　contact breaker　コンタクトブレーカ，断続器【点火装置の一種】
　contact breaker cam　断続器カム
　contact breaker lever　断続子，ブレーカレバー
　contact cracking　接触分解
　contact erosion　接触侵食
　contact error　接触誤差
　contact face　当たり面，接触面
　contact factor　コンタクト係数
　contact interval　接触率，かみ合い率
　contact maker　接触装置
　contact point　接点，断続器接点
　contact pressure　接触圧力
　contact ratio　接触比，かみ合い率
　contact resistance　接触抵抗
　contact stress　接触応力
　contact surface　接触面
　contact system　接触式
　contact thermal resistance　接触熱抵抗
　contactless switch 無接点スイッチ
contactor 接触器，接触片
　contactor control system　接点制御システム
contain 動含む，入っている
container コンテナ，容器
　container refrigerator　コンテナ用冷凍機
containment 原子炉格納容器，封じ込め，保持，抑制
contaminant 汚染物質
contaminate 動汚す，汚染する
contaminated 形汚染された，汚れた
　contaminated washings　汚濁洗浄水
　contamination of the sea　海洋汚濁
content 量，含有量，容量，容積，内容，目次
　関 sulfur content：硫黄分
context 前後関係
contiguous 形隣接する，近接する
contingent survey 臨時検査
continual 形断続的
continue 動続く・ける
continuity 連続性，導通〈電気〉
　continuity equation　連続の式
　continuity test　導通試験
continuous 形連続的な
　continuous action controller　連続動作調整器
　continuous beam　連続ばり
　continuous bleed type　連続放出型
　continuous blow　連続ブロー
　continuous combustion　連続燃焼
　continuous cruising power　連続定格出力
　continuous current　DC，直流
　continuous diagram　連続線図
　continuous distillation　連続蒸留法
　continuous duty　連続使用
　continuous indicator　連続指圧器
　continuous injection system　連続噴射方式
　continuous line　実線
　continuous liner　一体ライナ
　continuous load　連続負荷
　continuous maximum power　連続最大出力
　continuous operation　連続運転
　continuous output　連続出力
　continuous power　連続出力
　continuous purifying　連続清浄
　continuous rating　連続定格
　continuous running　連続運転
　continuous service output　常用出力

continuous spectrum 連続スペクトル
continuous survey 継続検査
continuous weld 連続溶接
continuous work 連続動作
continuously 副連続的に
contour 輪郭
contour gauge 輪郭ゲージ
contouring control 輪郭制御【NC工作機関連用語】
contra- 接頭「逆」の意
contra flow 逆流
contra flow condenser 逆流復水器
contra propeller コントラプロペラ
Contra Rotating Propeller CRP, コントラロテーティングプロペラ【二重反転プロペラ】
contra rudder コントラかじ【かじ柱の前縁に上下逆の捻れをつけたかじ】
contra turning propeller 反動プロペラ
contract 名契約 動収縮する
contract demand 契約電力
contracted 形収縮した
contracted method 偏縮製図法
contracted vein 縮流
contraction 収縮, 縮流, 絞り
contraction coefficient 収縮係数 同 coefficient of contraction
contraction of area 絞り【引張り試験の測定項目の一つ】
contraction percentage 縮み率【引張り試験の測定項目の一つ】
contraction scale 縮尺
contrast 名コントラスト, 対照, 対比 動対比する, 対照させる
contrate 形横歯の
contrate gear フェースギア, 横歯車
contribute 動寄与する, 貢献する, 一因となる
contributor 誘因, 一因
control 名コントロール, 制御, 管理, 調節, 操縦, 規制, 調整 動コントロールする, 制御する, 操作する, 指揮する

control action 制御動作
control air compressor 制御空気圧縮機
control air dehumidifier 制御空気除湿機
control air pipe 制御用空気管
control air reservoir 制御空気だめ
control board 制御盤
control circuit 制御回路
control combustion 制御燃焼
control computer 制御コンピュータ
control console コントロールコンソール, 制御盤
control desk 制御卓, 制御盤
control device 制御装置
control engineering 制御工学
control equipment 制御装置
control input 制御入力, 操作量
control lever コントロールレバー, 操作レバー
control mechanism 制御機構
control of air-fuel ratio 空気比制御
control of oil discharge 油の排出規制
control oil pipe 制御用油圧管
control oil pump 制御用油ポンプ
control panel 制御盤
control platform 運転台
control point 制御点
control rack 燃料調節ラック
control relay 制御継電器
control rod 制御棒, 燃料調節棒
control room 制御室
control signal 制御信号
control spring 調整ばね
control station 制御場所
control surface 検査面
control switch 制御スイッチ
control system 制御システム, 制御装置, 制御系, 制御方式, 操縦装置, 管理システム
control system failure 制御系統の故障
control theory 制御理論
control transformer 制御変圧器
control unit 制御装置

control valve 制御弁, 調節弁
control winding 制御巻線
control with fixed setpoint 定値制御
controllable 形制御可能な
Controllable Pitch Propeller CPP, 可変ピッチプロペラ
controlled 形制御された
controlled medium 制御媒体
controlled object (system) 制御対象
controlled variable 制御変数, 制御量
controller コントローラ, 制御器
controller lag 調整器遅れ
controller response 制御器応答
controlling 制御
controlling circuit 制御回路
controlling device (equipment) 制御装置
controlling element 調節部, 制御要素
controlling force 制御力
controlling gear 操縦装置
controlling magnet 制御磁石
controlling magnetic field 制御磁界
controlling means 調節部
controlling mechanism 制御機構
controlling spindle 調節棒
controlling torque 制御トルク
controlling valve 制御弁, 操縦弁
controlling variable 操作量
convection 対流
convection boiler 対流形ボイラ
convection current 対流
convection heat transfer 対流熱伝達
convection heater 対流加熱器
convection superheater 対流過熱器, 接触過熱器
convective 形対流の, 伝達力のある
convective acceleration 対流加速度
convective heat transfer 対流熱伝達
convective heating surface 対流伝熱面
convector 対流放熱器
convenient 形便利な, 都合の良い
conveniently 副便利に, 好都合に
convention 条約, 慣例, 慣習, 規約, 規定, 規則, 会議
conventional 形平凡な, 通常型の, 慣例の, 従来の, 伝統的な
converge 動収束する, 集中する
convergence 収束, 集中
convergent 形集中的な, 収束性の
convergent divergent nozzle 中細ノズル
convergent nozzle 先細ノズル, 絞りノズル
convergent pipe 細まり管
converging lens 収束レンズ, 集光レンズ
conversion 変換, 改造, 転換, 発電
関 energy conversion：エネルギ変換
conversion table 換算表
convert 動変える, 転換する, 変換する, 変形する・させる
convert A into B 動AをBに変える
converter コンバータ, 変換器, 転炉, 転換炉, 変流器
converter transformer 整流器用変圧器
convex 形凸面の 名凸形表面
convex cam 凸面カム
convex lens 凸レンズ
convex mirror 凸面鏡
convex roller bearing 凸面ころ軸受
convexity ratio 膨らみ率
convey 動運搬する, 伝達する
conveyer (conveyor) コンベア
conveyer system コンベアシステム
cool 形冷たい 動冷やす, 冷却する
coolant クーラント, 冷媒, 冷却剤, 冷却液, 焼入れ液
coolant tank 冷却水タンク
coolant temperature 冷媒温度
cooler クーラ, 冷却器, 冷房機
cooling 冷却 関 slow cooling：徐冷
cooling air duct 導風板
cooling apparatus 冷却装置
cooling area 冷却面
cooling blade 冷却式羽根
cooling capability 冷却性能
cooling capacity 冷却能力
cooling coil 冷却コイル, 冷却管

cooling effect　冷却効果
cooling fan　冷却ファン
cooling fin　冷却ひれ
cooling fluid flow　冷却液循環量
cooling fresh water　冷却清水
cooling fresh water cooler　清水冷却器
cooling fresh water expansion tank　冷却清水膨張タンク
cooling fresh water pump　冷却清水ポンプ
cooling jacket　冷却ジャケット
cooling junction　冷接点
cooling liquid　冷却液
cooling loss　冷却損失
cooling method　冷却方法
cooling pipe　冷却管
cooling rate　冷却速度
cooling sea water pump　冷却海水ポンプ
cooling spray　クーリングスプレイ【水の蒸発潜熱を利用する噴水】
cooling stress　冷却応力
cooling surface　冷却面
cooling system　冷却系統, 冷却装置
cooling tower　冷却塔
cooling velocity　冷却速度
cooling water　冷却水
cooling water consumption　冷却水消費量
cooling water flow　冷却水循環量
cooling water manifold　冷却水管寄せ
cooling water pump　冷却水ポンプ
cooperage store　小道具倉庫
Cooperative Fuel Research engine　CFR機関
coordinate　形同等の, 等位の, 座標の 名同格, 座標 動同格にする, 調和させる
coordinate axis　座標軸
coordinate plane　座標平面
coordinate system　座標系
COP (**Coefficient Of Performance**)　動作係数, 成績係数
copolymer　共重合体

copolymerization　共重合
copper　銅　記Cu
copper alloy　銅合金
copper base alloy　銅基合金
copper hammer　銅ハンマ
copper loss　銅損【銅線に負荷を掛けたときに生ずる損失】
copper pipe　銅管
copper tube cutter　チューブカッタ
copper wire　銅線
copy　名コピー, 複写, 模写　動コピーする, 複写する, 模写する
copy drawing　複写図
copy machine　コピー機
copying　すり合せ, ならい削り
copying control　ならい制御
cord　コード, ひも
cordage　索具類
cordage store　索具類格納庫
core　コア, 鉄心, 核, 芯, 中子, 中心, 炉心
core bar　心金, 芯
core current　鉄損電流
core diameter of thread　ねじの谷の径
core loss　鉄損
core material　心材
core type　コア型, 内鉄型
core wire　心線
cored steel　中空鋼
cork　コルク
corkscrew rule　右ねじの法則
corn part　コーンパート
corner joint　かど継手
corona discharge　コロナ放電
correct　形正しい, 正確な, 適切な 動訂正する, 直す, 修正する
correct value　正常値, 正しい値
corrected mass flow　修正流量
corrected power　修正出力
corrected speed　修正速度
correcting moment　修正モーメント
correcting signal　修正信号
correction　修正, 補正
correction curve　修正曲線
correction factor　修正率, 補正率
correction factor for current rating

電流定格の補正係数
correction fluid　修正液
corrective　訂正の，修正の
　corrective action　修正動作，修正処理，補正処理
　corrective maintenance　事後保守
　corrective process　修正加工
correctly　副正しく，正確に
correctness　正当性
corrector　修正器
　corrector magnet　修正用磁石
correlation　相関，相関関係，関連性，相関性，対比，関連，相関作用
　correlation coefficient　相関係数
correspond　動相当する，対応する，一致する，匹敵する
corresponding　形一致する，対応する，それ相応の，類似の
　corresponding pressure　相当圧
　corresponding speed　対応速度
correspondingly　副準じて，対応して，同様に，それ相応に
corridor　通路
corrode　動腐食する
corrosion　腐食
　corrosion allowance (margin)　腐食しろ
　corrosion control　防食措置
　corrosion cracking　腐食割れ
　corrosion embrittlement　腐食脆化
　corrosion erosion　浸食
　corrosion fatigue　腐食疲労
　corrosion inhibitor　腐食防止剤
　corrosion limit　腐食限界
　corrosion margin　腐食しろ
　corrosion pit　腐食孔
　corrosion prevention　防食
　corrosion preventive compound　防食剤
　corrosion preventive paint　さび止め塗料
　corrosion product　腐食生成物
　corrosion proof　防食
　corrosion protection　防食
　corrosion protective covering　防食被膜
　corrosion resistance　耐食性
　corrosion resistant aluminium alloy　耐食アルミニウム合金
　corrosion resistant material　耐食性材料
　corrosion resistant protected cover　防食保護被膜
　corrosion resisting alloy　耐食合金
　corrosion test　腐食試験
corrosive　形腐食性の　名腐食剤
　corrosive elements　腐食性成分
　corrosive substance　腐食性物質
　corrosive wear　腐食摩耗
corrosiveness　腐食性
corrugated　形波形の
　corrugated expansion joint　波形継手
　corrugated furnace　波形炉筒
　corrugated sheet　波[形]板
corrugation　起伏，しわ
corundum　コランダム，鋼玉
cosine　コサイン，余弦
cotangent　余接
cotter　コッタ【平形のくさびの一種】
　cotter joint　コッタ継手
　cotter pin　脱出止めピン，割ピン
cotton　コットン，綿，木綿
　cotton canvas　綿キャンバス
　cotton cloth　[木]綿布
Couette flow　クエット流れ
could　助「can」の過去形，～ということもあり得る
couldn't　～のはずがない
coulomb　クーロン【電気量の単位】
　Coulomb force　クーロン力
　Coulomb friction　クーロン摩擦
　Coulomb's law　クーロンの法則
count　動数える
　名計算，放射線測定の単位
counter　カウンタ，計数[管]，計数器　形反対の，逆の
counter-　接頭「反，逆，対応，副」の意
　counter-act　中和する，打ち消す
　counter-action　反作用，中和作用
　counter-balance　つり合わせ，つり合いおもり，平衡力

counter-balance valve　カウンタバランス弁【圧力制御弁の一種】
counter-bore　皿穴, 端ぐり【ボルトなどの頭を沈めるために穴の口を広げ平らな底にすること】
counter-boring　もみ下げ
counter-clockwise　形／副反時計回りの(に)
counter-clockwise rotation　反時計回り
counter-current　対向流, 逆流, 向流
counter-current breaking　逆転制御
counter-current heat exchange　向流熱交換
counter-electromotive force　逆起電力
counter-flow　向流, 逆流
counter-flow heat exchanger　向流形熱交換器, 逆流形熱交換器, 対向流形熱交換器
counter-flow heat transmission　逆流伝熱
counter-pressure brake　背圧ブレーキ
counter-scale　カウンタスケール
counter-shaft　中間軸, 仲介軸
counter-sink　皿穴【皿状の穴】
counter-sinking　皿もみ, 皿座ぐり, 皿穴
counter-voltage　逆電圧
counter-weight　つり合いおもり, バランスウェイト

counting　カウント, 計数, 計算
counting rate meter　比率計

couple　名偶力【大きさが等しく, 向きが反対の一対の力】
動つなぐ, 連結する, 結合する
couple of force　偶力
couple unbalance　偶不つり合い

coupler　カップラー, 連結器, 結合器, 接続器

coupling　カップリング, 継手, 軸継手, 結合, 連結器
coupling agent　結合剤
coupling bolt　継手ボルト
coupling device　連結装置
coupling flange　継手フランジ
coupling rod　連結棒

course　方向, 針路, 経過
covalency　共有結合
covalent　形共有結合の
covalent bond　共有結合

cover　名カバー, おおう物, 表紙, ふた　動おおう, ふたをする
cover glass　保護ガラス

coverage　適用範囲, 有効範囲, 被覆率, 記載内容, 担保範囲
coverage diagram　有効範囲図

covered electrode　被覆電極, 被覆アーク溶接棒

covered wire　被覆線

covering　被覆, 被覆剤
covering plate　かぶせ板

CPP（**Controllable Pitch Propeller**）可変ピッチプロペラ

CPU（**Central Processing Unit**）中央処理装置

crab winch　移動ウインチ

crack　名クラック, 亀裂, 割れ
動［熱］分解する, 亀裂が入る, すきまができる
crack arrester　割れ止め
crack detection　探傷試験
crack growth　亀裂成長
crack initiation　亀裂発生
crack propagation　亀裂伝播
crack stopper　クラックストッパ【亀裂の成長を防止するためにつける穴】
crack test　割れ試験

cracked edge　割れ口
cracked gasoline　分解ガソリン
cracked naphtha　分解ナフサ
cracker　分解装置
cracking　名クラッキング, 亀裂, 割れ, 分解蒸留, 熱分解
形活発な, 速い
cracking coil　分解コイル
cracking distillation　分解蒸留
cracking gas　分解ガス
cracking load　ひび割れ荷重
cracking pressure　クラッキング圧

【リリーフ弁などで流量が認められる圧力】
cramp［**iron**］ かすがい,締付け金具
crane クレーン
crank クランク,回転腕【往復運動を回転運動に,また回転運動を往復運動に変える装置】
 crank angle　クランク角
 crank arm　クランクアーム,クランク腕
 crank-case　クランクケース,クランク室　⊜ crank chamber
 crank-case compression　クランク室掃気
 crank-case explosion　クランクケース爆発
 crank-case scavenging　クランク室掃気　⊜ crank-case compression
 crank chamber　クランクケース
 crank effort　クランク回転力
 crank effort diagram　クランク回転力線図
 crank journal　クランクジャーナル
 crank mechanism　クランク機構
 crank-pin　クランクピン
 crank-pin bearing　クランクピン軸受
 crank-pin bolt　クランクピンボルト,連接棒ボルト
 crank-pin metal　クランクピンメタル
 crank radius　クランク半径
 crank-shaft　クランクシャフト,クランク軸　⊜ crank axle
 crank-shaft bearing　クランク軸受
 crank-shaft deflection　クランクデフレクション,クランク腕開閉量
 crank-shaft distortion dial gauge　クランクデフレクションゲージ
 crank-shaft gear　クランク軸歯車　⊜ crank-shaft pinion
 crank-shaft sprocket　クランク軸スプロケット
 crank-throw　クランクスロー,クランクアーム有効長さ
 crank-web　クランク腕
 crank-web deflection　クランクウェブデフレクション
cranking motor　スタータ,始動電動機
crash　形応急の　名クラッシュ,故障,衝突,破壊,崩壊　動砕ける,衝突する,壊す・れる
 crash astern　クラッシュアスターン,緊急逆転
 crash astern stop　緊急逆転停止
 crash back　全力後進
 crash stop and astern test　急速停止後進試験
 crash stopping　急速停止
crater　クレータ,穴,くぼみ
 crater crack　クレータ割れ
 crater wear　クレータ摩耗
crawling　クローリング【電動機始動時の異常現象】
crazing　クレージング【ひび割れの一種】
creaming　クリーミング【コロイド分散相が不均一になる現象】
create　動作る,生み出す,創造する,引き起こす
creation　創造,発生,生成
creep　クリープ【材料に一定温度で一定荷重が加わった状態で時間の経過と共にひずみが増大する現象】
 creep damage　クリープ損傷
 creep ductility　クリープ延性
 creep elongation　クリープ伸び
 creep failure　クリープ破損
 creep fatigue　クリープ疲労
 creep forging　クリープ鍛造
 creep fracture　クリープ破壊
 creep limit　クリープ限度
 creep rate　クリープ速度,クリープ率
 creep rupture test　クリープ破断試験
 creep strain　クリープひずみ
 creep strength　クリープ強さ
 creep test　クリープ試験
creepage distance for insulation　沿面距離【漏電が起きる最短距離】
creosote oil　クレオソート油【防腐

剤，殺虫剤などに用いる】
crest 頂上，みね
  crest clearance 山頂すきま
  crest of thread ねじ山の長
  crest working reverse voltage ピーク動作逆電力
crevice corrosion すきま腐食
crew 乗組員，船員
  crew member 乗組員
crisis management 危機管理
criteria 「criterion」の複数形
criterion 標準，条件，基準
  criterion numeral 標準数
critical 形臨界の，限界の，危機の，重大な，批判的な
  critical accident 臨界事故
  critical compression ratio 臨界圧縮比
  critical cooling rate (velocity) 臨界冷却速度【焼入れ硬化する最小冷却速度】
  critical damping 臨界減衰
  critical damping constant 臨界減衰定数
  critical density 臨界密度
  critical diameter 臨界直径【焼入れの良否を決める鋼材の大きさ】
  critical dimension 臨界寸法
  critical flow 臨界流
  critical frequency 危険振動数
  Critical Heat Flux CHF，バーンアウト熱流束，限界熱流束
  critical load 臨界荷重，危険荷重
  critical Mach number 臨界マッハ数
  critical number of revolution 危険回転数
  critical point 臨界点
  critical point tester 変態点測定器
  critical pressure 臨界圧力
  critical quality 限界クオリティ
  critical radius 臨界半径
  critical resistance 臨界抵抗
  critical revolution 危険回転数
  critical Reynolds number 臨界レイノルズ数
  critical section 危険断面
  critical slope 臨界こう配
  critical speed 臨界速度，危険速度
  critical state 臨界状態
  critical temperature 臨界温度
  critical torsional vibration 危険ねじり振動
  critical velocity 臨界速度，危険速度
  critical wave length 臨界周波数
criticality 臨界
crocodiling クロコダイリング【熱間加工の際に生じる亀状の亀裂】
cross 形横の，交差した，横切った，反対の
動横切る，交差する，逆らう
  cross-ampereturn 交差アンペア回路
  cross-bar クロスバー，横棒
  cross-board 切換盤
  cross-butt 継手交差部
  cross-compound engine 並列二段膨張機関
  cross-compound steam turbine クロス形蒸気タービン
  cross-compound turbine 並列形タービン，横並び高低圧タービン
  cross-current 横流，逆流
  cross-cut chisel えぼしたがね
  cross-cut file 両丸やすり
  cross-cut saw 横挽きのこ
  cross-draft carburetor 横向き気化器
  cross-feed 横送り
  cross-fin type heat exchanger クロスフィン型熱交換器
  cross-flow 交差流，直交流
  cross-flow fan 横流ファン
  cross-flow heat exchanger 直交流型熱交換器
  cross-flow scavenging 横断掃気
  cross-head クロスヘッド【ピストン棒と連接棒を結ぶ部分】
  cross-head bearing クロスヘッド軸受
  cross-head guide クロスヘッドガイド
  cross-head pin クロスヘッドピン，ガジョンピン，ピストンピン
  cross-head pin bearing クロスヘッ

ドピン軸受
cross-head shoe　クロスヘッドシュー【クロスヘッドにつける滑り金】
cross-head [type] engine　クロスヘッド型機関
cross-joint　クロス継手, 十字継手
cross-lap joint　重ね継手
cross-magnetization　交差磁化
cross-mark　クロスマーク【プロペラに生じる傷の一種】
cross-over　交点, 交差
cross-over frequency　折れ点振動数, 分割周波数, 交差周波数
cross-recessed screw　十字穴付きねじ
cross-scavenging　横断掃気, 横掃気
cross-section　[横]断面
cross-section area　断面積
cross-sectional view　断面図
cross-shaft　たすき棒
cross-talk　クロストーク【回路に生じる不具合の一種】
cross-valve　三方弁
cross-ventilation　通風
cross web　クロスウェブ
crossed belt　交差ベルト
crossed card　クロス示圧図
crossed helical gear　ねじ歯車
crosswise　形反対の, 交差した, 横方向の　副反対に, 交差して
crown　クラウン, トップ, 頂部
crown gear　クラウン歯車, 冠歯車
crown nut　菊ナット
crown wheel　クラウン歯車
crown plate　天井板, 冠板
crowning　クラウニング【歯車加工の一種】
CRP (Contra Rotating Propeller)　コントロロテーティングプロペラ【二重反転プロペラ】
CRT (Cathode-Ray Tube)　ブラウン管, 陰極線管
crucible　るつぼ
crucible furnace　るつぼ炉
crucible induction furnace　るつぼ形誘導炉
crude　形天然のままの, 自然の, 加工していない
crude copper　粗銅
crude oil　原油
crude oil tanker　原油タンカ
crude still　原油スチル【蒸留装置の一種】
cruel　形悲惨な
cruise　動巡航する
cruising　巡航
cruising power　巡航出力
cruising speed　巡航速力, 経済速度
cruising turbine　巡航タービン
crus　動押しつぶす, 粉砕する　名クラッシュ, 粉砕
crusher　粉砕機械
crushing　破砕, 圧壊, 粉砕
crushing strength　破砕強さ
crushing stress　破砕応力
crushing test　破砕試験
crust　クラスト, 堅くなった表面
cryoelectronics　[極]低温電子工学
cryogenic　形極低温の
cryology　低温研究, 氷雪学
cryometer　低温計
cryopump　低温ポンプ
cryostat　クライオスタット【断熱容器】
crystal　クリスタル, 結晶
crystal control alloy　結晶制御合金
crystal grain　結晶粒
crystal lattice　結晶格子
crystal oscillator　水晶発振器
crystal oven　クリスタルオーブン【温度制御炉の一種】
crystalline structure　結晶組織
crystallization　結晶, 結晶化
crystallize　結晶化する, 具体化する
CT (Computed Tomography)　コンピュータ断層撮影
CT (Current Transformer)　変流器
C-type screw clamp　シャコ万力
cube　立方体, 三乗, 立法, 正六面体
cubic　形体積の, 3次の, 等軸晶系の
cubic meter　立方メートル

cubic lattice　立方格子
cubic root　立方根，三乗根
cubical dilatation　体積ひずみ
cubical expansion　体積膨張
cubical expansion coefficient　体積膨張係数
cubicle　キュービクル，小部屋
cumulative　形累積の，集積した，次第に増加する
cumulative fatigue failure　累積疲労破壊
cup　カップ，くぼみ
cup chuck　ベルチャック
cup grease　カップグリース
cup leather　Uパッキン
cup packing　カップパッキン
cup-shaped　形カップ状の
cupola　キューポラ，溶銑炉【銑鉄を溶融する炉】
cupping test　カップテスト【材料の絞り性能試験の一種】
cupro nickel　白銅【耐食性，高温強さの大きい銅合金】
cure　名硬化　動硬化する・させる
curie　キュリー【放射能の単位】
curing　硬化
curing time　硬化時間，成形時間
curl　カール，渦，回転
curling　カーリング【縁を丸める加工法】
current　形今の，最新の
名流れ，電流　関 direct current：直流／alternating current：交流
current [carrying] capacity　電流容量
current carrying conductor　通電導体
current carrying part　導電部
current coil　電流コイル
current control　電流調整
current density　電流密度
current [flow]　電流
current gain　電流利得
current limit relay　限流継電器
current limiter　電流制限器
current limiting fuse　限流ヒューズ
current limiting reactor　限流リアクタ
current meter　流速計
current plate　整流板
current relay　電流継電器
current track　流路
Current Transformer　CT，計器用変流器
current variation　電流変動
currently　副今は，広く，一般に
Curtis stage　カーチス段
Curtis turbine　カーチスタービン【衝動タービンの一種】
curvature　曲率，屈曲，曲がり
curve　名カーブ，曲線　動曲げる
curve of center of buoyancy　浮心曲線
curve of extinction　減衰曲線
curve of metacenter　メタセンタ曲線
curve of stability　復原力曲線
curved beam　曲がり梁(ばり)
curved line　曲線
curved pipe　曲がり管
curved surface　曲面
curvilinear motion　曲線運動
cushion　クッション
cushion rubber　緩衝ゴム
cushion valve　緩衝弁
cushioning　緩衝作用
cushioning material　緩衝材
cut　動切る，切断する　名切断
関 finishing cut：仕上げ削り
cut across　動鎖交する
cut in [on]　動割り込む
cut-in pressure　カットイン圧力
cut-off　切断，締切り，遮断
cut-off of injection　噴射の切れ
cut-off port　逃し穴
cut-off ratio　締切比
cut-off tool　突切りバイト
cut-off valve　締切弁
cut open test　切開試験
cut-out　遮断装置，安全開閉器
cut-out governing　締切調速法
cut-out switch　遮断スイッチ
cut tap　切削タップ
cutlery　刃物類
cutter　カッタ，刃，切断機，切削

工具,刃物
cutting 切断,切削,切削加工
  cutting action 切削作用
  cutting allowance 削りしろ
  cutting amount 切削量
  cutting angle 削り角
  cutting area 切削面積
  cutting chip 削りくず
  cutting edge 切り口
  cutting edge chipping チッピング
  cutting efficiency 切削効率
  cutting fluid 切削油剤
  cutting force 切削力,切削抵抗
  cutting load 切削荷重
  cutting motion 切削作動
  cutting nipper ニッパ【切断工具の一種】
  cutting off 突切り
  cutting off tool 突切りバイト
  cutting oil 切削油
  cutting performance 切削性能
  cutting plane 切断面
  cutting pliers ペンチ,プライヤ
  cutting pressure 切削圧力
  cutting process 切削工程,切削方法
  cutting resistance 切削抵抗
  cutting sequence 切断順序
  cutting speed 切削速度
  cutting stroke 削り行程
  cutting tip 切断火口
  cutting tool バイト,切削工具
  cutting work 切削加工
cutwater 水切り
CVD (Chemical Vapor Depositions) 化学蒸着法
cyanidation シアン化,青化法【表面硬化法の一種】
cycle 名サイクル,周期,循環 動循環する
  cycle diagram サイクル線図
  cycle efficiency サイクル効率
  cycle life サイクル寿命
cyclic irregularity 回転不整率
  cyclic irregularity of injection 不整噴射
cyclic load 繰返し荷重

cyclic process 循環過程
cyclic stress 繰返し応力
  cyclic stress-strain diagram 繰返し応力‐ひずみ線図
cycloconverter サイクロコンバータ【電力変換器の一種】
cyclograph サイクログラフ【電磁誘導試験法の一種】
cyclohexanone シクロヘキサノン【溶剤の一種】
cycloid サイクロイド【一直線上の円が回転するとき,円周上の一点が描く軌跡】
  cycloid gear サイクロイド歯車
  cycloid tooth サイクロイド歯形
cycloidal cam サイクロイドカム
cyclone サイクロン【集塵機の一種/低気圧の総称】
  cyclone burner サイクロンバーナ
  cyclone combustor サイクロン燃焼器
  cyclone dust collector サイクロン集塵機
  cyclone separator サイクロン分離器
  cyclone type air cleaner 遠心分離式空気清浄器
cyclotron サイクロトロン【粒子加速装置】
cylinder シリンダ,円筒,円柱,ボンベ
  cylinder arrangement シリンダ配列
  cylinder bank シリンダ列
  cylinder barrel シリンダ胴
  cylinder block シリンダブロック
  cylinder body シリンダ本体
  cylinder bore シリンダ内径
  cylinder bore gauge ボアゲージ【シリンダ内径測定器】
  cylinder bush シリンダライナ
  cylinder capacity 総行程容積
  cylinder casing シリンダケーシング
  cylinder column シリンダ柱
  cylinder constant シリンダ定数
  cylinder cover シリンダカバー
  cylinder cover bolt シリンダヘッ

ドボルト
cylinder cowling　導風板
cylinder cushioning　シリンダクッション
cylinder cut off test　シリンダカットオフ試験
cylinder diameter　シリンダ直径
cylinder distance　シリンダピッチ
cylinder face　シリンダ面
cylinder gas　ボンベガス
cylinder head　シリンダヘッド
cylinder head bolt　シリンダヘッドボルト
cylinder head cover　シリンダヘッドカバー
cylinder head gasket　シリンダヘッドガスケット
cylinder injection of fuel　筒内噴射
cylinder injection system　シリンダ噴射方式
cylinder jacket　シリンダジャケット
cylinder liner　シリンダライナ
cylinder lubrication　シリンダ潤滑
cylinder offset　シリンダオフセット
cylinder oil　シリンダ油
cylinder oil lubricating pump　シリンダ注油ポンプ
cylinder [oil] lubricator　シリンダ注油器
cylinder output　シリンダ出力
cylinder ratio　シリンダ比
cylinder volume　シリンダ容積
cylinder wall　シリンダ壁
cylindrical　形円柱の，円筒形の
cylindrical bearing　真円軸受
cylindrical boiler　丸ボイラ
cylindrical cam　円筒カム
cylindrical coil　円筒コイル
cylindrical coordinate system　円筒座標系
cylindrical friction wheel　円筒摩擦車
cylindrical gauge　円筒ゲージ【限界ゲージの一種】
cylindrical gauge glasses　円筒形ゲージグラス
cylindrical roller bearing　円筒ころ軸受
cylindrical rotor　円筒形ロータ，円筒形回転子
cylindrical shell　円筒シェル，円筒胴
cylindrical spring　筒型スプリング
cylindrical tank　円筒形タンク

D-d

D slide valve D形すべり弁
D-A (Digital-Analog) converter D-A変換器
daily 形毎日の, 日常の 副日々, 常に
 daily tank　小出しタンク
Dalton's law　ダルトンの法則
dam　ダム, 堰(せき)
damage　名ダメージ, 損傷, 損害, 被害　動傷む, 破損する, 損なう, 損傷を与える
 damage function　損傷関数
damaging stress　損傷応力
damp　動弱める, 湿らす　名湿気, 水分
 damp proof machine　耐湿型電機
dampen　動弱める, 湿らせる
damped natural frequency　減衰固有振動数, 減衰自然周波数
damped oscillation (vibration)　減衰振動, 減衰動揺
damped wave　減衰電波
damper　ダンパ, 緩衝器, 制動器, 通風調節装置
 damper gear　ダンパ装置
 damper winding　制動巻線
damping　ダンピング, エネルギ分散, 減衰, 制動
 damping capacity　減衰能力
 damping coefficient　減衰係数
 damping coil　制動コイル
 damping constant　減衰定数
 damping curve　減衰曲線
 damping device　制動装置
 damping factor　減衰率
 damping force　減衰力, 制動力
 damping magnet　制動磁石
 damping oil　制動油
 damping ratio　減衰比
 damping resistance　制動抵抗
 damping ring　制動環
 damping vane　制動羽根
danger　危険

 danger signal　危険信号
dangerous　形危険な
 dangerous cargo　危険貨物
 dangerous section　危険断面
Daniel cell　ダニエル電池
dark current　暗電流
dark flame　暗炎
darken　動暗くする・なる
dash　名ダッシュ, 注入, 衝突, 突進　動ぶつける, 突進(入)する
 dash board　ダッシュボード, 計器盤, 仕切り板
 dash plate　波よけ板
 dash pot　ダッシュポット, 流体緩衝器
data　データ, 資料, 資材
 data communication　データ通信
 data logger　データロガー【計測値などの物理量を継続的に記録する装置】
 data processing system　データ処理システム
date stamp　日付印
dating machine　日付印字機
datum　データ, 資料, 基準点
 datum level　基準面
 datum line　基準線
davit　ダビット【ボートや貨物の上げ下ろし装置】
daylight　昼間, 日光, 日中
dB (decibel)　デシベル【音の強さや電力比を表す単位】
DC (Direct Current)　直流
 DC amplifier　直流増幅器
 DC generator　直流発電機
 DC machine　直流機
 DC motor　直流電動機
 DC potentiometer　直流電位差計
 DC thyristor variable motor　DCサイリスタ可変モータ
De Laval turbine　デラバルタービン
dead　形デッドな, 動きのない, 機能の停止した, 完全な

㊥ go dead：故障する
dead angle　死角
dead beat　無振揺
dead calm　静穏【とても穏やかな状態】
dead center　死点
dead earth　完全接地
dead load　静荷重，死荷重
dead point　死点
dead slow　極微速
dead soft steel　極軟鋼
dead space　空所
dead time　むだ時間，不動作時間
dead weight　静荷重，死荷重，おもり，自重，載貨重量（D/W）
dead weight safety valve　おもり安全弁
dead weight tonnage　載貨重量トン数
dead zone　不感帯
deaerating feed heater　脱気給水加熱器
deaeration　脱気【水中の酸素を除去すること】
deaerator　脱気器，空気分離器
deal　图マツ材，取引
㊥ a deal of 〜：かなりの〜
働扱う，分配する，加える
deal with　働扱う，処理する，対処する
dearth　不足，欠乏
debooster cylinder　減圧シリンダ
debug　图デバッグ，修正
働デバッグする【コンピュータのプログラムを検査し，その誤りを修正すること】
debugging　デバッグ，デバッギング，手直し，虫取り
decade counter　十進カウンタ
decanter　デカンタ【ガラス瓶の容器】
decarburization　脱炭
decatron　デカトロン，計数放電管
decay　图減衰，崩壊　働崩壊する，減衰する，朽ちる，腐る
decay heat　崩壊熱

decelerate　働減速する・させる
deceleration　デセレレーション，減速，減速度
㊉ acceleration：加速，加速度
deceleration flow　減速流
deceleration valve　デセレレーション弁【流量を減速させる弁】
decelerator　減速器
decelerometer　減速計
dechlorination　脱塩素
deci-　接頭「十分の一」の意
decibel　デシベル，dB【音の強さや電力比を表す単位】
decide　働決定する，決心する
decimal　图十進法，小数
形十進数の，小数の
㊥ first decimal place：小数第1位
decimal notation　十進法，十進法表記
decimal point　小数点
decision　決定，判断
deck　デッキ，甲板
deck crane　甲板クレーン
deck department　甲板部
deck machinery　甲板機械
declare　働明らかにする，表す，明言する
declared power　定格出力
declared speed　定格回転数
declination　偏角，傾斜，磁針偏差，回旋
declinometer　偏角計
declivity　勾配，下り勾配
declutch　働クラッチを切る
decoloring（decolorization）　脱色
decolorize　働脱色する，漂白する
decolorizer　脱色剤
decompose　働分解する，分析する，腐敗する
decomposition　分解，変質，溶解
decomposition combustion　分解燃焼
decomposition explosion　分解爆発
decomposition of forces　力の分解
decomposition point　分解点
decomposition pressure　分解圧
decomposition voltage　分解電圧

decompression 減圧
  decompression device　減圧装置
  decompression valve　減圧弁
decontaminate 動汚染を除く
decontamination 汚染除去
decorate 動飾る
decorative plating　装飾めっき
decoupling 減結合, 非干渉化
decrease 動減少する, 減る 名減少
decremeter 減衰計, 減幅計
dedendum 歯元, 歯元のたけ
  dedendum angle　歯元角
  dedendum circle　歯元円
deduce 動推定する, 割り出す
dedusting 脱塵, 除塵
dedustor 脱塵装置
deenergize 動通電しない
deep 形深い
  deep notch test　ディープノッチ試験
  deep slot motor　深溝電動機
  deep slot squirrel cage motor　深溝かご型電動機
  deep tank　深水タンク
defeat 動無効にする
defect 欠陥, 欠点, 不足, きず
defective 形欠陥のある, 不完全な
defense in depth　多重防護
define 動定義する
definition 定義, 限定
deflagration デフラグレーション, 爆燃, 暴燃
deflect 動たわむ, そらす
deflecting plate　そらせ板
deflection デフレクション, たわみ, ゆがみ, 偏向, 偏差
  deflection angle　たわみ角, 偏向角, 転向角
  deflection coil　偏向コイル
  deflection curve　たわみ曲線
  deflection gauge　たわみ計
  deflection method　偏位法
  deflection plate　偏向板
  deflection wheel　そらせ車
deflectometer　たわみ計
deflector そらせ板, 水切りリング【ポンプの軸受シール】

defoaming agent　あわ止め剤
deform 動変形させる
deformation 変形, ひずみ
  関 mode of deformation：変形モード
  deformation heat　変形熱
  deformation processing　塑性加工
deformeter 応力計, ひずみ測定器
deforming 変形性
  deforming stress　変形応力
deformity 変形
defrost 名デフロスト, 除霜
  動霜を取る, 解凍する
  defrost system　除霜装置
defrosting 除霜, 解凍
deg. 略 degree
degas 動脱気する, ガスを抜く
degasification 脱気, 脱ガス
degassing 脱ガス, 脱気
degauss 動消磁する
degradation 劣化, 粉化
  関 size reduction：粉砕
degrade 動劣化する, 低下する
degrease 動脱脂する
degreasing 脱脂
degree 度, 角度, 程度, 温度, 度合い
  degree Celsius　摂氏温度
  degree Fahrenheit　華氏温度
  degree of burn-up　燃焼度
  degree of confidence　信頼度
  degree of constant volume　等容度
  degree of dispersion　分散度
  degree of dissociation　解離度
  degree of dryness　乾き度
  degree of elasticity　弾性率
  degree of explosion　最高圧力比
  degree of finish　仕上げ程度
  degree of freedom　自由度
  degree of hardness　硬度
  degree of plasticity　可塑度
  degree of reaction　反動度
  degree of reliability　信頼度
  degree of saturation　飽和度
  degree of stability　安定度
  degree of subcooling　サブクール度, 過冷度
  degree of superheat　過熱度

dehumidification 72

degree of supersaturation　過飽和度
degree of undercooling　過冷度
degree of uniformity　均等度
degree of vacuum　真空度
degree ruler　角度計
dehumidification　除湿, 減湿
dehumidification machine　除湿機
dehumidifier　除湿器
dehumidify　動除湿する
dehydrating agent　脱水剤
dehydration　脱水
dehydrator　脱水機, 脱水剤, 乾燥機
deicing　氷結防止
delamination　剥(はく)離, 表層剥離, 層間剥離
delay　名遅延, 遅れ
　動遅らせる, 延ばす
delay time　遅れ時間
delayed fracture　遅れ破壊
delete　名デリート, 削除, 除去, 消去　動削除する
deliver　動放出する, 噴出する, 引き渡す, 手放す, 届ける, 産出する
Delivered HorsePower　DHP, 伝達馬力, 伝達出力
delivered output　伝達出力
delivery　出口, 吐出, 放出
　動放出する
delivery head　吐出し水頭
delivery pipe　出口管, [燃料]噴射管
delivery pressure　出口圧, 吐出圧力
delivery pump　流出ポンプ
　関 variable delivery pump：可変流出ポンプ
delivery ratio　給気比
delivery stroke　送出し行程
delivery valve　吐出弁, 出口弁
delta　δ(デルタ)
delta connection　デルタ結線, Δ結線
　関 star connection：Y(星形)結線
delta delta connection　三角三角結線
delta function　δ(デルタ) 関数
delta iron　δ(デルタ) 鉄
delta metal　デルタメタル【四六黄銅の一種】
delta star connection　三角星形結線
delta wing　デルタ翼
demagnetization　脱磁, 消磁, 減磁
demagnetization ampere-turn　減磁アンペア回数
demagnetize　動消磁する
demagnetizer　消磁器, 脱磁器
demagnetizing effect　減磁作用
demagnetizing factor　減磁率
demagnetizing field　減磁界
demand　名デマンド, 要求
　動要求する, 必要とする
demineralization　脱塩
demineralizer　純水装置, 脱塩装置
demister　デミスタ【乾燥器の一種】
demodulation　復調
demodulator　復調器
demolition　破壊, 取り壊し, 解体
demonstrate　動説明する, 証明する, 明らかにする
demulsibility　抗乳化性, 抗乳化度
demulsifier　抗乳化剤, 乳化破壊剤
demulsifying agent　抗乳化剤
denaturant　変性剤
denaturation　変性, 変質
denatured alcohol　変性アルコール
dendrite　デンドライト, 樹枝状結晶
denitration　脱硝
denitrification　脱硝, 脱窒素
denominator　分母
dense　形濃い
dense matrix　密行列
denser　形「dense」の比較級
densimeter　密度計, 比重計
densitometer　濃度計
density　濃度, 密度【単位体積あたりの質量】
density meter　密度計
deny　動否定する, 打ち消す
deodorant　脱臭剤, 防臭剤
deodorization　脱臭
deoxidation　脱酸[素]
deoxidization　還元, 脱酸
deoxidize　動脱酸素する, 還元する
deoxidized copper　脱酸銅

deoxidizer 脱酸剤
deoxidizing agent 脱酸剤
depart 動出発する,外れる
departure 出発,東西距
  Departure from Nucleate Boiling DNB,核沸騰限界
depend 動依存する
  depend [up] on 〜 動〜によって決まる,〜による
dependence 依存,従属
dependent 形依存する,左右される,従属の,依存性の
deplete 動減少させる,消耗させる,使い果たす,空にする,減らす
depletion 減少,空乏,消耗,劣化,枯渇,除去,破壊
  depletion layer 空乏層
depolarization 減極,消極,脱分極
depolarizer 減極剤,消極剤
deposit 名デポジット,堆積物,沈殿物,付着した燃焼生成物
  動堆積する,沈殿する,付着する
deposited metal 溶着金属
deposition 堆積[物],沈殿[物],析出,蒸着
  deposition efficiency 溶着率
  deposition rate 溶着速度
depress 動押し下げる,下に押す
depressant 降下剤,抑制剤,低下剤
depression 減圧,くぼみ,機能低下,沈下,降下,低下
depressor bar type recorder 打点式記録計
deprive 動奪う,与えない
  deprive A of B AからBを奪う
depth 深さ
  depth gauge 深さゲージ【穴などの深さを測る測定器】
  depth of water 水深
derivation 誘導,導出,微分
derivative 導関数,微分,誘導体
  derivative action 微分動作
  derivative action time 微分時間
  derivative control 微分制御
  Derivative control action D動作,微分動作
  derivative element 微分要素
derive from 動由来する
derived unit 組立単位
derusting 脱錆,除錆,さび除去法
desalinization 脱塩,除塩
desalting 脱塩,淡水化
describe 動説明する,記述する,描く,描写する,述べる
describing circle ころがり円
description 記述,品目,銘柄
deserve 動〜に値する
desiccant 脱湿剤
desiccating agent 乾燥剤
desiccator デシケータ,乾燥器
design 動設計する
  名デザイン,設計,計画,目的
  design condition 設計条件
  design drawing 設計図
  design point 設計点
  design pressure 設計圧力
  design specification 設計仕様書
  design thickness 設計厚さ
designate 動指定する,任命する
designation 記号
designed speed 計画速力
designer 設計者
desirable 形望ましい
desire 動望む,願う
  名要求,願望,欲求,要望
desired 形望まれた,要求された,適切な
  desired value 目標値
  desired angle 目標(希望)角度
desorption 脱着,脱離
despite 前〜にもかかわらず
destroy 動破壊する,損なう
destruction 破壊
destructive 形破壊的な,有害な
  destructive distillation 乾留
  destructive inspection 破壊検査
  destructive test 破壊試験
desulfurization 脱硫
  desulfurization equipment 脱硫装置
desuperheated steam 緩熱蒸気
desuperheater 過熱低減器
detach 動引き離す,取り外す

detail 名詳細，細目，細部
　関 in detail：詳細に
　動詳しく述べる，列挙する
　detail design　詳細設計
　detail drawing　詳細図
detailed 形詳細な
detect 動検出する，検知する
detectable 形検知可能な
detecting element　検出部
detection　探知，発見，検波
　関 early detection：早期発見
　detection limit　検出限界
detector　検出器，検知器，探知器
　detector demodulator　復調器
detention　貯留，滞留
detergency　洗浄性，洗浄力
detergent　洗剤，洗浄剤，界面活性剤
　detergent dispersant　清浄分散剤
deteriorate 動劣化する，低下する
deterioration　劣化，変質，低下
　関 aged deterioration：経年変化，経年劣化
determinant　行列式
determination　決定
determine 動決める，決定する
detonation　デトネーション，異常爆発，爆燃，瞬間的燃焼
deuterium　重水素
develop 動発生する，発達する，開発する，展開する，明らかになる
developed area ratio　展開面積比
development　開発，発達，展開，現像
　development drawing　展開図
deviation　偏差，制御偏差，偏り
　関 standard deviation：標準偏差
　deviation angle　偏向角
　deviation from circular form　真円度
deviator　偏差成分
deviatoric stress　偏差応力
device　装置，機器，器具，素子
　関 control device：制御装置
devise 動工夫する，考案する
dew　露
　dew point　露点
　dew point hygrometer　露点湿度計
　dew retardation　防湿，防露

dewater 動脱水する，排水する
　名脱水機
dewaterer　脱水機
dewing　結露
dezincification　脱亜鉛現象
　dezincification corrosion　脱亜鉛腐食
DHP (Delivered HorsePower)　伝達馬力
di- 接頭「2つの」「二重の」の意
diagnosis　診断
　関 failure diagnosis：故障診断
diagnostic system　診断システム
diagonal 形対角線の，斜めの
　名対角線
　diagonal flow pump　斜流ポンプ
　diagonal line　斜線，対角線
　diagonal pitch　斜めピッチ
diagram　図，図表，曲線，線図
　diagram efficiency　線図効率
　diagram factor　線図係数
　diagram work　線図仕事
diagram [m] atic sketch　略図
dial　ダイヤル，文字盤，目盛盤
　dial ca[l]liper　ダイヤルキャリパ
　dial gauge　ダイヤルゲージ
　dial indicator　ダイヤルインジケータ
　dial thermometer　ダイヤル式温度計，指針形温度計
dialing system　ダイヤル方式
diamagnetic 形反磁性の
　diamagnetic material (substance)　反磁性体
diamagnetism　反磁性
diameter　直径
　diameter ratio　内外径比
diametral clearance　直径すきま
diametral pitch　直径ピッチ
diamond　ダイヤモンド
　diamond coil　ひし形巻線
　diamond indenter　ダイヤモンド圧子，ダイヤモンド圧子硬度計
diaphragm　ダイアフラム，隔膜，膜板，仕切り板，振動板，絞り
　diaphragm manometer　ダイアフラム圧力計
　diaphragm meter　ダイアフラム流

量計
diaphragm plate　ダイアフラム板
diaphragm pump　膜ポンプ
diaphragm type pressure gauge　ダイアフラム圧力計
diaphragm vacuum gauge　隔膜真空計
diaphragm valve　ダイアフラム弁

diatomaceous earth heat insulating material　けい藻土保温材

die　ダイ，ダイス，打ち型，金型
die casting　ダイカスト【金型に溶融金属を圧入して鋳物を作る方法】
die casting alloy　ダイカスト合金
die forging　型鍛造
die quenching　ダイクエンチング，プレス焼入れ

dielectric　形誘電性の　名誘電体
dielectric breakdown　絶縁破壊
dielectric constant　誘電率
dielectric heating　誘電加熱
dielectric loss　誘電損
dielectric material　誘電材料
dielectric strain　誘電ひずみ
dielectric strength　絶縁耐力

dies　ダイス

diesel　ディーゼル[機関]
diesel boat　ディーゼル船
diesel cycle　ディーゼルサイクル
diesel engine　ディーゼル機関
　関 gasoline engine：ガソリン機関
diesel generator　ディーゼル発電機
diesel index　ディーゼル指数
diesel knock　ディーゼルノック
diesel oil　ディーゼル油

differ　動異なる，違う
differ from　動～と異なる

difference　差，違い，相違

different　形違う，異なる，別の

differential　形差動の，特異な，区別の，差のある，微分の
名差動装置，差，微分
differential accumulator　差動アキュムレータ
differential accuracy　精度差

differential action　微分動作
differential aeration　気曝差，通気差
differential amplifier　差動増幅器
differential block　差動滑車
differential brake　差動ブレーキ
differential calculus　微分学，微分法
differential capacitance　微分キャパシタンス
differential circuit　微分回路
differential compound motor　差動複巻電動機
differential connection　差動接続
differential control　微分制御
differential controlling valve　差動制御弁
differential cylinder　差動シリンダ
differential equation　微分方程式
differential flow meter　差圧式流量計
differential gear　差動歯車，差動歯車装置
differential generator　差動発信器
differential governor　差動調速器
differential manometer　示差マノメータ，差圧計
differential measurement　示差測定法
differential method　差動法，示差法，微分法
differential micrometer　差動マイクロメータ
differential motion　差動運動
differential piston　差動ピストン
differential plunger pump　差動プランジャポンプ
differential pressure　圧力差，差圧
differential pressure gauge　差圧計
differential pressure liquid level gauge　差圧液面計
differential pressure switch　差圧スイッチ
differential pressure type flowmeter　差圧式流量計
differential pulley　差動滑車
differential relay　差動継電器
differential screw　差動ねじ
differential transformer　差動トランス，差動変圧器

differential valve　差動バルブ
differential winding　差動巻線
differentially　副差動的に
differentiate　動区別する，差別する，微分する
differentiating circuit　微分回路
differentiation　差別，差別化，微分，微分回路，微分法，分化，区別，分別，差異
differentiation circuit　微分回路
differently　副別に，それとは違って，異なって
difficult　形難しい，困難な
difficulty　困難，難しさ
diffract　動回折する，分散する，分解する
diffraction　回折
diffractometer　回折計
diffuse　形拡散した，散乱性の　動拡散する・させる，発散する・させる
diffuse reflection　拡散反射
diffuser　ディフューザ，拡散器，散布器，案内羽根
diffuser efficiency　ディフューザ効率，拡散効率
diffuser pump　拡散ポンプ
diffuser vane　ディフューザ案内羽根，ディフューザ翼
diffusion　拡散
diffusion burning (combustion)　拡散燃焼
diffusion equation　拡散方程式
diffusion factor　拡散係数
diffusion pump　拡散ポンプ
diffusion velocity　拡散速度
diffusive　形広がる，拡散性の
diffusive burning　拡散燃焼
diffusivity　拡散係数，拡散率，温度伝導率，拡散性
dig　動掘る
digest　動消化する，整理する，要約する，温浸する
digit　ディジット，桁，数字
digital　形ディジタルの，計数形の，数字で表示する
名ディジタル，計数形
digital circuit　ディジタル回路
digital computer　ディジタルコンピュータ，ディジタル計算機
digital control　ディジタル制御
digital instrument　ディジタル計器，数字式計器
digital meter　ディジタル計器
digital representation　ディジタル表現
digital signal　ディジタル信号
Digital-to-Analog converter　DA変換器
digitize　動ディジタル化する，数字で表す，計数表示する
digitron　ディジトロン，蛍光表示管
dilatation　膨張，拡張，膨張率
dilation　膨張，拡張，伸び，拡大
dilatometer　膨張計
dilution　希釈，希釈度
dilution flow point　希釈流動点
dilution law　希釈律
dilution ratio　希釈率
dim　形かすんだ，薄暗い，かすかな　動かすむ
dimension　寸法，次元，大きさ
dimension line　寸法線
dimensional　形寸法の
dimensional analysis　次元解析
dimensional tolerance　寸法公差，寸法許容差
dimensionless　無次元
dimensionless expression　無次元式
dimensionless number　無次元数
diminish　動減る・らす，縮小する・させる
diminished pressure　減圧
dimmer switch　調光スイッチ
diode　ダイオード，二極体
関 light emitting diode：発光ダイオード
dioxide　二酸化，二酸化物，過酸化物
関 carbon dioxide：二酸化炭素
dioxin　ダイオキシン【猛毒の化学物質】
dip　傾斜，傾斜角，沈下，くぼみ，

たるみ 動浸す,沈下する,沈む,下がる,入れる
dip brazing　どぶ付け
dip feed lubrication　油浴潤滑
dipmeter　傾斜計
dipole　ダイポール,双極子
dipole moment　双極子モーメント
dipper　ひしゃく
dipstick　計量棒,検油棒
direct　形直接の,まっすぐな
動向ける,注ぐ,指導する,指図する,指示する,案内する
direct acting engine　直動機関【ピストン運動をクランク機構で直接回転運動に変える機関】
direct acting governor　直動調速機
direct axis　直軸
direct bilge suction　独立ビルジ吸込【機関室より直接ビルジを排出できるビルジ吸込】
direct connection　直結
direct contact heat exchanger　直接接触式熱交換器
direct control　直接制御
関 indirect control：間接制御
direct coupled　形直結した
direct coupled turbine　直結タービン
Direct Current　DC,直流
関 alternating current：AC,交流
direct digital control　直接計数制御
direct drive　直結駆動
direct expansion system　直接膨張式
direct expansive refrigeration　直接膨張式冷凍
direct illumination　直接照明
direct injection type combustion chamber　直接噴射式燃焼室
direct load　直接荷重
direct measurement　直接測定
関 inverse proportion：反比例
direct proportion　正比例
direct reading　直読式
direct reversing　自己逆転【軸系に逆転装置を設けることなくカム軸を移動することにより機関の回転を変えること】

direct short　直接短絡,直接接地
direct spring loaded safety valve　ばね安全弁
direct transfer type heat exchanger　熱通過式熱交換器
direct transmission　直接伝動
directed line segment　有向線分
direction　方向,指示,指図
direction finder　方向探知器
direction indicator　方向指示器
direction of rotation　回転方向
directional control valve　方向制御弁,切換え弁【流体の流れを制御する弁】
directly　副直接に
directly coupled turbine　直結形タービン
director　ディレクタ,監督,導波器,管理者
director valve　振分け弁【蓄圧式燃料噴射用の弁】
dirt　泥,汚れ,ほこり,ごみ
dirty　形汚れた　動汚す,汚染する
dirty oil　汚油
dirty tank　ダーティタンク,汚油タンク
dis-　頭動詞につけて「反対の動作」を示す,名詞・形容詞につけて「不～」「非～」の意
disable　動無力にする,出来なくさせる,使用不可にする,機能を無効にする
disadvantage　短所,不便さ,不利,欠点　反 advantage：長所,有利
disagree　動食い違う,一致しない
disappear　動消す,消える
disappoint　動がっかりする・させる
disapprove　動不賛成である,非とする,認可(承認)しない
disassemble　動分解する,解体する
disassembly　分解,取外し
disc　ディスク,円板,(清浄機の)分離板
disc and drum turbine　円板胴タービン
disc area　全円面積【プロペラが

回転する円の面積】
- disc brake　円板ブレーキ
- disc cam　円板カム
- disc clutch　円板クラッチ
- disc coil　円板コイル
- disc friction　円板摩擦
- disc piston　円板ピストン
- disc rotor　円板回転子，円板羽根車
- disc type　ディスク形，円板形
- disc valve　円板弁
- disc wheel　翼車〈タービン〉

discard　動捨てる，放棄する

discharge　動放電する，排出する，放出する　名放電，排出，放出，吐出，流出，流量
- discharge angle　流出角
- discharge coefficient　流量係数
- discharge current　放電電流，放出流
- discharge cut-off voltage　放電終止電圧
- discharge direction　流量方向
- discharge head　吐出しヘッド
- discharge pipe　吐出管，吐出し管
- discharge plug　放電点火プラグ
- discharge port　吐出口
- discharge pressure　吐出圧力
- discharge regulator　流量調整装置
- discharge side　吐出側
- discharge tube　放電管
- discharge valve　吐出弁【流体を吐き出す側の弁】
- discharge voltage　放電電圧

discharged battery　放電した電池

discipline　名訓練，規律　動訓練する

discolor　動変色する・させる

discomfort index　不快指数

disconnect　動外す，分離する

disconnecting switch　ディスコン，断路器【高電圧回路に用いるスイッチ】

discontinuity　不連続，欠陥

discontinuous　形不連続の
- discontinuous action　不連続動作
- discontinuous combustion　息づき燃焼
- discontinuous control　不連続制御
- discontinuous flow　不連続流

discord　不一致

discover　動発見する

discrimination　区別，差別，識別，弁別，分解能，相違点，感度限界

discriminator　弁別器，選別器

disengage　動外す，離す，解放する，クラッチを切る

disequilibrium　不平衡，不均衡

dished end　さら形鏡板

disinfection　消毒，殺菌

disk　回 disc

dilocation　転位，食違い
- dislocation hardening　転位強化

dislodge　動取り外す，除去する

dismantle　動取り除く，分解する

dismount　動分解する，取り外す，降りる，降ろす

dismounting　分解，解体

dispense　動調合する，施す，配分する

dispersal　分散，散布

dispersant　名分散剤　形分散性の

disperse　動分散する

dispersed　形分散した
- dispersed control system　分散制御方式
- dispersed flow　分散流

dispersion　分散，分光，散乱，ばらつき
- dispersion strengthened alloy　分散強化合金【耐クリープ，耐熱衝撃に強い材料】

displace　動置き換える，移す，動かす，～に取って代わる

displaced card　引き伸ばし指圧図

displaced volume　排水容量【船が排除した水の容量】

displaced weight　排水重量【船が排除した水の重量】

displacement　変位，偏位，置換，取替え，押しのけ量，排水量，排気量，排除
- displacement compressor　容積形圧縮機

displacement curve　排水曲線，排気曲線
displacement diagram　変位線図
displacement function　変位関数
displacement plating　置換めっき
displacement thickness　排除厚さ
displacement tonnage　排水トン数【満載時に船が排除する水の重さ】
displacement work　排除仕事
display　動表示する　名ディスプレイ，表示，表示装置
display control switch　表示制御スイッチ
display information　表示情報
display lamp panel　表示ランプ盤
disposal　処理，処分
disregard　動無視する，軽視する
disrupt　動中断させる，破壊する
disruptive　形破壊的な，分裂性の
disruptive voltage　破壊電圧
dissimilar　形似ていない
dissipate　動放散する・させる
dissipation　散逸，消失，放散，損失
dissipation factor　損失係数
dissipative heat loss　放熱損失
dissociation　分離，解離
dissociation pressure　解離圧
dissolution　融解，溶解
dissolve　動溶かす，溶解する
dissolved acetylene　溶解アセチレン
dissolved gas　溶解ガス，溶存ガス
dissolved oxygen　溶存酸素
　関 concentration of dissolved oxygen：溶存酸素濃度
dissolved solid　溶解固形物
distance　距離，間隔，程度
distance piece　ディスタンスピース【間隔を保つために挿入する物体】
distant　形遠く離れた
distant control　遠隔制御
distemper　水性塗料
distill　動蒸留する
distillate　蒸留[物]，留[出]分
distillate fuel　留出燃料油

distillation　蒸留，蒸留作用
　関 vacuum distillation：真空蒸留／cracking distillation：分解蒸留
distillation gas　蒸留ガス
distilled water　蒸留水
distiller　蒸留器
distilling condenser　蒸留復水器
distilling plant　造水装置
distinct　形明確な，異なる，別個の
distinctive　形独特な，特徴的な
distinguish　動区別する，分類する，特徴づける
distort　動ゆがめる，変形させる，ひずませる，ねじる
distorted pattern　歪模様
distortion　ねじれ，ゆがみ，変形
distortion factor　ゆがみ率
distortion tolerance　許容ゆがみ
distress　遭難，災難，困難
distress signal　遭難信号
distribute　動分配する，分類する，散布する，配布する，分布する
distributed defect　分布欠陥
distributed header　分配管寄せ
distributed inductance　分布インダクタンス
distributed load　分布荷重
　関 concentrated load：集中荷重
distributing valve　分配弁
distribution　分配，配分，配電，配水，分布
distribution board　分電盤，配電盤
distribution box　分電箱
distribution function　分布関数
distribution line　配電線
distribution panel　分電盤，配電盤
distribution system　配電方式
distributor　分配器，配電器
district heating　地域暖房
disturb　動かき乱す，阻害する
disturbance　乱れ，外乱，妨害，障害
dither　ディザ，震え，振動
dive　動潜る　名潜水
diver　潜水夫
divergence　発散，拡散，多様化，偏差，分岐，逸脱，相違

divergence line　発散線
divergent　形発散性の，放射状の，分岐している
　divergent current (flow)　広がり流れ
　divergent nozzle　末広ノズル【圧力を減らすため後方の広がったノズル】
　divergent pipe　広がり管
diverging duct　末広ダクト
diverging lens　発散レンズ
divert　動転送する，迂回させる，そらす
divide　動分割する，分類する
　divide sleeve bearing　割り軸受
divided chamber type combustion chamber　副室式燃焼室
divided circumferential pitch　周割りピッチ
divided combustion chamber　副室式燃焼室
divided flow turbine　分流タービン
divider　ディバイダ，仕切り，分割器，除算器，割算器
division　割算，課，部，隔壁，仕切り，分割，分配，部分，部門，区画，区分，分裂，目盛
　division plate　分離板，仕切板【タービンのノズルを取り付ける板】
DNB (**Departure from Nucleate Boiling**)　核沸騰限界
dock　ドック，船渠
dockyard　造船所
document　動記録する，文書化する　名文書，書類
dog　回し金，つかみ道具
　dog clutch　かみ合いクラッチ
dolly　当て盤，台車，トロッコ
dome　ドーム，丸天井
domestic　形国内(産)の
done　動「do」の過去分詞
donkey　形補助の
　donkey boiler　補助ボイラ
　donkey engine　補機
　donkey pump　補助ポンプ
donor　ドナー【半導体に混入して自由電子を増加させる不純物】
Doppler effect　ドップラー効果【観測者と波源の相対運動のため，観測される波動の長さが変化する現象】
Doppler flow meter　ドップラー流量計
dose　照射量，線量，吸収線量
　dose equivalent　線量当量
　dose rate　線量率
dosimeter　線量計
dot　ドット，点
dotted line　点線
double　形二重の，2倍の　名ダブル，2倍，複式
　double acting　形複動の，複動式の，両開きの，2倍の効力のある
　double acting engine　複動機関【ピストンの両側から圧力が交互に作用し動く機関】
　double bond　二重結合
　double bottom construction　二重船底構造
　double bottom tank　二重底タンク
　double cage　二重かご形
　double caliper　両用パス
　double cascade　二重翼列
　double compound　複軸
　double ended　形両端が同形の
　double ended boiler　両面ボイラ【焚口が両側にあるボイラ】
　double ended wrench　両口スパナ
　double ender　両頭船
　double expansion engine　二重膨張機関
　double flow turbine　複流タービン
　double fluid cell　二次電池
　double geared drive　二段歯車運転
　double header　複式管寄せ
　double helical gear　やまば歯車
　double hull construction　二重船こく構造
　double injector　複式インゼクタ
　double locating　二重位置決め
　double nut bolt　両ナットボルト【特殊ボルトの一種】

double pipe condenser 二重管凝縮器
double pipe heat exchanger 二重管熱交換器
double precision 倍精度
double reduction gear 二段減速装置
double reheat 二段再熱
double scale 二重目盛
double seat valve 両座弁, 複座弁【上下に弁座のある弁】
double sided impeller 両側吸込み羽根車
double squirrel-cage (induction) motor 二重かご型（誘導）電動機
double suction pump 両吸込みポンプ
double threaded screw 二条ねじ【2本のねじ山の巻きついているねじ】
double throw switch 双投スイッチ
double tube type heat exchanger 二重管型熱交換器
double wire system 二線式【2本の電線による配電方式】
dovetail ばち形, あり継【鳩の尾のような形の溝とほぞを組み合わせて接合すること】
dovetail groove あり溝
dovetail joint あり継手
Dow metal ダウメタル【85％以上のMgを含む合金の商品名】
dowel 合わせ釘, だぼ【2つの部品をつなぐ釘】
dowel pin ドエルピン, 位置合せピン
down 動降ろす 副下に
形ダウンして 名下降
down to 動〜まで下がって(減って)
down stroke 下り行程【ピストンが下向きに動く行程】
down wash 洗流
downcast header 降水管管寄せ
downcast pipe 降水管【水管ボイラでボイラ水を下降させる管】
downcast ventilator 給気通風筒【船室や船倉に吸気する筒】
downcomer ダウンカマー, 降水管, 下降管
downward 副下向きに
形下向きの
反 upward：上向きに, 上向きの
draft 名ドラフト, 下図, 設計図, 通風, 喫水
動下図を描く, 起草する
draft equipment 通風設備
draft gauge 通風計, 喫水計
draft gear 緩衝器, 引張装置
draft head 通風水頭
draft loss 通風損失
draft power 通風力
draft pressure 通風圧力
draft tube 吸出し管
drafter ドラフタ【製図器械】
drag 動引く
名引くこと, 抗力, 抵抗
drag coefficient 抗力係数
drain 動排水する
名排水, 水抜き, 排水管, 排出, 流出, ドレン【蒸気が復水した水】
drain board 水切板
drain cock ドレンコック
drain collecting tank ドレンコレクティングタンク
drain cooler ドレンクーラ
drain pipe ドレンパイプ, 排水管
drain plug ドレンプラグ, ドレン栓
drain tank ドレンタンク
drain trap ドレントラップ【ドレン除去器】
drain valve ドレン弁
drainage 排水, 排水装置
drainage pump 排水ポンプ
dramatically 副劇的に, 著しく
draught 「draft」に同じ
draw 動製図する, 誘因する, 吸引する, 引く, 引っ張る, 引き出す, 取り出す, 描く 名引くこと
draw card 引伸ばし指圧図【シリンダ内の圧力を描いた図】
draw curve (diagram) 手引き線図
draw from 動引き抜く, 抜き取る
draw into 動引き込む
drawing 製図, 図面, 引抜き, 引

抜き加工, 絞り加工
drawing instrument　製図器具
drawing number　図面番号
drawing rate　絞り率
drawn　動「draw」の過去分詞
形引かれた
　drawn tube　引抜管【継ぎ目なしの鋼管】
dredging pump　ドレッジポンプ【遠心ポンプの一種】
drencher for fire protection　ドレンチャ設備【防火設備の一種】
dresser coupling　ドレッサ形管継手
dressing　仕上げ, 散布, 装飾
dribble　動垂れる, したたる
dribbling　後だれ
drift　動漂う
名ドリフト, 漂流, 横流れ, 移動
　drift angle　偏流角, 偏角
　drift pin　ドリフトピン, 打込みピン【穴合わせに用いるピン】
drill　訓練, 穴あけ機, きり
　drill socket　きりソケット
driller　穴あけ機, 穴あけ工
drilling　穴あけ
　drilling machine　ボール盤
drinking water pump　飲料水ポンプ
drip　動ぽたぽた落ちる
名しずく, 水滴
　drip feed lubrication　滴下潤滑
　drip proof type　防水型, 防滴形
drive　動運転する, 駆動する
名ドライブ, 装置, 駆動, 駆動装置, 伝動, 励振, 運転
関 belt drive：ベルト駆動／direct drive：直結駆動
　drive shaft　駆動軸, 伝動軸
　drive unit　駆動ユニット(装置)
driven　動「drive」の過去分詞形
名従車, 従動歯車
　～ driven　～によって動く
　　関 diesel driven generator：ディーゼル(駆動)発電機／motor driven blower：電動送風機
　driven axle　動軸
　driven gear　従動歯車
　driven pulley　供車, 従車, 従動車
　driven machinery　駆動機械
　driven shaft　従軸, 従動軸, 被駆動軸
　driven wheel　供車, 従車, 従動車
driver　ドライバ, ねじ回し
driving　形駆動の, 推進の, 運転の
名駆動, 運転, 推進, 操縦
　driving action　駆動作用
　driving arrangement　駆動装置
　driving belt　駆動ベルト
　driving coil　励振巻線
　driving cylinder　駆動シリンダ
　driving device　駆動装置
　driving face　圧力面〈プロペラ〉
　driving fit　たたきばめ
　driving force　推進力
　driving gear　駆動歯車, 駆動装置
　driving key　打込みキー
　driving mechanism　伝動機構, 駆動機構
　driving motor　駆動電動機
　driving pulley　原車, 伝動プーリ【従車に運動・動力を伝える車】
　driving shaft　駆動軸
　driving source　駆動源
　driving surface　圧力面
　driving torque　駆動トルク
　driving wheel　駆動輪, 原動車
drop　名落下, 降下
動落ちる・とす, 下がる・げる
　drop feed oiling　滴下給油
　drop lubrication　滴下注油
　drop test　落下試験【物体を落下させ強度を試す試験】
droplet　小滴, 水滴, 飛沫, 液滴
dropping point　滴点
drum　ドラム, 胴【中空円筒の容器】
　drum cam　円筒カム
　drum clutch　ドラムクラッチ
　drum controller　ドラム制御器
　drum level　ドラム水位
　drum rotor　ドラムロータ, 胴形回転子, 筒形羽根車
　drum stage　胴段〈タービン〉

drum winding　鼓状巻
dry　形乾燥した，乾式の，潤滑油を使用しない，無水の，乾性の　動乾く，乾かす
　dry air　乾燥空気
　dry air pump　乾式空気ポンプ【混合気用ポンプ】
　dry back boiler　乾燃式ボイラ
　dry battery　乾電池
　dry bearing　ドライ軸受
　dry bulb temperature　乾球温度　関 wet bulb temperature：湿球温度
　dry cell　乾電池
　dry chemicals　ドライケミカル
　dry combustion boiler　乾燃室ボイラ【燃焼室の周囲に水のないボイラ】
　dry combustion gas　乾き燃焼ガス
　dry compression　乾き圧縮
　dry condition　乾燥状態
　dry corrosion　乾食
　dry distillation　乾留
　dry dock　ドライドック，乾ドック
　dry element cell　乾電池
　dry friction　乾燥摩擦
　dry gas meter　乾式ガスメータ
　dry ice　ドライアイス
　dry lapping　乾式ラップ仕上げ
　dry liner　乾式ライナ
　dry out　ドライアウト
　dry point　乾点
　dry process　乾式【ボイラ保存法】
　dry saturated steam　乾き飽和蒸気
　dry saturated steam line　乾き飽和蒸気線
　dry sump lubrication　乾式潤滑
　dry type transformer　乾式変圧器
　dry vapor　乾き蒸気
dryer　ドライヤ，乾燥器，乾燥剤
drying　乾燥
　drying agent　乾燥剤
dryness　乾燥，乾き度
　dryness fraction　乾き度，乾燥率
dual　形二重の，2 の
　dual combustion cycle　二重燃焼サイクル，複合サイクル

　dual ignition　二重点火
　dual pump　複式ポンプ
　dual reduction drive　二段減速装置
duality　二重性
duct　ダクト，煙道，通気管，管路，送水管【矩形断面の流体通路】
ductile　形延性のある
　ductile cast iron　ダクタイル鋳鉄，可鍛鋳鉄，球状黒鉛鋳鉄
　ductile failure　延性破壊
　ductile fracture　延性破壊，延性破面
　ductile material　延性材料
　ductile rupture　延性破壊
　ductile steel　延性鋼
ductility　延性，変形能
　ductility test　延性試験
　ductility transition temperature　延性遷移温度
ductilometer　伸び計
due　形正当な，しかるべき
　due to ～　前～が原因で，～のために
dummy　名ダミー，模型，偽物，見本　形模型の，模造の，見掛けの
　dummy coil　ダミーコイル，遊びコイル
　dummy gauge　ダミーゲージ
　dummy piston　つり合いピストン
　dummy ring　ダミーリング【バランスピストンの気密装置に用いられるリング】
dump bolt　押込みボルト
dump test　縦圧試験，負荷遮断試験
duplex　形複式の，二重の，2 倍の
　duplex bearing　組合せ軸受
　duplex burner　複式バーナ
　duplex pump　複式ポンプ【2 個のシリンダが相補し動作するポンプ】
　duplex winding　二重巻
durability　耐久性
durable year　耐用年数
duralumin　ジュラルミン【軽くて丈夫なアルミニウム合金】
duration　継続[時間]，持続時間，所要時間，持続，期間

during ～　前～の間
durometer　デュロメータ，硬度計
duroscope　デュロスコープ【硬度計の一種】
dust　ダスト，ちり，ほこり
  dust and mist collection　集塵
  dust collection　集塵
  dust collector　集塵器
  dust explosion　粉塵爆発
  dust proof　形ごみよけの，防塵の　名防塵，ちり止め
  dust-proof glasses　防塵眼鏡
  dust respirator　防塵マスク
dusty　ほこりっぽい
duty　義務，任務，職務，関税
  duty cycle　動作周期比率
  duty ratio　デューティ比，動作比
dwell　ドウェル，ねじ面の平行部
dye　名染料，染色，色素　動色をつける，しみ込む，染色する
  dye check　染色探傷
  dye penetrant　染色浸透剤，染色浸透探傷
  dye penetrant test　染料浸透試験，浸透探傷試験
dyeing　染色，浸染
dynamic　形動力の，動的な，力学的，動力学的
  dynamic balance　動的つり合い，動的平衡
  dynamic braking　発電制動
  dynamic characteristic　動特性
  dynamic damper　ダイナミックダンパ，動吸収器【金属ばねを使った減衰装置】
  dynamic electricity　動電気
  dynamic fatigue　動的疲れ
  dynamic friction　動摩擦
  dynamic lift　揚力
  dynamic load　動荷重
  dynamic precision　動的精度
  dynamic pressure　動圧，動圧力
  dynamic sensitivity　動作感度
  dynamic stability　動的安定性，力学的安定
  dynamic stress　動応力
  dynamic test　動的試験【材料に衝撃を加えて試験すること】
  dynamic unbalance　動的不つり合い
  dynamic vibration reducer　動吸振器
dynamical　形ダイナミックな，活動的な
  dynamical balancing　動的つり合い
  dynamical friction　動摩擦
  dynamical pressure　動圧，動圧力
  dynamical stability　動的復原力
  dynamical stress　動応力
dynamics　力学，動力学
  dynamics of machinery　機械力学
dynamite　ダイナマイト
dynamo　発電機
  dynamo engine　発電機用機関
  dynamo oil　ダイナモ油
  dynamo type tachometer　発電機型回転計
dynamometer　動力計【原動機の動力を測る計器】
dynamometric power　正味出力
dyne　ダイン【力の単位】

E-e

each 形各々の，各　副それぞれ
　each end　各端
　each occasion　毎回，その都度
　each of　～のそれぞれ
　each other　互いに
　each revolution　回転毎
　each side　両側
　each time　いつも，毎回，～するごとに
ear plug　耳栓
earlier　「early」の比較級
early　副早く　形初期の，早期の
　early detection　早期発見
　early failure　初期故障
earmuff　イヤマフ，耳覆い
earth　名アース接地，地球，地面　動接地する
　earth connection　接地
　earth current　接地電流
　earth detector　検漏器
　earth fault　地絡
　earth lamp　アースランプ，検漏灯
　earth magnetism　地磁気
　earth metal　土金属
　earth resistance meter　接地抵抗計
earthing　接地
earthquake proof　耐震
ease　容易さ　動和らげる，取り除く，緩和する，容易にする
easement curve　緩和曲線
easily　副容易に，簡単に
easiness　容易さ
easing gear　弁上げ装置
easy　形やさしい，容易な，簡単な
ebonite　エボナイト，硬質ゴム
ebulliometer　沸点[測定]計
ebullition　沸騰
ECA（Emission Control Area）　排出規制海域
eccentric　形偏心の　名偏心
　eccentric angle　偏心角
　eccentric arm　偏心距離
　eccentric load　偏心荷重
　eccentric pivot　偏心接触点
　eccentric radius　偏心距離
　eccentric ring set　偏心環
　eccentric rod　偏心棒
　eccentric sleeve　偏心スリーブ
　eccentric throw　偏心距離
eccentrically　副偏心して
　eccentrically-located　形偏在する
eccentricity　偏心[率]，離心率
echo　エコー，こだま，反響
　echo sounder　音響測深器
economic　形経済の
economical　形経済的な，節約する
economizer　エコノマイザ，節炭器，給水予熱器【燃焼ガスの余熱で給水を予熱する装置】
eddy　名エディ，渦，渦巻き　動渦を巻く
　eddy current　渦電流
　eddy current braking　渦電流制動
　eddy current examination　渦流探傷検査
　eddy current loss　渦電流損
　eddy current test　渦流探傷検査
　eddy loss　渦損失
　eddy making resistance　渦抵抗
　eddy resistance　渦抵抗
　eddy viscosity　渦粘性，渦粘度
edge　エッジ，端，刃先，縁，周辺
　edge cam　側面カム
　edge condition　周辺条件
　edge preparation　開先加工
　edge strip　へり目板，縁目板【鋼板を結合するための板】
edition　～版
eductor　エダクタ，放出器
　関 ejector
effect　名効果，結果，影響　動～をもたらす，引き起こす
　関 in effect：実質的に，事実上，実際には
effective　形効果的な，有効な
　effective breadth　有効幅

effective clearance volume 有効すきま容積
effective compression ratio 有効圧縮比
effective cross current 有効横流
effective current 実効電流
effective cylinder volume 有効シリンダ容積
effective diameter 有効直径
effective efficiency 有効効率【熱機関の有効仕事と理論仕事の比】
effective electric power 有効電力
effective head 有効落差, 有効水頭
Effective HorsePower EHP, 有効馬力, 有効出力
effective impedance 実効インピーダンス
effective output 有効出力
effective pitch 有効ピッチ〈プロペラ〉
effective power 有効動力, 有効電力
effective pressure 有効圧力
effective pull 有効張力
effective range 有効範囲
effective resistance 有効抵抗
effective sectional area 有効断面積
effective slip 有効失脚〈プロペラ〉
effective strain 有効ひずみ
effective stroke 有効行程
effective temperature 有効温度
effective tension 有効張力
effective thermal efficiency 有効熱効率【熱機関に供給された熱量と有効仕事の比】
effective thread 有効ねじ
effective throat thickness 有効のど厚
effective value 有効値, 実効値
effective voltage 実効電圧
effective wake 有効伴流
effective work 有効仕事
effectively 副効果的に, 効率よく
effectiveness 効果, 有効性, 効率
effervescence 沸騰, あわ立ち, 発泡
effervescent steel リムド鋼
efficiency 効率, 能力
efficiency by input-output test 実測効率
efficiency of boiler ボイラの効率
efficiency ratio 効率比, 測定能率
efficiency test 効率試験
efficient 形効果的な, 効率的な
efficiently 副能率的に, 効果的に
efflux angle 流出角
effort 作用力, 努力, 労力
effusion 流出, 噴出
effusion cooling しみ出し冷却
e.g. 例えば
EHP (Effective HorsePower) 有効馬力, 有効出力
eigenvalue 固有値
eigenvector 固有ベクトル
either 形どちらかの, いずれか一方の 代どちらか一方
関 on either side (of): 〜(の)両側に, 〜(の)左右に
either A or B 接 A かまたは B
eject 動排出する
ejector エジェクタ, エゼクタ, 排出器, 放出器 同 eductor
ejector condenser エゼクタ復水器
elabolate 形入念な, 精巧な
elapsed time 経過時間
elastance 弾性, エラスタンス【静電容量の逆数】
elastic 形弾性の, 弾力のある, しなやかな, 伸縮自在の
elastic axis 弾性軸
elastic body 弾性体
elastic break-down 弾性破損
elastic collision 弾性衝突
elastic constant 弾性定数
elastic coupling 弾性継手
elastic curve 弾性曲線
elastic deformation 弾性変形
elastic energy 弾性エネルギ
elastic failure 弾性破損
elastic force 弾性力
elastic hardness 弾性硬さ
elastic hysteresis 弾性履歴, 弾性ヒステリシス
elastic inflation 弾性膨張
elastic limit 弾性限度

elastic medium　弾性媒質
elastic modulus　弾性係数, 弾性率
elastic oscillation　弾性振動
elastic range　弾性域
elastic ratio　弾性比
elastic recovery　弾性回復
elastic region　弾性域
elastic scattering　弾性散乱
elastic sleeve bearing　弾性スリーブ軸受
elastic stability　弾性安定
elastic strain　弾性ひずみ
elastic strain energy (resilience)　弾性ひずみエネルギ
elastic stress　弾性応力
elastic surface　弾性曲面
elastic vibration　弾性振動

elasticity　弾性, 弾力, 伸縮
elasticplastic fracture　弾塑性破壊
elastoplasticity　弾塑性
elbow　エルボ, L字継手
electric　形電気の, 電動の【「電気で動く」「電気で生ずる」の意】

electric apparatus　電気装置
electric appliance　電気器具
electric arc welding　アーク溶接
electric boiler　電気ボイラ
electric brake　電気式ブレーキ
electric brazing　電気ろう付
electric cable　電線
electric capacity　電気容量
electric charge　電荷
electric circuit　電気回路
electric coil　電気コイル
electric conduction　電気伝導, 電導
electric conductivity　導電率
electric conductor　導電体
electric control　電気制御
electric controller　電気式調節計
electric cradle dynamometer　電気動力計
electric current　電流
electric discharge　放電
electric discharge machining　放電加工
electric drill　電気ドリル

electric drive　電気駆動
electric dust collector　電気集塵装置
electric dynamometer　電気動力計
electric energy　電気エネルギ, 電力量
electric engineering　電気工学
electric eye　光電池, 光電管
electric fan　扇風機
electric field　電界, 電場
electric flux density　電束密度
electric force　電気力
electric furnace　電気炉
electric generator　発電機
electric heater　電気ヒータ, 電熱器
electric heating　電熱, 電気加熱
electric ignition engine　電気点火機関
electric indicator　電気式指圧器, 電気式表示器
electric installation　電気設備
electric insulation　電気絶縁性
electric load　電気負荷
electric machine　電気機械
electric meter　電気計器
electric micrometer　電気マイクロメータ【測長器の一種】
electric motor　モータ, 電動機
electric painting　静電塗装
electric potential　電位
electric potential difference　電位差
electric power　電力
electric power supply　電源
electric precipitator　電気集塵装置
electric pressure　電圧
electric property　電気的性質
electric propulsion ship　電気推進船
electric protection　電気防食
electric pulse motor　電気パルスモータ
electric resistance　電気抵抗
electric shock　感電, 電撃
electric signal　電気信号
electric soldering iron　電気はんだごて
electric spark　電気火花
electric spark machining　放電加工

electrical

electric steel 電気鋼
electric tachometer 電気式回転計
electric thermostat 電気サーモスタット
electric water heater 電気温水器
electric welded pipe 電縫管
electric welding 電気溶接
electric wire 電線
electric work 電気工事

electrical 形電気の, 電気的な【「電気に関する」の意】
electrical angle 電気角
electrical circuit 電気回路
electrical conduction 電気伝導, 電導
electrical control 電気制御
electrical energy 電気エネルギ
electrical engineering 電気工学
electrical equipment 電気機器, 電気設備, 電気装置
electrical field 電界
electrical fire 電気火災
electrical indicator 電気式インジケータ
electrical machining 電気加工
electrical neutral axis 電気的中性軸
electrical potential 電位
electrical power 電力
electrical signal 電気信号
electrical system 電気システム, 電気系統
electrical tachometer 電気式回転計
electrical tape 絶縁テープ
electrical transmission gear 電気式伝動装置
electrical unit 電気単位
electrical wire 電線
electrical wiring 電気配線
electrical work 電気仕事

electrically 副電気的に
electrically driven 電動

electrician 電気技師, 電気工
electricity 電気, 電力, 電気学
electrification 帯電, 感電, 電化
electrified body 帯電体
electrify 動電化する・させる, 電力を供給する, 帯電させる, 充電する

electro- 接頭「電気」の意
electroanalysis 電解分析
electrochemical 形電気化学的な
electrochemical corrosion 電気化学的腐食, 電食
electrochemical discharge machining 電解放電加工
electrochemical equivalent 電気化学当量
electrochemical machining 電解加工

electrochemistry 電気化学
electroconductive glass 電導性ガラス
electrocute 動感電死させる
electrode 溶接棒, 電極
electrode holder 電極ホルダ, 溶接棒ホルダ【溶接棒支持器】
electrode potential 電極電位
electrode tip 電極チップ

electrodeposition 電着, 電着塗装
electrodynamic force 電流力
electrodynamic pump 電磁ポンプ
electroforming 電鋳, 電気鋳造
electrogalvanizing 電気亜鉛メッキ
electrohydraulic 形電気流体式の, 電気水力学の
electrohydraulic control 電気油圧制御
electrohydraulic dissociation 電離
electrohydraulic pulse motor 電気-油圧式パルスモータ
electrohydraulic servo valve 電気-油圧式サーボ弁
electrohydraulic steering gear 電動油圧舵取機

electrokinetic potential 界面動電位
electroless plating 無電解めっき, 化学めっき
electrolysis 電気分解, 電解
electrolyte 電解液, 電解質
electrolytic 形電解質の
electrolytic capacitor (condenser) 電界コンデンサ
electrolytic corrosion 電食
electrolytic dissociation 電解
electrolytic etching 電気侵食

electrostatic

electrolytic picking　電解酸洗い
electrolytic solution　電解質溶液
electromagnet　電磁石
electromagnetic　形電磁石の，電磁気の
　electromagnetic attraction relay　電磁吸引形継電器
　electromagnetic brake　電磁ブレーキ
　electromagnetic chuck　電磁チャック【電磁石を利用した保持装置】
　electromagnetic clutch　電磁クラッチ
　electromagnetic contactor　電磁接触器
　electromagnetic controller　電磁制御器
　electromagnetic coupling　電磁継手
　electromagnetic crack detector　電磁気的探傷機
　electromagnetic damping　電磁制動，電磁制振
　electromagnetic deflection　電磁偏向
　electromagnetic energy　電磁エネルギ
　electromagnetic examination　渦流探傷検査
　electromagnetic field　電磁界
　electromagnetic flowmeter　電磁流量計
　electromagnetic force　電磁気力
　electromagnetic ground detector　電磁検漏器
　electromagnetic induction　電磁誘導
　関 the principle of electromagnetic induction：電磁誘導の原理
　electromagnetic membrane thickness gauge　電磁式膜厚計
　electromagnetic oscillograph　電磁オシログラフ
　electromagnetic relay　電磁継電器
　electromagnetic survey　電磁探査
　electromagnetic switch　電磁スイッチ
　electromagnetic unit　電磁単位【電流の磁気効果をもとにして作った単位系】
　electromagnetic vibrograph　電磁振動計
　electromagnetic wave　電磁波
electromagnetism　電磁気，電磁気学
electrometallurgy　電気冶金[学]
electrometer　電位計，電位差計
electromotive　形起電の，電動の
　Electro Motive Force　EMF，起電力
electron　エレクトロン，電子
　electron affinity　電子親和力
　electron beam welding　電子ビーム溶接
　electron bombardment　電子衝撃
　electron diffraction　電子回折
　electron flow　電子流
　electron lens　電子レンズ
　electron microscope　電子顕微鏡
　electron volt　エレクトロンボルト，電子ボルト【記号：eV】
electronic　形電子工学の，電子の
　electronic automatic-balancing thermometer　電子式自動平衡温度計
　electronic circuit　電子回路
　electronic control　電子制御
　electronic engineering　電子工学
　electronic mail　電子メール，電子郵便
　electronic voltmeter　電子式電圧計
electronics　エレクトロニクス，電子工学
electropainting　電気塗装，電着塗装
electroplating　電気めっき
electropneumatic converter　電空変換器
electropneumatic operated control valve　電気空気式調節弁
electroscope　検電器
electrospark machining　放電加工
electrostatic　形静電気の
　electrostatic capacity　静電容量
　electrostatic charge　静電荷
　electrostatic force　静電力
　electrostatic induction　静電誘導
　electrostatic painting　静電塗装
　electrostatic potential　静電位

electrostatic precipitator　静電集塵器, 電気集塵装置
electrostatic unit　静電単位【略: esu】
electrostatics　静電学, 静電気学
electrothermal　形電気と熱に関する, 熱発電の
electrothermal instrument　熱電形計器
electrothermal voltmeter　熱電式電圧計
electrothermic type　熱電型
element　エレメント, 要素, 元素, 素子, 成分
elementary　形初歩の, 初等の, 基本の, 単純な, 元素の
elementary analysis　元素分析
elementary color　原色
elementary particle　素粒子
elementary reaction　基礎反応
elementary stream　流線
elevate　動上げる, 上昇させる
elevated temperature　高温
elevation　エレベーション, 標高, 側面図, 仰角, 正面図, 立面図
elevation angle　仰角
elevator　エレベータ　同lift
eliminate　動除去する, 消去する
elimination　削除, 除去, 消去, 排除
elimination method　消去法
elimination of labor　省力化
eliminator　エリミネータ, 交流整流器, 排除器, 分離器
ellipse　楕円, 長円
elliptic　形楕円形の
elliptic gear　楕円歯車
elliptical　形楕円形の
elliptical blade　楕円羽根
elliptical valve diagram　楕円弁線図
elongate　動長くする, 引き伸ばす, 伸びる　形細長い, 伸長した
elongation　伸び, 伸び率, 伸長
elongation after fracture　破断伸び
elongation-load diagram　伸び-荷重線図
elongation percentage　伸び率

elongational strain rate　伸びひずみ速度
else　副さもなければ
elsewhere　副他の部分(場所)に(へ)
elution　溶出, 溶離
embed　動埋め込む, 埋める
embeddability　埋没性, 埋込み性
embedded temperature detector　埋込温度計
embrittlement　脆(ぜい)性, 脆化
embrittlement characteristic　脆化特性
embrittlement cracking　脆化割れ
emerge　動現れる, 明らかになる
emergence　発生, 出現
emergency　名緊急事態, 非常時, 危機, 緊急　形緊急の
　関 in an emergency：非常の場合には
emergency bilge pump　非常用ビルジポンプ
emergency brake　非常ブレーキ
emergency call　非常呼出し
emergency cutout gear　危急遮断装置
emergency equipment　非常用装置
emergency exit　非常口
emergency fire pump　非常用消火ポンプ
emergency generator (dynamo)　非常用発電機
emergency lighting　非常灯
emergency measures　応急対策, 緊急措置
emergency propulsion plant　非常用推進装置
emergency safety device　危急安全装置
emergency stairs　非常階段
emergency switch　非常スイッチ
emergency valve　非常弁
emergency vent　非常吐出口
emergency work　応急作業
emery　エメリ, 金剛砂
emery cloth　布やすり
emery wheel　砥石車
EMF (**Electro Motive Force**)　起電力
emission　排出物, 放射, 放出, 放出物

emission angle　放射角
Emission Control Area　ECA，排出規制海域
emission standard　排出基準
emissive power　放射度，放射能
emissivity　放射率
emit　動放つ，放出する，出す
emittance　エミッタンス，発散度，放射率
emitter　エミッタ，放射体
emitting surface　放射面
emphasize　動強調する
empirical　形経験の，実験の
　empirical equation　実験式
employ　動使用する，雇う
　employ for　〜に用いる
employment　使用，雇用
empty　形空の　動空にする
emulsification　乳化
emulsified fuel　エマルジョン燃料
emulsified water　乳化水
emulsifier　乳化剤，乳化機
emulsify　動乳化する
emulsifying agent　乳化剤
emulsifying oil　乳化性油
emulsion　エマルジョン，乳濁液
　emulsion breaker　乳化破壊剤
　emulsion fuel　エマルジョン燃料
　emulsion stabilizer　乳化安定剤
　emulsion tube　ブリード管
　emulsion type lubricants　乳化性潤滑油
emulsoid　乳濁質
enable　動可能にする，容易にする
enamel　エナメル，琺瑯（ほうろう）
　enamel paint　エナメル塗料
enameled wire　エナメル線
enclose　動囲む，密閉する，含む
enclosed compressor　密閉型圧縮機
enclosed space　密閉空間
enclosed type　閉鎖形
encoder　暗号器，符号変換器
encoding　符号化，コード化
encounter　動遭遇する，直面する
encourage　動激励する，奨励する
end　エンド，端，境界，停止，目標，終わり　動終わる
end cam　エンドカム，端面カム
end gas　末端ガス
end lap　横縁
end plate　鏡板，端板
end play　軸方向の遊び
end point　終点，乾点
end ring　端絡環
end thrust　軸端スラスト
end view　端面図
endanger　動危険にさらす
endless belt　継目なしベルト
endothermic reaction　吸熱反応
endurance　持続性，耐久性
　関 year of endurance：耐用年数
　endurance limit　疲れ限度,耐久限度
　endurance ratio　耐久比
　endurance strength　疲れ強さ
　endurance test　耐久試験
endure　動耐える，我慢する
energize　動励磁する，通電する
energy　エネルギ
　関 kinetic energy：運動エネルギ
　energy conservation　省エネルギ
　energy conservation law　エネルギ保存則
　　同 law of conservation of energy
　energy conversion　エネルギ変換
　energy conversion efficiency　エネルギ変換効率
　energy diagram　エネルギ線図
　energy efficiency　エネルギ効率
　energy fluctuation　エネルギ変動率
　energy level　エネルギ準位
　energy loss　エネルギ損失
　energy of activation　活性化エネルギ
　energy of motion　運動エネルギ
　energy of nature　自然エネルギ
　energy of rupture　破壊エネルギ
　energy saving　省エネルギ
enforce　動強化する，補強する
enforcement　強制，施行
engage　動かみ合う，連動する
engaging　連結，結合
engine　エンジン，機関，発動機
　engine alarm panel　機関警報盤

engine bearing　エンジン軸受
engine bed　機関台
engine brake　エンジンブレーキ
engine casing　機関室囲壁
engine department　機関部
engine efficiency　エンジン効率, 機関効率
engine fittings　機関取付物
engine frame　架構
engine load　機関負荷
engine monitoring system　エンジン監視システム
engine oil　エンジン油
engine output　エンジン出力
engine performance　エンジン性能
engine power　機関[軸]出力
engine rating　エンジン定格
engine remote control system　主機関遠隔操縦装置
engine room　機関室
engine speed　エンジンスピード
engine telegraph　エンジンテレグラフ
engine telegraph logger　エンジンテレグラフロガー
engine torque　エンジントルク, 軸トルク
engine trial　機関試運転
engineer　機関士, エンジニア, 技師
engineer's logbook　機関日誌
engineering　エンジニアリング, 技術, 工学, 工学技術
　㊥ mechanical engineering：機械工学
　engineering development　技術開発
　engineering heat transfer　伝熱工学
　engineering material　工学材料
　engineering unit　工学単位
Engler viscosimeter　エングラー粘度計
enhance　動強化する, 高める
enhanced　形強化された
enhancement　改良, 拡張, 強調, 強化, 向上, 増強, 増大, 促進
enlarge　動拡(増)大する
enlarged character　拡大文字
enlargement　拡大, 増大

enlargement loss　拡大損失
enough　形十分な
　…enough to 〜　…なので〜できる, 〜できるほど…
enrich　動高める, 強化する, 濃縮する
enrichment　濃縮, 濃縮度
ensemble average　アンサンブル平均, 集団平均
ensure　動確実にする, 保証する
enter　動入力する, 記入する
enterprise　企業
enthalpy　エンタルピ, 熱含量
　enthalpy-entropy diagram　H-S線図, エンタルピ-エントロピ線図
entire　形全体の
entirely　副完全に, すっかり, 全く
entity　実在, 本体, 存在
　㊥ separate entity：独立した存在
entrainment　気水共発, 飛沫同伴
　entrainment limitation　飛散限界
entrance　入口, 入場, 水切部
　entrance length　助走距離
　entrance region　合流域
entropy　エントロピ
　entropy chart (diagram)　エントロピ線図
　entropy function　エントロピ関数
entry　入ること, 記入, 項目
　entry angle　入射角
envelope　包絡線, 包晶, 外被, 封筒
environment　環境, 状況, 周囲
environmental assessment　環境アセスメント, 環境影響評価, 環境審査
environmental pollution　公害, 環境汚染
enzyme　酵素
epicyclic　形周転円の
　epicyclic gear　遊星歯車[装置]
　epicyclic reduction gear　遊星歯車減速装置
epicycloid　外[転]サイクロイド
epitrochoid　外[転]トロコイド
epoxy resin　エポキシ樹脂【耐熱性合成樹脂の一種】
Eq. (Equation)　方程式, 等式

equal 形等しい, 同等の, 一様な 動～に等しい, ～に匹敵する
  equal angle steel 等辺山形鋼
equality 同等, 等式
equalization 等価, 等化, 均等化
equalize 動等しくする, 一様にする
equalizer イコライザ, 等価器, 等化器, 均圧管, 均圧線, 均圧母線, 平衡装置, つり合い装置
equalizing 等化, つり合い, 均等, 均圧, 同等
  equalizing bus-bar 均圧母線
  equalizing pipe 均圧管
  equalizing piston つり合いピストン
  equalizing ring 均圧環
  equalizing spring つり合いばね
equally 副同様に, 等しく
equate 動一致する, 同等とみなす
equation 式, 方程式, 等式, 反応式, 平衡, 均一化, 均等化, 誤差
  equation of continuity 連続の式
  equation of motion 運動方程式
  equation of state 状態[方程]式
equilateral triangle 正三角形
equilibrant 平衡力
equilibrate 動つり合う
equilibria 「equilibrium」の複数形
equilibrium つり合い, 平衡
  equilibrium diagram 平衡状態図
  equilibrium distillation 平衡蒸留
  equilibrium moisture 平衡水分
  equilibrium of forces 力のつり合い
  equilibrium phase diagram 平衡相状態図
  equilibrium point 平衡点
  equilibrium state 平衡状態
  equilibrium valve つり合い弁
equip 動装備する, 備える
equipment 機器, 装置, 設備, 器具
  equipment drawing 装置図
equipotential 形等電位の
  equipotential surface 等ポテンシャル面, 等電位面, 等磁位面
equivalence 当価, 等価, 同等, 同量
  equivalence ratio 当量比, 等価比
equivalent 形等価の, 同価の, 同等の, 同量の, 対等の
名当量, 等価, 相当, 対等, 等量
  equivalent bending moment 相当曲げモーメント
  equivalent circuit 等価回路
  equivalent conductivity 相当導電率
  equivalent cross section 等価断面
  equivalent diameter 相当直径, 等価直径
  equivalent disc 相当円板
  equivalent dynamical system 相当振動系
  equivalent evaporation 相当蒸発量, 換算蒸発量
  equivalent impedance 等価インピーダンス
  equivalent length 相当長さ, 等価長さ
  equivalent mass 相当質量
  equivalent mean effective pressure 相当平均有効圧力
  equivalent moment of inertia 相当慣性モーメント
  equivalent nozzle area 相当ノズル面積
  equivalent radius 相当半径
  equivalent ratio 等量比
  equivalent resistance 等価抵抗
  equivalent round 等価円
  equivalent shaft horsepower 相当軸馬力
  equivalent sine wave 等価正弦波
  equivalent spray angle 等噴霧角
  equivalent spur gear 相当平歯車
  equivalent strain 相当ひずみ
  equivalent stress 相当応力
  equivalent temperature 相当温度
  equivalent torsional stiffness 等価ねじりこわさ
  equivalent twisting moment 相当ねじりモーメント
  equivalent value 当量
erase 動消去する, 削除する
erection stress 組立応力, 架設応力
ergonomics 人間工学
Ericsson cycle エリクソンサイクル

erosion エロージョン，侵食
 erosion corrosion　摩耗腐食,潰食,侵食腐食
 erosion shield　浸食保護
err　動誤る
error　エラー,誤り,誤差,偏差,差
 error function　誤差関数
escape　動漏れる，流出する，逃げる，避難する　名逃がし，逃げ口
 escape scuttle　逃げ口
 escape valve　逃し弁
especially　形特に，特別に，とりわけ
essential　形重要な，不可欠の，根本的な，本質的な
 関 be essential to：〜にとって絶対必要である
essentially　副本質的に，本来
establish　動設立する，制定する
estimate　動見積もる
estimated value　推定値
estimation　見積り
 estimation drawing　見積図
etc.（**et cetra**）　〜等，その他
etch　エッチング処理を行う
 etch test　腐食試験
etching　エッチング，腐食
 etching reagent　腐食剤
ethane　エタン
ethanol　エタノール
ether　エーテル
ethyl alcohol　エチルアルコール
ethylene　エチレン
 ethylene glycol　エチレングリコール
Euler number　オイラー数
Euler's equation　オイラーの式【長い柱の座屈強さを表す式】
Euler's formula　オイラーの公式
eutectic　共晶
 eutectic alloy　共晶合金
 eutectic point　共晶点
 eutectic reaction　共晶反応
eutectoid　共析
 eutectoid steel　共析鋼，共晶鋼
 eutectoid structure　共析組織
evacuate　動避難する・させる
evacuation　避難

evaluation　評価
 evaluation function　評価関数
 evaluation system　評価システム
evaporability　蒸発性
evaporate　動蒸発する
evaporating combustion　蒸発燃焼
evaporating pressure　蒸発圧力
evaporating temperature　蒸発温度
evaporation　蒸発，蒸着，気化
 evaporation curve　蒸発曲線
 evaporation rate　蒸発率，蒸発速度
 evaporation rate of heating surface　伝熱面蒸発率，伝熱面熱負荷
 evaporation ratio　蒸発倍数
evaporative　形蒸発の，気化の
 evaporative capacity　蒸発容量
 evaporative combustion　蒸発燃焼
 evaporative condenser　蒸発凝縮器，蒸発復水器
 evaporative cooler　蒸発冷却器
 evaporative factor　蒸発倍数，蒸発係数【ボイラにおいて燃料1kgあたりに発生する蒸気量】
 evaporative gas　蒸発ガス
 evaporative power　蒸発力，蒸発能
 evaporative test　蒸発試験
evaporator　エバポレータ，蒸発器
 evaporator section　蒸発部
 evaporator tube　蒸発管
evaporimeter　蒸発計
even　副…でさえ，さらに，それどころか
 even if（though）〜　接たとえ〜であっても，たとえ〜としても
 even number　偶数
 even with　〜と同等で，〜としても
event　出来事　関 in the event of [that]：[万一] 〜の場合には
eventual　形最終的な
eventually　副結局は，ついに
ever　副いつか，今までに，どんな時でも，今までで，いったい
every　形どの…も，毎…，…ごとに
 every A hours　A時間ごとに
 every A minutes　A分ごとに
 every single　一つ一つの

every time 毎回，〜するたびに
evidence 名証明，証拠，しるし 動証明する，明示する
exact 形正確な，精密な，完全な
　exact differential 完全微分
exactly 副正確に，精密に
exaggeration 誇張
examination 検査，試験，調査
examine 動検査する，試験する，調べる，調査する
example 例，事例，例題，実例，標本，見本　関 for example：例えば
exceed 動超える，上回る
exceeding 形非常な，過度な，異常な
exceedingly 副非常に，とても
excel 動勝る，卓越する，優れる
excellent 形優れた，すばらしい
except 動除く　前〜以外は，〜を除いて
　except [for] 〜　前〜以外は，〜を除いて
exception 例外，除外　関 with the exception of：〜を除いて，〜以外に
excess 過剰，過多，超過，余分
　excess air　過剰空気
　excess air coefficient　空気過剰係数
　excess air factor　空気過剰率
　excess air ratio　空気過剰係数，空気過剰率，空気比，過剰空気率
　excess oxygen　過剰酸素
　excess pressure　過圧，過剰圧力
excessive 形過度の，過大な，極端な
　excessive air　過剰空気
　excessive heating　過熱
　excessive wear　過度の摩耗
excessively 副過度に，はなはだしく
exchange 名交換　動交換する
　exchange action　交換作用
　exchange loss　交換損失
exchanger 交換器，交換剤
excitation 励磁，励振，励起，加振
excite 動励磁する
exciter 励磁機，励振器
exciting current 励磁電流
exclude 動除外する

exclusion 排除，除外
exclusive 形独占的な，唯一の
excursion 行程，偏位
exducer 誘導羽根
executable 形実行可能な
execute 動実行する，実施する
exercise 名練習　動練習する
exergy エクセルギ，有効エネルギ
exert 動及ぼす，働かせる
exfoliation 剥(はく)離
　exfoliation corrosion　剥脱腐食，層状腐食
exhale 動吐き出す，発散する
exhaust 名排気，排気ガス，排出　動排出する，使い果たす
　exhaust brake　排気ブレーキ
　exhaust cam　排気カム
　exhaust emission　排気物質，排気排出物，排気(出)ガス
　exhaust emission regulation　排出ガス規制
　exhaust energy　排気エネルギ
　exhaust fan　排気送風機
　exhaust fume　排(出)ガス
　exhaust gas　排気ガス，排ガス
　exhaust gas analyzer　排気ガス分析計
　exhaust gas boiler　排ガスボイラ
　exhaust gas cleaning device　排気ガス浄化装置
　exhaust gas economizer　排ガスエコノマイザ
　exhaust gas turbine　排気ガスタービン
　exhaust heat recovery　排熱回収
　exhaust interference of pressure　排気干渉
　exhaust loss　排気損失
　exhaust manifold　排気マニホールド
　exhaust pipe　排気管
　exhaust port　排気口
　exhaust pressure　排気圧
　exhaust scrubber　排気清浄器
　exhaust silencer　排気消音器
　exhaust sound　排気音
　exhaust steam turbine　排気タービン

exhaust stroke 排気行程
exhaust temperature 排気温度
exhaust turbine 排気タービン
exhaust turbocharger 排気タービン過給機
exhaust valve 排気弁
exhaust velocity 排気速度

exhausting 排気作用
exhaustion 排気，排出，消耗，[極度]疲労，枯渇
exhaustive test 徹底的な試験
exhibit 動示す，展示する，提出する
exist 動存在する，実在する
existing 形既存の，現在の，存在する
existing condition 現状，状況
exit 出口
exit angle 出口角，流出角
exit loss 流出損失
exit velocity 流出速度
exothermic change 発熱変化
exothermic reaction 発熱反応
expand 動拡張する，展開する，拡大する，膨張する・させる
expanded area ratio 展開面積比
expanded blade area 羽根展開面積
expander エキスパンダ，拡管器
expansion 膨張，展開
expansion bend 伸縮ベンド
expansion coefficient 膨張係数，膨張率
expansion cooling 膨張冷却
expansion efficiency 膨張効率
expansion engine 膨張機関
expansion index 膨張指数
expansion joint 膨張継手，伸縮継手
expansion loop 伸縮ループ
expansion plan 展開図
expansion rate 膨張率
expansion ratio 伸縮率，膨張比
expansion stroke 膨張行程
expansion tank 膨張タンク
expansion valve 膨張弁
expect 動予期する，期待する
expected value 期待値
expel 動排出する，吐き出す
expense 経(出)費，犠牲

expensive 形高価な
experience 名経験 動経験する
experiment 名実験，試験 動実験をする
experimental 形実験の
experimental formula 実験式
experimentally 副実験的に
expert 熟練者，専門家
expire 動切れる，吐きだす，満了する
explain 動説明する
explanation drawing 説明図
explode 動爆発する
exploit 動開発する，利用する
explore 動調査する
explosion 爆発，破裂
explosion chamber 爆発室
explosion door 爆発戸
explosion limit 爆発限界
explosion mixture limits 爆発限界
explosion pressure 爆発圧力
explosion proof type 防爆型
explosion ratio 爆発比
explosion relief valve 爆発安全弁
explosion stroke 爆発行程
explosion temperature 爆発温度
explosion test 爆発試験
explosive 形爆発性の 名爆発物，爆薬
explosive limits 爆発限界
explosive welding 爆発圧接
exponent 指数，指数部，べき指数
exponential 形指数の 名指数，指数関数
exponential function 指数関数
exponential law 指数法則
exponentiation べき乗，指数化，べき算
expose 動さらす，陳列する，公表する，感光する・させる，暴露する
exposed surface 露出面
exposure 被爆，露出，照射
express 動表す，表現する 名急行，速達
expression 式，表現
extend 動伸ばす，延ばす，広げる，

延長する，拡張する
extended 形拡張した
extensibility 伸び性，伸展性
extension 伸び，伸張，拡張，拡大，延長，範囲，限度，内線
  extension line 寸法補助線
  extension scale 伸び尺
  extension spring 引張りばね
extensional rigidity 伸び剛性
extensive 形広い，広大な
  extensive property 示量変数，示量性，示量的性質
  extensive quantity 示容量，外延量
extensively 副広く
extensometer 伸び計
extent 程度，範囲，大きさ，広がり，広さ　関 to a certain content：ある程度まで
exterior angle 外角
exterior product 外積
external 形外部の，外的な
  external characteristic curve 外部特性曲線
  external combustion engine 外燃機関
  external diameter 外径
  external energy 外部エネルギ
  external firing boiler 外だきボイラ
  external force 外力
  external load 外部負荷
  external loss 外部損失
  external pressure 外圧
  external resistance 外部抵抗
  external source 外部電源
  external thread 雄ねじ
  external work 外部仕事
extinction 消火，消滅，減衰，終息
  extinction curve 減衰曲線
  extinction of arc 消弧
extinguish 動消す，消滅させる
extinguisher 消火器

extinguishing appliance 消火装置
extra 形余分な，極上の，特別な
  extra feed valve 補給弁
  extra hard steel 最硬鋼，極強鋼
  extra soft steel 極軟鋼
  extra thick 形極太の
extract 動抽出する，抜粋する，引き出す，[圧縮データを]解凍する
extraction 抽出
  extraction turbine 抽気タービン
extractive distillation 抽出蒸留
extranuclear electron 核外電子
extreme 形極端な，過度の，非常な，極限の 名極値，極度，極端
  関 in extreme cases：極端な場合
  extreme breadth 最大幅，全幅
  extreme elevation 最大仰角
  extreme fiber stress 縁応力【材料の外表面に生じる応力】
  extreme high vacuum 超高真空
  extreme length 全長
  extreme pressure 極圧
  extreme pressure agent (additive) 極圧剤，極圧添加剤
  extreme pressure grease 極圧グリース
  extreme pressure lubricating oil 極圧潤滑油
extremely 副極端に，きわめて
extrude 動押し出す，突き出る，成形する
extruding 押出し加工
extrusion 押出し
eye plate アイプレート【フックなどを掛けるための穴の開いた鉄板】
eye sight のぞき穴
eyebolt アイボルト
eyehole 目穴，のぞき穴
eyelet はと目，小穴
eyenut アイナット

F-f

fabricate 動製作する，組み立てる，建造する
fabrication 製造，製作，組立て，成形加工，二次加工，加工法
 fabrication drawing　製作図
face 名フェース，面，表面，顔　動直面する，向かう
 face angle　面角，すくい角，歯先円すい角
 face bar　面材
 face bend　表曲げ
 face cam　正面カム
 face cavitation　圧力面キャビテーション
 face centered cubic lattice　面心立方格子
 face gear　フェースギア，正面歯車
 face of tooth　歯末の面
 face pitch　圧力面ピッチ
 face plate　面板，銘板
 face seal　端面シール
 face width　歯幅
facility 設備，施設，能力
facing フェーシング，表面仕上げ，座ぐり
 facing ring　座金
 facing surface　接面
 facing up　すり合わせ
facsimile ファクシミリ，ファックス
fact 事実
factor ファクタ，係数，要因，要素，因数，因子，素，率，指数
 factor of evaporation　蒸発係数
 factor of safety　安全率
factory 工場
 Factory Automation　FA，ファクトリオートメーション，工場自動化
 factory test　工場試験
fade 動[色が]あせる，薄れる，消える，変色する
 名フェード【ブレーキの利きが一時的に低下する現象】
fading 衰退，減退，フェーディング【受信電波が変化する現象】
Fahrenheit 形華氏の
 名華氏温度，華氏温度計
 Fahrenheit degree　華氏温度
 Fahrenheit temperature scale　華氏温度目盛
 Fahrenheit thermometer　華氏温度計
fail 名失敗，故障
 動失敗する，故障する，停止する
 fail safe　フェールセイフ，事故時安全
failure 故障，損傷，破壊，破損，破断，失敗，不足
 failure diagnosis　故障診断
 failure mode　故障モード
 failure period　故障期間
 failure probability　破損確率
 failure rate　故障率
 failure strength　破損強度
 failure stress　破壊応力
fair 形公正な，公平な，正当な，きれいな，かなりの
fairly 副かなり，公正に
fairness limit 許容ひずみ量
fall 動落ちる，下がる，倒れる
 名落下，降下，転倒，吊綱
 fall of potential　電位降下
 fall off　低下，悪化，下落
falling ball viscometer 落球粘度計
falling film 流下膜式
falling object 落下物
false 形間違った，仮の，虚偽の
 false alarm　誤報
 false indication　誤示
 false reading　誤測
familiar 形精通している，使いやすい
 関 be familiar with：精通している，なじんだ
familiarize 動馴じむ
fan ファン，扇風機，送風機
 fan motor　ファンモータ，送風電

動機
fang bolt 鬼ボルト【基礎ボルトの一種】
fanning action かきまぜ作用
far 副遠く,大いに 形遠い,遠いほうの
 関 too far:極端に,余計に
 far infrared ray (radiation) 遠赤外線
Farad ファラッド【静電容量の単位:F】
Faraday's law ファラデーの法則
fashion 方法,形状
 関 in a opposite fashion:逆(反対)に,逆の方法で
fast 形早い,速い,抵抗性の,耐性の,固着した 副早く
 fast breeder reactor 高速増殖炉
 fast neutron 高速中性子
 fast relay 速度継電器
fasten 動締め付ける,固定する
fastener ファスナ,留め具,締め具
fastening bolt 締付けボルト
fastening torque 締付けトルク
fatal 形致命的な,重大な,致死の
 fatal accident 死亡事故,人身事故
 fatal electric shock 感電死
fatigue 名疲労,疲れ
 動疲れさせる,疲労させる
 fatigue corrosion 腐食疲れ
 fatigue crack 疲労き裂
 fatigue deformation 疲労変形
 fatigue failure (breakdown, fracture) 疲労破壊,疲れ破損
 fatigue life 疲労寿命,疲れ寿命
 fatigue limit 疲労限度
 fatigue notch factor 切欠き係数,疲れ切欠き係数
 fatigue ratio 疲れ限度比,疲れ率
 fatigue strength 疲労強度,疲れ強さ
 fatigue test 疲労試験
fatty 形脂肪質の
 fatty acid 脂肪酸
faucet コック,蛇口,給水栓
 faucet joint 印ろう継手【管継手の一種】
fault 欠陥,傷,故障,障害,漏電
 fault circuit 故障回路
 fault diagnosis 故障診断
 Fault Tree Analysis FTA,フォールトツリー解析,故障樹解析【信頼度評価法の一種】
faultily 副誤って,不完全に
faulty 形欠陥のある,誤った,欠点のある,不完全な
 faulty circuit 故障回路
favor 名好意,支持
 動賛成する,好む
favorable 形好意的な,好都合な,賛成の,順調な,友好的な,有利な
faying surface 接合面
feather key フェザキー,滑りキー,案内キー
feature 名特徴,特色,特質,機能,性能,機構
 動重大に扱う,呼び物とする
fed 動「feed」の過去,過去分詞
feed 動供給する
 名送り,供給,給水
 feed back 動フィードバックする
 feed-back 名フィードバック,帰還
 feed-back control フィードバック制御
 feed check valve 給水逆止弁
 feed-forward control フィードフォワード制御
 feed-function F機能,送り機能
 feed-gear 送り装置
 feed-motion 送り運動
 feed-valve 給水弁
 feed-water 給水
 feed-water check valve 給水逆止弁
 feed-water control 給水制御
 feed-water heater 給水加熱器
 feed-water pump 給水ポンプ
 feed-water regulator 給水調整器
 feed-water softener 給水軟化装置
 feed-water stop valve 給水止め弁
 feed-water system 給水系,給水方式
 feed-water tank 給水タンク
 feed water temperature 給水温度
feeder フィーダ,供給装置
 feeder panel 給電盤

feeler フィーラ, すきまゲージ, 触手
 feeler gauge　すきまゲージ
 feeler pin　感じ針

feet フィート【「foot」の複数形】

felt フェルト, 毛せん

female 形雌の 名雌
 female screw　雌ねじ
 関 male screw：雄ねじ

ferric 形鉄の, 第二鉄の

ferrite フェライト【α鉄の金属組織上の名称】

ferro- 「鉄の」「第一鉄の」の意を表す連結形
 ferroalloy　合金鉄, 鉄合金
 ferrochrome　フェロクロム, クロム鉄
 ferroconcrete　鉄筋コンクリート
 ferromagnetic material (substance)　強磁性体
 ferromanganese　フェロマンガン
 ferronitride　窒化鉄

ferrous 形鉄の, 第一鉄の
 ferrous alloy　合金鉄
 ferrous chloride　塩化第一鉄
 ferrous metal　鉄合金, 鉄金属
 反 nonferrous metal：非鉄金属

ferrule 口輪

FET (Field Effect Transistor) 電界効果トランジスタ

few 形ほとんどない
 関 a few：少数の

fiber ファイバ, 繊維, 合成繊維
 fiber grease　ファイバグリース【グリースの一種】
 Fiber Reinforced Plastics　FRP, 繊維強化プラスチック
 fiber scope　ファイバスコープ【内視鏡の一種】

fibrous structure 繊維状組織

field 場, 界磁, 磁界
 関 magnetic field：磁界, 磁場
 field circuit　界磁回路
 field coil　界磁コイル
 field control　フィールド制御, 界磁制御
 field core　界磁鉄心
 field current　界磁電流
 Field Effect Transistor　FET, 電界効果トランジスタ
 field intensity　電界(磁界)の強さ
 field ion　電解イオン
 field magnet　界磁石, 界磁
 field magnet coil　界磁巻線
 field ohmic loss　界磁銅損
 field pole　界磁極
 field regulator　界磁調整器
 field resistance　界磁抵抗
 field rheostat　界磁加減抵抗器, 界磁調整器, 界磁抵抗器
 field system　界磁
 field winding　界磁巻線

Fig. (Figure) 図
 関 as shown in Figure A：図A参照, 図Aで示されるように

fight a fire 動消火にあたる

figure 名図[形], 数字, 挿絵, 桁 動計算する, 図(絵)に表す

filament フィラメント, 細糸【電球, 真空管などの発熱部の細い線】

file 動ファイルする, やすりをかける 名やすり

filiform corrosion 糸状腐食

filing やすり仕上げ

fill 動満たす, いっぱいにする 名充填
 fill factor　充填率, 充填比
 fill with　動～で一杯になる

filled thermometer 封入式温度計

filler フィラ, 充てん剤, 詰め物, 混ぜ物
 filler metal　溶加剤, 埋め金

fillet フィレット, すみ肉, 面取り
 fillet radius　歯底の丸み
 fillet weld　すみ肉溶接

filling フィリング, 補充, 充填, 注入, 詰め物, 盛り金
 filling chock　詰め材
 filling pipe　注入管

film 名フィルム, 薄膜 動薄膜でおおう, 撮影する
 film bearing　膜軸受
 film boiling　膜沸騰

finned

film cooling 膜冷却
film integrated circuit 膜IC, 膜集積回路
film lubrication 油膜潤滑
film thickness 膜厚, 油膜厚さ
film-wise 形膜状の
filter 動ろ過する, 浸透する 名フィルタ, ろ過器, 平滑回路 関 air filter：エアーフィルタ
filter aid ろ過助剤, ろ過促進剤
filter cartridge ろ過材
filter cloth ろ布
filter dust collector ろ過集塵装置
filter element フィルタエレメント, フィルタ素子【ろ[過]材のこと】
filter medium ろ[過]材
filter paper ろ紙, こし紙
filter press フィルタプレス, 圧ろ器【ろ過器の一種】
filterability ろ過性
filteration ろ過
filtering 形フィルタの, ろ過の 名フィルタリング
filtering-type smoke meter ろ過式排気濃度計
fin フィン, 鋳ばり, ひれ, 鍔(つば), 垂直安定板
fin efficiency フィン効率
fin plate heat exchanger フィン付形熱交換器
fin tube フィン付管
final 形最後の, 最終の
final annealing 仕上げ焼なまし
final controlling element 調節部, 操作部, 最終制御要素
final plan 完成図
final pressure 終圧
final temperature 最終温度
final velocity 終速
finally 副最後に, ようやく, ついに
find 動見つける, わかる
finder ファインダ, 距離計, 探知器, 測定器
fine 形微細な, 細い, 良質の, 鋭い, 精巧な, 希薄な 副うまく
fine adjustment 微調整, 微動装置, 細密調整
fine ceramics ファインセラミックス【高性能のセラミックス製品】
fine chemical ファインケミカル【高純度の化学製品】
fine finishing 精密仕上げ
fine line 細線
fine polymer ファインポリマ, 機能性高分子
fine screw thread 細目ねじ
fineness ファインネス, 詳細, 精細度, 純度, 微粉度, 粉末度, 粒度, 細かさ
finer 形「fine」の比較級
fines 微粉
finger 指
finger gauge フィンガゲージ【スラスト軸のすきまを計測する装置】
finger pin closer フィンガピンクロージャ【圧力容器に使われる継手法】
finish 動終える, 仕上げる 名仕上げ
finish machining 仕上げ加工
finish mark (symbol) 仕上げ記号
finished 形完成した, 仕上がった
finished bolt 仕上げボルト
finished product 完成品
finished size 仕上げ寸法
finished steel 仕上げ鋼
finished surface 仕上げ面
finishing 仕上げ, 仕上げ削り, 処理
finishing allowance 削りしろ, 仕上げしろ
finishing cut 仕上げ削り
finishing tool 仕上げバイト
finite 形有限の, 制限された
finite difference method 有限差分法
finite element method 有限要素法
finite strain 有限ひずみ
finned 形フィン付の, ひれをもった
finned radiator フィン付放熱器
finned surface フィン付伝熱面
finned tube フィン付管

finned tube heat exchanger　フィン付管式熱交換器

fire　名ファイア，火，炎，火災，燃焼　動燃やす，点火する

　fire alarm　火災報知器，火災警報装置

　fire and bilge pump　消防ビルジポンプ

　fire area　防火区画

　fire axe　消防おの

　fire brick　耐火れんが

　fire clay　耐火粘土

　fire control　防火，消火

　fire detector　火炎探知器

　fire disaster　火災

　fire door　防火扉

　fire drill　防火訓練，消防訓練

　fire escape　火災避難装置

　fire exit　非常口

　fire extinguisher　消火器

　fire fighting　消火，消防

　fire fighting equipment　消火設備，消防器具

　fire foam　消火あわ，あわ消火剤

　fire hose　消火ホース

　fire hydrant (plug)　消火栓

　fire point　燃焼点，着火点，発火点

　fire preventing separation　防火区画

　fire protection　防火

　fire pump　消火(防)ポンプ

　fire resisting construction　耐火構造

　fire resisting division　耐火区画

　fire resisting material　耐火材料

　fire resistive　形耐火性の，難燃性の

　fire surface　火面

　fire tube　煙管

　fire tube boiler　煙管ボイラ

　fire wall　防火壁

fired boiler　燃焼ボイラ

fired heater　加熱炉

fireproof　名防火，耐火　形防火性の，耐火性の

　fireproof material　耐火材料

　fireproof paint　防火塗料

firework fuel　助燃剤

firing　ファイアリング，発火，着火，燃焼，焼成，発射

　firing circuit　点弧回路

　firing order　点火順序

　firing pressure　燃焼圧力

　firing rate　燃焼速度

firm　動固める，安定させる　形頑丈な，安定した

firmly　副堅く，堅固に，しっかりと

first　形第一の，最初の　副第一に，最初に

　first angle projection (system)　第一角法

　first law of motion　運動の第一法則，慣性の法則

　first law of thermodynamics　熱力学第一法則

　first order lag element　一次遅れ要素

　first stage　第１段

firstly　副まず第一に

fish　魚

　fish eye　銀点【溶接金属の破断面に現れる円形の欠陥】

　fish plate　継目板，添板

　fish scale　フィッシュスケール【うろこ状破面】

　fish tailing　フィシュテーリング【軸受に生じる尾びれ状の割れ】

fission　分裂

　関 nuclear fission：核分裂

fit　動取り付ける，据え付ける，設ける，一致する，適合する　名はめあい，適合，ぴったり合うこと，嵌合　形適当な　関 pressure fit：圧力ばめ

　fit quality　はめあい等級，はめりの等級

　fit tolerance　はめあい公差

　fit with　動～に一致する，取り付ける

fitting　はめあい，適合性，すり合わせ，調整，取付け，仕上げ，継手，付属品

　fitting out basin　艤装岸壁

　fitting shop　仕上げ工場，組立工場

　fitting strip　はさみ板

fittings　取付部品，付属品，艤装品

fix　動固定する，修理する，定める

名調整,修理
fixed 形固定した,固定の,一定の,凝固した,不揮発性の
fixed acid 不揮発性酸
fixed alkali 不揮発性アルカリ
fixed beam 固定梁(はり)
fixed blade 固定羽根,静翼
fixed capacitor 固定コンデンサ
fixed coil 固定コイル
fixed command control 定値制御
fixed cycle 固定サイクル,定周期
fixed delivery pump 定容量形ポンプ
fixed displacement pump 定容量形ポンプ
fixed element 固定部
fixed end 固定端
  関 free end:自由端
fixed end moment 固定端モーメント
fixed guide vane 固定案内羽根
fixed load 固定荷重
fixed moment 固定モーメント
fixed oil 固定油,不揮発性油
Fixed Pitch Propeller FPP,固定ピッチプロペラ
fixed point 固定点,定点
fixed pulley 定滑車
fixed resistor 固定抵抗器
fixed stay 固定揺れ止め
fixed stroke pump 不変行程ポンプ
fixed support 固定支点
fixed type 固定式
fixed value 固定値,一定値
fixed value control 定値制御
fixed vane 固定羽根
fixed way 固定台
fixing moment 固定モーメント
fixity 固定,連続性
fixture 固定具,備品,取付け設備,取付け器具
flag 旗
flake 名フレーク,薄片,白点 動薄片にはがす
flaking フレーキング,はく離
flame 名フレーム,炎,火炎 動炎を出す,燃え上がる
flame annealing 火炎焼なまし
flame arrester 火炎防止装置
flame cone 白心,炎心
flame detector 火炎検出器
flame eye フレームアイ
flame failure detector 消炎検出器
flame front 火炎面
flame hardening 火炎焼入れ【酸素アセチレンなどの火炎によって焼入れする表面硬化法】
flame impingement 熱衝撃
flame machining 炎切断,ガス切断
flame piloting baffle 保炎板
flame plate 炎板
flame plating 炎めっき
flame proof 形耐炎性の
flame propagation 火炎伝播,延焼
flame quenching 消炎
flame radiation 火炎放射
flame reaction 炎色反応
flame retardant material 難燃性材料
flame temperature 火炎温度
flame tube 煙管
flame velocity 火炎速度,火炎伝播速度
flammability 燃焼性,引火性
flammability limit 燃焼限界,可燃限界
flammable 形可燃性の
flammable gas 可燃性ガス
flammable liquid 可燃性液体,引火性液体
flammable mixture 可燃性混合物
flammable substance 助燃剤
flange フランジ,鍔(つば)
flange coupling フランジカップリング,フランジ継手
flange joint フランジ継手,つば継手
flange nut フランジナット,フランジ付ナット,つば付ナット
flanged connection フランジ継手(接続)
flanged valve フランジ付弁
flank 逃げ面,側面
flank wear フランク摩耗,逃げ

面摩耗
flap 名フラップ, 防火板, はね蓋
動たたく, はためく
  flap valve　蝶(ちょう)形弁
flapper　フラッパ
  flapper valve　フラップ弁, 蝶(ちょう)形弁
flare groove joint　フレア継手
flareback　後炎
flareless type pipe joint　食込み式管継手
flash 名フラッシュ, 閃光, ひらめき, 鋳ばり
動ぴかっと光る, ぱっと燃える
  flash back　フラッシュバック, 逆火, さか火
  flash back arrester　逆火防止器
  flash boiler　フラッシュボイラ【水を噴霧状にして吹き込み, 直ちに蒸発させるボイラ】
  flash distillation　フラッシュ蒸留
  flash evaporation　フラッシュ蒸発
  flash freeze　急速冷凍
  flash gas　フラッシュガス
  flash memory　フラッシュメモリ【コンピュータ内でデータの消去・書き込みができ, 電源を切っても内容が消えないメモリ】
  flash over　フラッシュオーバ, 爆燃現象
  flash point　引火点, 発火点
  flash set　急速硬化
  flash steam　熱水混合蒸気
  flash tester　引火点試験器
flashing　点滅, 閃光
  flashing point　引火点
  flashing relay　断続継電器
  flashing signal　点滅信号
flashlight　懐中電灯
flat 形平らな
  flat bar　平鋼
  flat belt　平ベルト
    関 V belt：Vベルト
  flat chisel　平たがね
  flat compound[ed] generator　平複巻発電機
  flat key　平キー
  flat plate　平板
  flat scavenge　フラット掃気
  flat seated valve　平座弁, 平面座弁
  flat slide valve　平滑弁
  flat spring　板ばね
  flat surface　平面
    関 curved surface：曲面
  flat valve　平弁
flatcar　台車
flattened tube　扁平管
flattening　フラットニング, 展伸加工, 平坦化, 平面加工, 扁平率
flaw　割目, 傷, 欠陥
  flaw detector　探傷器
flee 動逃げる
Fleming's left hand rule　フレミングの左手の法則
Fleming's right hand rule　フレミングの右手の法則
flex 動曲げる, 収縮させる
flexibility　フレキシビリティ, たわみ性, 柔軟性, 融通性
  flexibility factor　たわみ率
flexibilizer　可塑剤
flexible 形たわみやすい, 柔軟な, 弾力的な, 曲げやすい, 融通の利く
  flexible bearing　たわみ軸受
  flexible cord　柔軟電線
  flexible coupling　たわみ継手, 弾性継手【原動機軸と被駆動軸を固定せず, 緩みを与えて回転を伝える継手】
  flexible joint　たわみ継手
  flexible shaft　たわみ軸
  flexible shaft coupling　たわみ軸継手
flexural rigidity　曲げこわさ【曲げ荷重に対する変形抵抗】
flexure　屈曲, たわみ, 曲げ
flicker 名フリッカ, ちらつき
動明滅する
  flicker relay　フリッカ継電器
flip flop circuit　フリップフロップ回路
float 動浮かぶ, 浮く
名フロート, 浮き, 浮子, 浮揚物

float carburetor　フロート気化器
float chamber　フロート室
float gauge　フロートゲージ
float ring　遊動リング【軸封部の漏止めの一種】
float switch　フロートスイッチ
float type area flow meter　フロート形面積流量計, ロータメータ
float type liquid level gauge　フロート液面計
float valve　フロート弁
floatation　浮遊　関 flotation
floater　いかだ, 救命具
floating　形浮いている, 浮動の, 移動性の, 動的な
　floating action　浮動動作
　floating axle　浮動軸
　floating body　浮体, 浮揚体
　floating bush　浮筒, 浮動ブッシュ
　floating dock　浮きドック
　floating frame bearing　浮動軸受
　floating head　遊動頭, 浮動ヘッド
　floating ring　浮動リング
flocculant　凝集剤
flocculation　フロキュレーション, 凝結, 凝集, 凝集沈降【コロイド粒子が沈殿する現象】
flon gas　フロンガス
flood　名洪水, 浸水, 充満　動あふれ出る
flooded evaporator　満液式蒸発器
flooding　フラッディング, あふれ, 浸水
　flooding pipe　張水管
　flooding valve　張水弁
floor　床, 階
　floor plan　間取り図
flotation　浮揚, 浮力　関 floatation
flow　名フロー, 流れ, 流量
　動流れる　関 current flow：電流
　flow back　動逆(環)流する
　flow boiling　対流沸騰, 強制対流沸騰
　flow chart　フローチャート, 流れ図
　flow coefficient　流量係数
　flow collecting valve　合流弁
　flow control valve　流量制御弁
　flow controller　流量調節器
　flow dividing valve　分流弁
　flow down　動流れ落ちる
　flow improver　流動性向上剤
　flow measurement　流量測定
　flow meter　流量計
　flow nozzle　フローノズル【管に取り付け圧力差から流量を測定する器具】
　flow of air　空気流
　flow path　流路
　flow pattern　流動様式
　flow rate　流量
　flow ratio control　流量比率制御
　flow resistance　流動抵抗
　flow stream　フローストリーム
　flow stress　変形抵抗, 変形応力, 流動応力, 流れ応力
fluctuate　動変動する, 上下する
fluctuating load　変動荷重, 変動負荷
fluctuation　変動, ゆらぎ
　関 velocity fluctuation：速度変動
flue　フリュ, 煙道, 排気筒, 炉筒
　flue gas　煙道ガス, 燃焼ガス
　flue gas analysis　煙道ガス分析
　flue gas denitration　排煙脱硝
　flue gas desulfurization　排煙脱硫
　flue gas treatment　排ガス処理
　flue tube　煙管
　flue [tube] boiler　炉筒ボイラ
fluid　名流体　形流体の
　関 liquid：液体／ solid：固体
　fluid catalyst　流動触媒
　fluid catalytic cracking　流動接触分解
　fluid clutch　流体クラッチ
　fluid coupling　流体継手
　fluid drive　流体駆動
　fluid dynamics　流体力学
　fluid erosion　流体浸食
　fluid film　境膜
　fluid friction　流体摩擦
　fluid inlet angle　流入角
　fluid lubrication　流体潤滑, 液体潤滑, 完全潤滑

fluidic

fluid machinery　流体機械
fluid mechanics　流体力学
fluid outlet angle　流出角
fluid pressure　流体圧力
fluid torque converter　流体トルクコンバータ

fluidic　形流動性の，流体の
fluidics　流体工学
fluidity　流動，流動性
　関 fluidity test：流動性試験
fluidization　流動化
fluorescence　蛍光
fluorescent　形蛍光性の
　fluorescent lamp　蛍光灯
　fluorescent material　蛍光物質
　fluorescent penetrant inspection　蛍光[浸透]探傷検査
fluoride　フッ化物
fluorine　フッ素　記F
fluorocarbon　フルオロカーボン，フッ化炭化水素【冷媒，潤滑剤，消火剤などに利用】
　fluorocarbon resin　フッ素樹脂
fluorometer　蛍光計
fluorometry　蛍光測定，蛍光分析
fluororubber　フッ素ゴム
flush　形同一平面の
　flush bolt　皿ボルト，皿頭ボルト
　flush deck　平甲板
flushing　フラッシング，洗浄
flute　名溝　動溝を彫る
fluted nut　溝付ナット
flutter　名フラッタ，羽ばたき
　動バタバタする・させる
　flutter valve　フラッタ弁
fluttering　フラッタリング【弁が小さく振動している状態】
flux　フラックス，束，流束，融剤
　flux density　磁束密度
fly　名飛ぶこと　動飛ぶ
　fly ash　フライアッシュ，飛散灰，微粉灰
　fly cutter　舞いカッタ
　fly wheel　フライホイール，はずみ車
FM (Frequency Modulation)　周波数変調

FO (Fuel Oil)　燃料油
foam　名あわ　動あわ立つ
　foam type fire extinguisher　あわ消火器
foaming　フォーミング，あわ立ち
　foaming tendency　あわ立ち性
focal　形焦点の
　focal distance (length)　焦点距離
　focal point　焦点
focus　名焦点，集束，中心，震源
　動焦点を合わせる，集中する
fog　霧
　fog lubrication　噴霧潤滑
foil　フォイル，箔，金属箔
　foil bearing　フォイル軸受
fold　動折りたたむ，折り曲げる
folding　形折りたたみ式の
　folding measure　折尺
follow　動従う，進む，続く
　関 as follows：以下のように
follower　従節
　follower rest　移動振れ止め
following　形次の，以下の，次に述べる
　名下記のもの，次に述べるもの
　following current　順流
　following edge　後縁
　following equation　次式，次の式
　following up element　追従部
　following wake　伴流
follow-up　形再度の，追跡の，引き続き行われる　名フォロアップ，追跡，探究，再調査
　follow-up control　追従(値)制御
　follow-up failure　追従不良
　follow-up system　追従装置
font　フォント【文字の書体・大きさ】
fool proof　フールプルーフ，絶対安全
foot　フート，足，フィート【長さの単位：1フィート＝12インチ（約30cm）】
　foot brake　フットブレーキ，足踏みブレーキ
　foot step bearing　うす軸受【臼の形をした縦軸用のスラスト軸受】

foot switch フートスイッチ, 足踏スイッチ
foot valve フート弁, 吸上げ弁【ポンプの吸込み管に設け逆流を防ぐ】
for 前~のために(の), ~に(の), ~の間, ~と引き換えに, ~の代わりに, ~に向かって, ~に対して, ~にとって, ~の理由で 接~というわけで, なぜなら
for all 前~にもかかわらず, ~があっても, ~を考慮して
for example 副例えば
force 動強いる, 強制する, 押し出す, 押し込む, 押し付ける 名力【単位：ニュートン, N】
関 magnetic lines of force：磁力線
force diagram 示圧図
force fit 圧力ばめ【しまりばめの一種】
force of friction 摩擦力
force of gravity 重力
force of inertia 慣性力
force of restitution 復元力
force pump 押上げポンプ
forced 形強制された, 強制的な
forced air cooling 強制空冷
forced circulation 強制循環
forced circulation boiler 強制循環ボイラ
forced convection 強制対流
forced convection boiling 強制対流沸騰
forced convection heat transfer 強制対流熱伝達
forced draft 押込み通風, 強制通風
forced draft fan 押込み送風機
forced frequency 強制振動数
forced head 押込み揚程
forced lubrication 強制潤滑
forced response 強制応答
forced through flow boiler 貫流ボイラ
forced ventilation 強制通風, 押込み換気
forced vibration (oscillation) 強制振動

forced vortex 強制渦
fore 形前の, 前方の, 船首の
fore cooler 予冷器
foreign 形外国の, 異国の, 海外の, 外来の, 無関係な, 異質の
foreign element 外的要素
foreign material 異物
foreign matter 異物, きょう雑物, 不純物
forepeak bulkhead 船首隔壁
forge 動鍛造する, 鍛える 名鍛造
forge welding 鍛接
forgeability 可鍛性, 鍛造性
forged 形鍛造の, 鍛造された
forged crank shaft 鍛造クランク軸
forged iron 鍛鉄, 鋼鉄
forged part 鍛造品
forged rotor 鍛造ロータ
forged steel 鍛鋼, 鍛造鋼
forging 鍛造, 鍛造品, 鍛練
forging hammer 鍛造ハンマ
fork フォーク, 分岐, 二又, くま手【軸端が二又になっている】
fork end 二又端
form 名フォーム, 形, 形態, 形状, 種類, 型, 形式 動形成する, 構成する, 生じる, 作る, 結成する, 鍛える, 組み立てる
関 in the form of：~状の, ~の形式で
form drag 形状抗力
form factor 形状因子, 形状係数
form resistance 形状抵抗
form rolling 転造
formability 加工性, 成形性
formal 形正式な, 正規の, 公式的な, 幾何学的な, 規則正しい, 形式的な
formaldehyde ホルムアルデヒド【合成樹脂の原料, 防腐剤として用いる】
format フォーマット, 型, 書式, 様式, 形式, 構成
formation 形成, 生成, 構成
関 scale formation：スケール形成
formed coil 形巻コイル
formed tool 成形工具

former 形前者の, 以前の
  関 the former：前者
formerly 副以前は
forming 成形, 形成
formula 公式, 式, 形式, 規格
  関 Euler's formula：オイラーの式
fortunately 運よく, 幸いにも
forward 動送る, 回送する 形前部の, 船首の, 前の, 急進的な, 進んだ 副前方へ, 先へ
  関 look forward to：期待する, 待つ
  forward bias　順方向バイアス
  forward current　順電流, 順方向電流
  forward curved vane　前曲羽根, 前向き羽根
  forward direction　順方向
  forward end　前端
  forward voltage　順(方向)電圧
fossil fuel　化石燃料
Foucault current　フーコー電流, 渦電流
foul 形汚れた, 不潔な 動汚す, 汚れる
  foul air　汚れた空気
fouling　汚れ, 付着, 堆積
  fouling factor　汚れ係数
found 動「find」の過去・過去分詞
foundation　基礎, 土台
  foundation bolt　基礎ボルト
  foundation drawing　基礎図
foundry　鋳造, 鋳物工場, 鋳造場
four cycle diesel engine　4サイクルディーゼル機関
four [stroke] cycle engine　四サイクル機関
Fourier　フーリエ【フランスの数学者・物理学者】
  Fourier analysis　フーリエ解析
  Fourier series　フーリエ級数
fourthly 副第四に
FPP (Fixed Pitch Propeller)　固定ピッチプロペラ
fraction　一部, 断片, 比, 小数, 部分, 分数, 留分, 端数, 精留
  関 a fraction of：僅かな, 何分の1かの／a fraction of a millionth：100万分の1
  fraction defective　不良率
  fraction of distillate　留分
fractional 形小部分の, 断片的な, 分数の, 分留の
  fractional bond　部分結合
  fractional distillation　分留
    関 distillation：蒸留
  fractional pitch winding　短節巻
fracture 名破壊, 破面, 破損 動折れる, 壊れる, 破損する
  fracture ductility　破断延性
  fracture energy　破壊エネルギ
  fracture experiment　破壊実験
  fracture load　破壊荷重
  fracture mechanics　破壊力学
  fracture point　破断点
  fracture strength　破壊力, 破壊強さ
  fracture stress　破壊応力
  fracture surface　破面
  fracture test　破壊(断)試験
  fracture toughness　破壊靭性
  fracture wear　破壊摩耗
fragility　脆弱, 脆(もろ)さ
frame 名フレーム, 架構, 骨格, 骨組, 枠 動組み立てる, 考案する, 枠にはめる
  frame antenna　枠形アンテナ
  frame work　骨組構造, 枠組, 骨組
framed structure　骨組構造
frankly speaking　率直に言えば
Frary metal　フラリーメタル【軸受合金の一種】
free 形自由な, 解放された, 無料の 動解放する
  free acid　遊離酸
  free alkali　遊離アルカリ
  free bend　自由曲げ
  free carbon　遊離炭素【鋳鉄に含まれる黒鉛】
  free convection　自由対流
  free cutting ability　快削性
  free cutting steel　快削鋼
  free electron　自由電子
  free end　自由端
    関 fixed end：固定端

friction

free energy　自由エネルギ
free forging　自由鍛造
free from　動〜がない，含まない，使われない
free machining material　快削材料
free oscillation　自由振動
free radical　フリーラジカル，遊離基【不対電子をもつ原子または分子】
free response　自由応答
free shear layer　自由せん断層
free stream　自由流
free streamline　自由流線
free support　自由支点
free vibration　自由振動
free vortex　自由渦
free working distance　作動距離
free water　遊離水，分離水
freely　副自由に，十分に
freeze　名フリーズ，氷結
　動凍結する
freeze drying　凍結乾燥
freezer　フリーザ，凍結器，冷凍庫
freezing　凍結，冷凍，氷結，凝固
freezing chamber　凍結室
freezing machine　冷凍機
freezing point　凝固点，氷点
freezing point depression　凝固点降下
French curve　雲形定規
Freon　フレオン【商標名：冷媒の一種】
frequency　周波数，振動数，頻度，回数　関 high (low) frequency：高(低)周波
frequency analysis　周波数分析
frequency characteristic　周波数特性
frequency circuit　周波数回路
frequency distribution table　度数分布表
frequency equation　振動数方程式
frequency meter　周波数計
Frequency Modulation　FM，周波数変調
frequency modulator　周波数変調器
frequency relay　周波数継電器
frequency response　周波数応答
frequency spectrum　振動数スペクトル
frequency transfer function　周波数伝達関数
frequency transformation　周波数変換
frequent　形たびたびの
frequent cause of　形〜の原因になることが多い
frequently　副しばしば，たびたび
fresh　形新鮮な，新しい
fresh air　新気，新しい空気
fresh water　清水
fresh water cooler　清水冷却器
fresh water pump　清水ポンプ
fresh water tank　清水タンク
freshly　副新たに，新しく，新鮮に
fret saw　糸のこ
fretting　フレッチング，擦過【こすれて摩耗する現象】
fretting corrosion　フレッチングコロージョン，擦過腐食，摺動腐食
fretting wear　フレッチング摩耗
friability　破砕性
friction　摩擦
　関 mechanical friction：機械摩擦
friction angle　摩擦角
friction brake　摩擦ブレーキ
friction buffer　摩擦緩衝器【摩擦力を利用したエネルギ吸収器】
friction circle　摩擦円
friction clutch　摩擦クラッチ
friction coefficient　摩擦係数
　同 coefficient of friction
friction contact interface　摩擦接触面
friction coupling　摩擦継手
friction damper　摩擦ダンパ【摩擦によって振動を減衰させる装置】
friction disc　摩擦円板
friction drag　摩擦抗力，摩擦抵抗
friction drive　摩擦駆動，摩擦伝動
friction dynamometer　摩擦動力計【動力計の一種】
friction force　摩擦力
friction gearing　摩擦駆動，摩擦伝動装置
friction head　摩擦水頭，摩擦損

frictional

失水頭
friction horsepower　摩擦馬力
friction loss　摩擦損失
friction loss of head　摩擦損失水頭
friction loss of pipe flow　管摩擦損失
friction modifiers　摩擦調整剤【潤滑油添加剤の一種】
friction of motion　動摩擦
friction of rest　静止摩擦
friction pressure welding　摩擦圧接
friction pulley　摩擦車
friction pump　摩擦ポンプ【回転ポンプの一種】
friction resistance　摩擦抵抗
friction tape　絶縁用テープ
friction torque　摩擦トルク
friction transmission　摩擦駆動
friction vibration　摩擦振動
friction wake　摩擦伴流
friction welding　摩擦圧接
friction wheel　摩擦車
friction work　摩擦仕事
frictional　形摩擦の
frictional damping　摩擦減衰
frictional dynamometer　摩擦動力計
frictional electricity　摩擦電気
frictional force　摩擦力
frictional heat　摩擦熱
frictional loss　摩擦損失
frictional resistance　摩擦抵抗
frictional wear　摩擦摩耗
frictionless　摩擦のない
fringe pattern　縞模様
from A into B　AからBへ
from A to B　AからBまで
from one end to the other　端から端まで
front　名前部, 前方　形前の, 前方の　関 in front of：～の前に
front bearing　前軸受
front clearance　前逃げ角
front column　前柱
front elevation　正面図
　関 plan：平面図
front face　正面
front fillet weld　全面すみ肉溶接

front glass　フロントグラス
front header　前部管寄せ
front rake　すくい角
front view　正面図
frost　名霜　動霜で覆う, 凍る
frosted　形つや消しの, 霜で覆われた
frosting　きさげ仕上げ, つや消し
froth　名あわ　動あわ立つ
frother　気泡剤
Froude dynamometer　フルード動力計
Froude number　フルード数
Froude water brake dynamometer　フルード型水動力計
frozen　形凍った, 冷凍された
frozen cargo ship　冷凍船
FRP (Fiber Reinforced Plastics)　強化プラスチック
FTA (Fault Tree Analysis)　フォールトツリー解析, 故障樹解析【信頼度評価法】
fuel　名燃料　動燃料を供給する
fuel additive　燃料添加剤
fuel air cycle　燃料空気サイクル
fuel air mixture　混合気
fuel air ratio　燃料空気比
fuel cam　燃料カム
fuel cell　燃料電池
fuel consumption　燃料消費, 燃料消費量
fuel consumption ratio　燃料消費率
fuel controlling handle　燃料加減ハンドル
fuel cut-off　燃料遮断
fuel economy　燃料節約, 燃費
fuel feed pump　燃料供給ポンプ
fuel gas　燃料ガス
fuel injection　燃料注入, 燃料噴射
fuel injection device　燃料噴射装置
fuel injection nozzle　燃料噴射ノズル
fuel injection pipe　燃料噴射管
fuel injection pump　燃料噴射ポンプ
fuel injection system　燃料噴射システム, 燃料噴射装置
fuel injection valve　燃料噴射弁

furnace

fuel injector　燃料噴射器
fuel inlet valve　燃料弁
fuel line　燃料管, 燃料系路
fuel nozzle　燃料ノズル
Fuel Oil　FO, 燃料油
fuel oil additive　燃料添加物
fuel oil filter　燃料油フィルタ, 燃料油こし器
fuel oil heater　FOヒータ, 燃料油加熱器
fuel [oil] pump　FOポンプ, 燃料油ポンプ
fuel oil service pump　燃料油供給ポンプ
fuel oil service tank　燃料油サービスタンク
fuel oil settling tank　燃料油セットリングタンク
fuel oil sludge tank　燃料油スラッジタンク【スラッジなど再生の可能性のない燃料油タンク】
fuel oil storage tank　燃料油貯蔵タンク
fuel ratio (rate)　燃料比
fuel spray　燃料噴霧
fuel supply system　燃料供給装置
fuel system　燃料装置, 燃料系統, 燃料供給装置
fuel tank　燃料タンク
fuel treatment　燃料処理
fuel valve　燃料弁
fugacity　フガシティ, 揮発性, 揮散力
fulcrum　支点
　fulcrum shaft　支点軸, 揺れ軸
full　[形]いっぱいの, 完全な, 十分な, 全部の, 最大の, 最高の
　[動]満たす
　full admission　全周噴射
　full annealing　完全焼なまし
　full automatic control　無人制御
　full bearing　全周軸受
　full depth tooth　並歯
　full fillet weld　全厚すみ肉溶接
　full lift safety valve　全揚程安全弁
　full line　実線
　full load　全負荷, 全荷重, 満載

full load current　全負荷電流
full load displacement　満載排水量
full load efficiency　全負荷効率
full load torque　全負荷トルク
full pitch winding　全節巻
full power　全出力
full pressure　全圧
full radiation　完全放射体, 黒体
full size　実物大, 現尺, 原寸
full speed　全速[力], 最高速力
full voltage starting　全電圧始動
full wave rectification　全波整流
fully　[副]十分に, 完全に
　fully developed flow　完全に発達した流れ
　fully insulated tool　絶縁工具
　fully open condition　全開状態
fume　ヒューム, 蒸気, 煙, 煙霧, 排気ガス
function　[名]機能, 働き, 役目, 関数
　[動]働く, 作用する, 機能する
　function design　機能設計
functional　[形]機能的な
　functional material　機能材料
functionally gradient material　傾斜機能材料
fundamental　[形]基礎の, 基本の, 基本的な
　[名]基礎, 基本, 原理, 原則
　(関) transfer function：伝達関数
　fundamental equation　基礎式, 基礎方程式
　fundamental experiment　基礎実験
　fundamental expression　基礎式
　fundamental period　基本周期
　fundamental unit　基本単位
　fundamental vector　基本ベクトル
fundamentalty　根本的に
funicular polygon　速力図, 索多角形
funnel　煙突, 漏斗
furnace　ファーネス, 炉, 炉筒, 燃焼室, 溶鉱炉
　furnace cooling　炉冷【金属材料の熱処理の一種】
　furnace pressure control system　炉内圧力制御装置

furnace tube　煙管
furnace wall　炉壁
furnish　動供給する
further　副さらに，もっと先に　形それ以上の，もっと先の
furthermore　副さらに
fuse　ヒューズ　動ヒューズを取り付ける，溶かす，融合させる
fuse box　ヒューズ箱
fusibility　可融性
fusible　形可溶性の，可融性の
fusible alloy　可融合金
fusible disconnecting switch　ヒューズ付断路器
fusible plug　溶栓，可溶栓
fusing current　溶断電流
fusing point　融点

fusion　融合，融解，溶融
　関 heat of fusion：融解熱
fusion line　融合線
fusion point　融点
fusion welding　融接
fusion zone　融合部，融解域，融解部
future　名未来，将来
　形未来の，将来の
fuzzy　形あいまいな，柔軟な，縮れた，ぼやけた
fuzzy control　ファジィ制御【人間の認識のあいまいな部分をコンピュータで処理する技術】
F-V change　FV変換，周波数-電圧変換

G-g

G mark Gマーク
gage 同 gauge
gain 名ゲイン, 増加, 利得【出力量の入力量に対する増幅度を表す】動得る, 増す, 加える, 受ける
  gain diagram　ゲイン線図
  gain margin　ゲインマージン, ゲイン余裕
gallery　ギャラリ, 広い通路, 回廊, 廊下
galley　ギャレ, 調理室
galling　ゴーリング, かじり傷, 摩損【摩擦熱で金属の一部が溶ける現象】
gallium　ガリウム　記 Ga
gallon　ガロン【液量の単位】
galvanic　形ガルバニ電気の, 直流電気の
  galvanic action　電食作用, 電池作用
  galvanic cell　ガルバニ電池
  galvanic corrosion　電気腐食, 接触腐食, 局部電池腐食, 異種金属接触腐食, 電気化学的腐食, ガルバニ腐食
  galvanic electricity　動電気
galvanization　直流通電, 亜鉛めっき
galvanize　動亜鉛めっきする, 電気を通す
galvanized　形亜鉛めっきされた
  galvanized iron　亜鉛めっき鉄
  galvanized sheet　トタン板
  galvanized steel pipe　亜鉛めっき鋼管
galvanizing　亜鉛めっき
galvanometer　検流計
gamma iron　ガンマ鉄, γ鉄【鋼の組織】
gamma ray　ガンマ線, γ線【放射線の一種】
gang　作業隊, 一団
  gang control　連動制御
  gang socket　連結ソケット
  gang switch　連結スイッチ

ganister sand　珪(けい)砂
gap　ギャップ, すきま, 相違, 割れ目
garter spring　ガータスプリング
gas　ガス, 気体, ガソリン
  gas analysis　ガス分析
  gas brazing　ガスろう付け
  gas burner　ガスバーナ
  gas calorimeter　ガスカロリメータ, ガス熱量計
  gas carburizing　ガス浸炭【メタンガスを用いた浸炭法】
  gas charging　ガス充てん
  gas chromatography　ガスクロマトグラフィ【ガスの成分測定器】
  gas cleaner　ガス洗浄器
  gas constant　ガス定数, 気体定数
  gas cutting　ガス切断
  gas cycle　ガスサイクル
  gas cycle engine　気体サイクル機関
  gas detector　ガス検知器
  gas engine　ガス機関
  gas exchange　ガス交換
  gas film　気膜
  gas firing　ガスだき
  gas furnace　ガス炉
  gas generator　ガス発生装置
  gas leaking　ガス漏れ
  gas liquid equilibrium　気液平衡
  gas liquid separation　気液分離
  gas liquid two phase flow　気液二相流
  gas mask　ガスマスク, 防毒面
  gas meter　ガスメータ, ガス量計
  gas nitriding　ガス窒化【アンモニアガス中での硬化法】
  gas oil　軽油
  gas permeability　通気性
  gas phase　気相
    関 liquid phase：液相／solid phase：固相
  gas pipe joint　ガス管継手
  gas pocket　ブローホール, ガスポケット, 過熱きず, 巣

gas pressure regulator ガス圧力調節器
gas pressure welding ガス圧接
gas producer ガス発生器
gas recovery ガス回収
gas reheater ガス再熱器
gas shield ガスシールド
gas stream ガス流
gas table ガス表
gas temperature ガス温度
gas thermometer 気体温度計
gas thread ガスねじ,管用ねじ
gas tight 気密
gas turbine ガスタービン
gas turbine cycle ガスタービンサイクル
gas valve ガス弁
gas welding ガス溶接

gaseous 形気体の,ガスの
gaseous bearing 気体軸受
gaseous fuel 気体燃料
 関 liquid fuel:液体燃料
gaseous mixture 混合気
gaseous phase 気相
gaseous substance ガス状物質

gasification ガス化,気化
gasifier 気化器,ガス化装置
gasket ガスケット【漏止め材の一種】
gasket factor ガスケット係数【ガスケット締付け圧と内圧との比】
gasket packing ガスケットパッキン

gasoline ガソリン 同 petrol
gasoline engine ガソリン機関
 関 diesel engine:ディーゼル機関

gastight 形気密の,ガス密閉構造の

gate ゲート,門,出入口,仕切り,[半導体の]制御電極
gate alloy ゲート合金
gate amplifier ゲート増幅器
gate pulse ゲートパルス
Gate Turn Off thyristor GTOサイリスタ,ゲートターンオフサイリスタ
gate valve ゲート弁,仕切弁

gateway ゲートウェイ,入口
gather 動集める,採取する

gauge 名ゲージ,計器,定規 動測る
gauge board 計器盤
gauge cock 検水コック
gauge factor ゲージファクタ,ゲージ率【ゲージ固有の定数】
gauge glass 水面計ガラス,液面計ガラス
gauge length 標点距離【引張り試験片の2標点間距離】
gauge mark 標点
gauge notch 切欠き流量計
gauge panel 計器板
gauge pressure ゲージ圧力【大気圧を基準とした圧力の大きさ】
 関 absolute pressure:絶対圧力

gauss ガウス【磁束密度の単位】
gauss meter ガウスメータ【磁界の強さを測る磁力計】

Gaussian distribution ガウス分布,正規分布

gauze ガーゼ,金網
gauze wire ゴーズワイヤ,細目金網

Gay-Lussac's law ゲイ-ルサックの法則【一定圧力下での気体の温度と体積の関係】

GB (gigabyte) ギガバイト

gear 名歯車,装置,伝動装置 動かみ合う,連動させる
gear box ギアボックス,歯車箱,変速機箱,変速機,変速装置
gear case ギアケース,歯車箱
gear change ギアチェンジ,変速
gear coupling ギアカップリング,歯車形軸継手
gear cutting machine 歯切盤
gear drive (driven) 歯車駆動
gear grease ギアグリース【グリースの一種】
gear grinder 歯車研削
gear hob 歯車用ホブ【切削工具の一種】
gear lever 変速てこ
gear motor ギアモータ,歯車モータ
gear oil ギア油
gear puller ギアプーラ

gear pump　歯車ポンプ
gear ratio　歯車比, 歯数比, ギア比
gear shifting lever　変速てこ
gear tester　歯車検査機
gear tooth　歯
gear tooth gauge　歯形ゲージ
gear tooth micrometer　歯厚マイクロメータ
gear tooth vernier caliper　歯形ノギス
gear train　歯車列【複数の歯車がつながり動力を伝える機構】
gear wheel　大歯車, 親歯車【1対の歯車のうち歯数の多い歯車】
geared　形 ギアのある
geared motor　ギアドモータ, 歯車付モータ
geared pump　歯車伝動ポンプ
geared turbine　歯車減速タービン
geared type shaft coupling　歯車形軸継手
gearing　ギア, 伝動装置, 歯車, 歯車装置, 歯車伝動装置
Geislinger coupling　ガイスリンガ継手【捩じり振動を吸収できる弾性継手】
gel　ゲル【コロイド溶液が流動性を失ったゼリー状態】
gel point　ゲル化点
gelling　ゲル化, ゼリー化
gene engineering　遺伝子工学
general　形 一般的な, 全般的な, 普遍的な, 全体的な
関 in general：一般に
general alarm　非常警報
general angle　一般角
general arrangement　全体配置図, 一般配置
general corrosion　全面腐食
general drawing　全体図, 一般図
general illumination　全般照明
general lighting　全般照明
general procedure　基本手順
general purpose　形 汎用の, 多目的の
General Service pump　GSポンプ, 雑用ポンプ
general solution　一般解
general survey　普通検査
general view　全体図
general work　雑工事
generalization　一般化
generalized　形 一般化した
generalized force　一般化力, 一般力
generally　副 一般に, 一般的に, 通常
generally speaking　一般的に言えば
generate　動 生じる, 発生させる, 起こす, 生み出す
generated output　発電電力
generating　発電, 発生, 生成, 創成
generating circle　ころがり円, 母円
generating line　母線, 導線
generating plant (station)　発電所
generating surface　伝熱面
generating system　発電システム
generating tube　蒸発管
generating voltmeter　発電機形電圧計
generation　発生, 生成
generator　ジェネレータ, 発電機, 発生器
generator withstand voltage test　発電機耐電圧試験
genuine part　純正部品
geometric (geometrical)　形 幾何学の, 幾何学上の, 幾何学的な
geometrical factor　幾何学的因子, 形態係数
geometrical moment of area　断面二次モーメント
geometrical moment of inertia　断面二次モーメント
geometrical tolerance　幾何公差
geometrically　副 幾何学的に
geometry　幾何学, 形状, 配置(列)
German silver　洋銀
germanium　ゲルマニウム　記 Ge
get　動 得る, 受ける, 手に入れる
get along　動 うまくいく
get away　動 取り去る
get rid of　動 取り除く, やめる
get through　動 通り抜ける

getter alloy ゲッター合金

GFRP (Glass Fiber Reinforced Plastics) ガラス繊維強化プラスチック

ghost line ゴースト線

giant tanker 巨大タンカ

gib headed key 頭付キー

giga byte GB, ギガバイト【記憶容量の単位で10億バイトに相当する】

gilled 形 ひれ付の
　gilled cooler　ひれ付冷却器
　gilled superheater　ひれ付過熱器
　gilled tube radiator　ひれ付管放熱器

gimbal joint 水平自在継手

gin 三叉

girder ガーダ, けた【船体を縦通するけた】

give 動 与える, 供給する
　give off　動 放出する, 発する
　give rise to　動 生み出す, 生じさせる

gland グランド, パッキン押え
　gland bush　グランドブッシュ
　gland exhaust condenser　グランド復水器【タービンのグランドからの漏れ蒸気を復水させる装置】
　gland exhaust fan　グランド排気ファン
　gland packing　グランドパッキン【軸密封用パッキン】
　gland steam　グランド蒸気

glass ガラス
　glass fiber　ガラス繊維
　Glass Fiber Reinforced Plastics GFRP, ガラス繊維強化プラスチック 関 CFRP:炭素繊維強化プラスチック
　glass paper　紙やすり
　glass thermometer　ガラス温度計, 棒状温度計
　glass water gauge　ガラス水面計
　glass wool　グラスウール, ガラス綿

glaze 光沢, つや 動 つやを出す

glide 動 滑る・らせる, 滑るように動く

gliding movement 平面運動, すべり運動

global 形 球状の, 球形の, 世界的な, 地球的な
　global environment　地球環境
　Global Positioning System　GPS, 全地球的位置決定システム
　global warming　地球温暖化

globe 球, 球面, 球体, 地球
　globe cam　球面カム
　globe valve　玉形弁

globule 小球, (液体の)小滴

gloss 光沢

glossmeter 光沢計

glove 手袋

glow 名 グロー, 白熱, 赤熱, 光, 輝き 動 白熱する, 赤熱する, 輝く, 光る, 燃える
　glow discharge tube　グロー放電管【定電圧放電管, ネオン管など】
　glow lamp　グローランプ, グロー電球【蛍光灯の点灯用ランプ】
　glow switch　グロースイッチ, 点灯管

glowing 形 赤熱した, 真っ赤な

glue 名 にかわ, 接着剤, のり 動 接着する, 貼る

glued 形 にかわ(のり)づけした, 接着(接合)した
　glued connection　接着接合
　glued joint　接着継手, 張合わせ継手

glueing 接着

glycerin グリセリン
　同 glycerine《英》

glycol グリコール【不凍液, 溶剤にもちいる】

GM メタセンタ高さ

GMT (Greenwich Mean Time) グリニッジ標準時

go 動 達する, 作動する, 続ける, 入れられる
　go back in　動 中に戻る
　go dead　動 故障する
　go on　動 継続する, 点灯する, 起こる
　go over　動 点検する, 見直す
　go wrong　動 故障する

goggles 保護めがね

gold 金 記Au
 gold point 金点【温度定点の一つ】
gong ゴング, どら, ベル
goniometer 角度計, 測角計
good 形優良な, 良好な, 望ましい, 正しい
 関in good condition：良好な状態で, よく整備されて
 good combustion 良好な燃焼
 good conductor 良導体
 good design Gマーク
gouge 穴たがね, 丸のみ
governing 調速
 governing device (gear) 調速装置
 governing equation 基礎方程式, 支配方程式
 governing plunger 調節プランジャ
 governing valve 蒸気加減弁
governor ガバナ, 調速機
 governor gear 調速機装置
 governor motor ガバナモータ, 調速機用モータ
 governor test 調速機試験, 負荷遮断試験
 governor valve 調速弁
GPS (Global Positioning System) 全地球的位置決定システム
grade グレード, 程度, 等級, 勾配, 傾斜
 grade of fit はめあい等級
gradient 傾度, 勾配, 傾斜
grading 格付け, 粒度
gradiometer グラジオメータ, 傾度測定器【物理量の勾配を測定する機器】
gradual 形緩やかな
gradually 副徐々に, しだいに
graduate 動目盛りをつける
graduation 目盛, 分類, 平滑化, 補整
 graduation line 目盛線
grain 粒, 粒子, 木目, 結晶粒
 関fine-grained：きめ細かい, 微細構造の
 grain boundary 粒界, 結晶粒界
 grain boundary corrosion 粒界腐食
 grain boundary crack 結晶粒界割れ
 grain boundary segregation 粒界偏析
 grain boundary sliding 粒界すべり
 grain growth 結晶粒の成長
 grain size 粒径, 粒度
gram グラム
 gram equivalent グラム当量
granular 形粒の, 粒状の
 granular graphite 粒状黒鉛
 granular structure 粒状組織
granularity 粒度, 細分性
graph グラフ, 図表
grapher 記録計器
graphic 形グラフの, 図の, 図形の, 名グラフィック, 図形, 図式, 図形記号【絵や図形の総称】
 graphic panel グラフィックパネル, 図示パネル
 graphic[al] analysis 図式解法
 graphic[al] calculation 図式計算
 graphic[al] method 図示法, 図解方式
 graphic[al] solution 図式解法
 graphic[al] statics 図式力学, 図解力学
graphical symbol 図示記号
 graphical symbol for electrical apparatus 電気用図記号
 graphical symbol for fluid power diagram 油圧・空気圧用図記号
graphics グラフィックス, 製図法
graphite グラファイト, 黒鉛
 graphite brush 黒鉛ブラシ
 graphite flake 片状黒鉛
 graphite grease グラファイトグリース
 graphite packing 黒鉛パッキン
 graphite paint 黒鉛ペイント
 graphite resistance 黒鉛抵抗
graphitization 黒鉛化
grasp 動握る, つかむ
grate 火格子
grating グレーティング, 格子
gravimeter 重力計, 比重計
gravimetric 形重量測定の, 重力測定の, 重力計の

gravimetric analysis　重量分析
gravimetric unit　重力単位
　関 SI：国際単位
gravitation　重力, 引力　関 the law of gravitation：引力の法則
gravitational　形重力の
　gravitational acceleration　重力の加速度【略記：g】
　gravitational field　重力場
　gravitational system of units　重力単位系
　gravitational unit　重力単位, 重量単位
　gravitational wave　重力波
gravity　グラビティ, 重力, 重量, 重さ, 比重, 引力
　関 specific gravity：比重
　gravity cell　重力電池
　gravity circulation　重力式潤滑【潤滑油給油方式の一つ】
　gravity control　重力制御
　gravity disc　グラビティディスク, 比重板
　gravity oiling　重力給油
　gravity settling　重力沈降
　gravity tank　重力タンク
gray　灰色　形灰色の
　gray cast iron　ねずみ鋳鉄
grease　グリース, 油脂【半固体の潤滑剤】
　grease cup　グリースカップ【グリースを注入する容器】
　grease gun　グリースガン【グリースを注入する装置】
　grease lubrication　グリース潤滑
　grease packing　グリースパッキン
great　形大きな, 多量の, 際立った, 重要な, 偉大な, 優れた
greater/greatest　「great」の比較級／最上級
greater coasting area　近海区域
greatest elongation　最大離角
greatly　副著しく, 非常に, 大いに
greenhouse effect gas　温室効果ガス
grey　「gray」に同じ
grid　グリッド, 格子

grid bearing　グリッド軸受
grime　名あか, 汚れ, すす　動汚す
grind　動研ぐ, 擦れる, 研削する　名研削
grinder　グラインダ, 研削盤, 研磨機
grinding　研削, 研磨, 粉砕
　grinding allowance　研削しろ
　grinding attachment　研削装置
　grinding oil　研削油
　grinding stone　砥石
　grinding wheel　砥石車
grip　名グリップ, つかみ　動しっかり握る, 締め付ける
grit　砕粒, 砂
　grit stone　天然砥石
gritty　形粒子を含んだ, 砂の入った
grommet　グロメット, 環索
groove　名グルーブ, 溝, 開先　動溝をつける
　groove angle　開先角度【二つの溶接部材の間に設けられた溝の角度】
　groove corrosion　溝腐食, 溝状腐食
　groove depth　開先深さ, 溝深さ
　groove head nut　溝ナット
　groove weld　グルーブ溶接, 開先溶接
grooved cam　溝カム
grooved friction wheel　溝付摩擦車
grooved pulley　溝車
grooving　グルービング, 溝状腐食
　grooving corrosion　溝型腐食
gross　形総計の, 全部の
　gross calorific value　総発熱量, 高位発熱量
　gross error　総合誤差
　gross head　総落差
　gross heating value　総発熱量
　gross pump head　実揚程
　Gross Tonnage　総トン数
　　関 net tonnage：純トン数
　gross weight　総重量
ground　名グランド, アース, 地面, 接地, 基礎　形地面の, 基礎の, 研いだ　同 earth：接地　動「grind」の過去・過去分詞, 座礁する, アー

スにつなぐ
ground circle　基礎円
ground circuit　接地回路
ground connection　アース接続，接地接続
ground detecting lamp　検漏ランプ
ground detector　アース検知器，検漏器
ground fault　アース事故，接地事故，地絡事故
ground joint　すり合わせ
ground potential　大地電位
ground speed　対地速度
ground state　基底状態
ground surface　研磨面
grounded circuit　接地回路
grounded system　アース系統，接地系統
grounding　アース，接地
grounding resistance　接地抵抗
group　名グループ，群，基　動分類する，まとめる
group valve　グループ弁，集合弁
grouping　グルーピング，グループ化，組分け，部分組立図
growth　成長，増加，発展
growth mark　貝殻状のしま模様
growth of cast iron　鋳鉄の成長
GSP（Group Starter Panel）　集合始動器盤
GTO（Gate Turn Off）thyristor　GTO サイリスタ，ゲートターンオフサイリスタ
guarantee　保証
guarantee engineer　保証技師
guarantee speed　保証速力
guarantee test　保証試験
guarantee work　補償工事
guard　名ガード，防護物，見張り，監視，警戒　動保護する，監視する，防ぐ
guard lamp　警報ランプ，危険防止灯
guard plate　保護板
guard ring　ガードリング，保護環
guard wire　保護線

guardian valve　中間弁
gudgeon pin　ガジオンピン，ピストンピン，クロスヘッドピン
gudgeon pin bearing　ピストンピン軸受
guidance　ガイダンス，誘導，案内，手引き，指導
guide　動案内する，導く　名ガイド，案内，手引[書]，規準，指針，誘導装置
guide apparatus　案内装置
guide bar　案内棒
guide bearing　案内軸受
guide blade　案内羽根
guide block　滑り金
guide bush　案内ブッシュ，案内リング，案内輪
guide column　案内支柱
guide disc　案内円板
guide face　案内面
guide pin　ガイドピン，案内ピン
guide piston　ガイドピストン
guide plate　案内板
guide pulley　案内車，案内ベルト車
guide ring　ガイドリング，案内リング
guide rod　ガイドロッド，案内棒
guide screw　親ねじ
guide shoe　滑り金
guide vane　ガイドベーン，案内羽根，案内翼
guide wheel　案内車
guided bend　型曲げ
guideline　ガイドライン，指針
guillotine rupture　ギロチン破断
gun metal　砲金【青銅の一種】
gunwale　ガンネル，舷縁
gusset plate　ガセット板，補強用板
gusset stay　ガセットステー，ガセット控え
guy　支え綱，控え鋼，支え線
guy block　控え滑車
guy wire　支え線，支鋼索
gypsum　石膏
gyration　旋回，回転
gyrocompass　ジャイロコンパス，

回転羅針儀
gyrometer ジャイロメータ【回転速度計】
gyropilot ジャイロパイロット，自動操縦装置，自動舵取り装置

gyroscope ジャイロ，回転儀
gyrostabilizer ジャイロスタビライザ，ジャイロ安定機【横揺れを防ぐ装置】

H-h

habit 習慣
hafnium ハフニウム 記Hf
hair 髪の毛
  hair crack ヘアクラック, 毛割れ【毛髪状の細い割れ】
  hair hygrometer 毛髪湿度計
  hair line ヘアライン, 毛線【細い実線】
  hair spring ひげぜんまい
half 名半分 形半分の
  half a turn 半回転
  half ahead 前進半速
  half astern 後進半速
  half bearing 半周軸受
  half built up crankshaft 半組立クランク軸
  half center 切欠きセンタ
  half cycle 半周期
  half life 半減期
  half load 半負荷【定格出力の1/2に相当する負荷】
  half nut 半割りナット
  half round 形半円形の, 半円筒の
  half round bar steel 半丸[棒]鋼
  half round file 半丸やすり
  half speed 半速
  half value period 半減期
  half wave length 半波長
  half wave rectification 半波整流
  half-way 副中間で, 途中まで, 半分だけ, 不十分に 形中間の, 不完全な, 中途半端な, 部分的な
  half width 半値幅
halide ハロゲン化物【ハロゲン元素 (F, Cl, Br, I) との化合物】
  Halide Torch ハロイドトーチ【フロンガス検知器】
Hall effect ホール効果【磁界内の半導体に起電力が生じる現象】
hall element ホール素子
halogen ハロゲン, 造塩元素
halogenation ハロゲン化, ハロゲン置換

halt 停(休)止 動停(休)止する
halyard ハリヤード, 動索
hammer 名ハンマ, 槌 動たたく
  hammer dressing ハンマ仕上げ
  hammer test ハンマテスト
  hammer welding 鍛接
hand 動手渡す
名ハンド, 手, 手動
関 on the other hand：一方, これに反して／by hand：手で, 手動で
  hand bilge pump 手動ビルジポンプ
  hand brake 手動ブレーキ
  hand expansion valve 手動膨張弁
  hand finishing 手仕上げ
  hand firing 手だき
  hand gear 手動装置
  hand generator 手回し発電機
  hand hole 手穴【作業口の一種】
  hand lamp 手さげ灯
  hand lubrication 手差し注油
  hand oiling 手差し給油
  hand operated 形手動の
  hand operation 手操作
  hand over 動引渡す
  hand pump 手押しポンプ
  hand rail ハンドレール, 手すり
  hand regulation 手動調整
  hand reset relay 手動復帰継電器
  hand steering gear 手動舵取り装置
  hand wheel ハンドル車
  hand work 手仕事
handicap ハンディキャップ, 不利な条件
handle 名ハンドル, 取っ手 動取り扱う, 握る, 処理する, 操縦(作)する, 触れる
  handle grip 握り
hang 動吊るす, 掛ける, ぶら下げる
hanger ハンガ, 吊り具
happen 動起こる, 発生する
harbor 港
hard 形硬い 反soft：柔らかい
  hard anodic oxide coating 硬質皮膜

hard carbide 硬質炭化物
hard chrome plating 硬質クロムめっき
hard copper wire 硬銅線
hard copy ハードコピー【磁気記録に対し，紙などにコンピュータのデータを印刷したもの】
hard drawn copper 硬銅
hard error ハードエラー【回復できないプログラムエラーやOSの故障の原因になったハードの欠陥】
hard facing 表面硬化
hard finish ハード仕上げ
hard glass 硬質ガラス
hard hat 安全帽
hard lead 硬鉛，アンチモン鉛
hard metal 硬質金属，超硬合金
hard rubber 硬質ゴム
hard sludge 硬質スラッジ
hard solder 硬ろう，硬質はんだ
hard steel 硬鋼，高炭素鋼
hard water 硬水
harden 動 硬くする，固める，硬水にする
hardenability 焼入れ性，硬化性
hardened 形 硬化された，焼入れされた
hardened steel 焼入れ鋼，硬化鋼
hardening 焼入れ，硬化
hardening agent 焼入れ剤
hardening crack 焼割れ
hardening depth (penetration) 焼入れ硬化深度
hardening temperature 焼入れ温度
hardly 副 ほとんど～ない
hardness 硬度，硬さ
hardness meter 硬度計
hardness number 硬度数，硬さ数
hardness of water 水の硬度
hardness test 硬度試験，硬さ試験
hardness tester 硬さ試験機
hardware ハードウエア，金物，金属器具
harmful 形 有害な
harmless 形 無害な，無傷の

harmonic 形 調和の，調和した
harmonic analysis 調和解析，調和分析，調波分析
harmonic analyzer 調波分析器，調和分析器
harmonic distortion 高調波ひずみ
harmonic motion 調和運動
harmonic ringing 同調信号
harmonic vibration 調和振動
Hastelloy ハステロイ【耐酸合金】
hatch ハッチ，船倉口
hatchet 手斧
hatching ハッチング，けば付け
hauling engine 巻上げ機関
have 動 持っている，経験する，有する
have to ～する必要がある，～しなければならない
hawser ホーサ，太綱，大索
hazard 名 ハザード，危険
動 危険にさらす
関 natural hazard：天災，自然災害／public hazard：公害
hazard analysis 危険解析
hazard prevention 危険防止
hazardous 形 危険な，有害な
hazardous material (substance) 危険物，有害物
HCFC (Hydro-Chloro-Fluoro-Carbon) ヒドロクロロフルオロカーボン
head ヘッド，水頭，落差【単位体積当たりのエネルギを長さの単位で表現したもの】
head loss 損失ヘッド，損失水頭
head meter ヘッドメータ【流量計の一種】
head of fluid 揚程
head of water 水頭，揚程
head pressure 上部圧力，頭部圧力，ヘッド圧力
head pump 水頭ポンプ
head tank ヘッドタンク，重力タンク
head test 圧力水密試験
head valve 頭弁，吐出弁
header ヘッダ，管寄せ

header plate　管寄せ板
header tank　圧力調整タンク，膨張タンク，集水タンク
hear　動 聞く
heard　動 「hear」の過去・過去分詞
hearing protection　（耳栓やイヤマフによる）聴覚保護
heart cam　ハートカム【ハート形をしたカム】
heat　動［加］熱する　名［加］熱，熱量【熱量の単位：joule】
　heat absorption rate of heating surface　伝熱面熱負荷【ボイラ本体の伝熱面の性能を表す】
　heat accumulator　蓄熱器
　heat affected zone　変質部，熱影響部
　heat attack　熱中症
　heat balance　ヒートバランス，熱勘定，熱精算，熱平衡，熱収支
　heat balance diagram　熱勘定線図
　heat budget　熱収支，熱勘定
　heat capacity　熱容量
　heat characteristics　熱特性
　heat conduction　熱伝導
　heat conductivity　熱伝導率
　heat consumption　熱消費量
　heat control　熱管理
　heat convection　熱対流
　heat curing　熱硬化
　heat cycle　熱サイクル
　heat damage　変質
　heat deterioration　加熱劣化
　heat distortion　加熱ひずみ，熱変形
　heat distribution　熱分配，熱分布
　heat drop　熱落差
　heat energy　熱エネルギ
　heat engine　熱機関
　heat equation　熱［伝導］方程式
　heat equivalent of work　仕事の熱当量
　heat exchange　熱交換
　heat exchanger　熱交換器
　heat exhaustion　熱疲労
　heat extraction coefficient　冷却係数
　heat flow　熱流
　heat flow rate　熱流，熱流量，熱流束

　heat flux　熱流，熱流束
　heat generating rate of combustion chamber　燃焼室熱発生率
　heat insulating material　断熱材，保温材，熱絶縁材
　heat insulation　保温，熱絶縁，断熱
　heat insulator　保温材，熱断熱材
　heat irreversible　形 熱非可逆性の
　heat load　熱負荷
　heat load in combustion chamber　燃焼室熱負荷
　heat loss　熱損失，損失熱
　heat management　熱管理
　heat medium　熱媒体
　heat of absorption　吸収熱
　heat of combustion　燃焼熱
　heat of compression　圧縮熱
　heat of condensation　凝縮熱
　heat of decomposition　分解熱
　heat of dilution　希釈熱
　heat of dissociation　解離熱
　heat of evaporation　蒸発熱，気化熱
　heat of formation　生成熱
　heat of friction　摩擦熱
　heat of fusion　融解熱
　heat of linkage　結合熱
　heat of liquid　液体熱
　heat of reaction　反応熱
　heat of solidification　凝固熱
　heat of solution　溶解熱
　heat of sublimation　昇華熱
　heat of transition　転移熱，変態熱
　heat of vaporization　蒸発熱，気化熱
　heat pipe　ヒートパイプ【熱伝導管の一種】
　heat pipe type heat exchanger　ヒートパイプ式熱交換器
　heat proof　耐熱，防熱
　heat pump　ヒートポンプ，熱ポンプ【機械的エネルギを用いて低温から高温へ熱を伝える装置】
　heat quantity　熱量
　heat radiation　熱放射，熱輻射
　heat rate　熱消費率，熱発生率
　heat ray　熱線
　heat recovery　熱回収

heat regenerator　蓄熱式熱交換器
heat regulator　温度調節器
heat release rate　熱発生率
heat resistance　耐熱度, 熱抵抗
heat resisting alloy　耐熱合金
heat resisting glass　耐熱ガラス
heat resisting material　耐熱材料
heat resisting property　耐熱性
heat resisting steel　耐熱鋼
heat retaining　保温処理
heat retaining property　保温性
heat reversible　形熱可逆性の
heat run test　耐熱試験
heat sensitive paint　示温塗料
heat sensor　熱センサ, 熱検出器
heat sink　熱シンク【温度上昇を抑える吸放熱装置】
heat source　熱源
heat stability　熱安定性
heat storage　蓄熱
heat storage tank　蓄熱槽
heat stress　熱応力
heat supply　熱供給
heat test　耐熱試験
heat transfer　伝熱, 熱伝達, 熱移動【固体と流体の間で熱が伝わる現象】
heat transfer by convection　熱対流, 対流熱伝達
heat transfer coefficient　伝熱係数, 熱伝達率, 熱伝達係数
heat transfer rate　熱伝達率, 伝熱率, 伝熱速度
heat transfer surface　伝熱面
heat transfer medium　[伝]熱媒体
heat transmission　熱伝達, 熱貫流
heat transmission coefficient　熱貫流率
heat treating steel　熱処理鋼
heat treatment　熱処理【焼なまし, 焼戻し, 焼ならしなど】
heat unit　熱単位
heat value　発熱量
heated　形加熱された
heater　ヒータ, 加熱器
heater circuit　電熱回路

heating　ヒーティング, 加熱, 暖房
heating and cooling devices　暖冷房装置
heating apparatus　暖房装置, 加熱装置
heating area　伝熱面積
heating boiler　暖房ボイラ
heating coil　加熱管, 加熱コイル
heating current　加熱電流
heating curve　加熱曲線
heating element　発熱体, 加熱素子
heating furnace　加熱炉
heating steam　加熱蒸気
heating surface　伝熱面, 加熱面
heating surface area　伝熱面積
heating system　加熱システム, 加熱系統, 暖房方式, 暖房装置
heating tube　伝熱管, 加熱管
heating unit　発熱体
heating value　発熱量
heating wire　電熱線
heavier　「heavy」の比較級
heavily　副大量に, 激しく, 重く
heaving　上下揺れ
heavy　形重い, 濃い, ひどい, 激しい　反light：軽い
heavy article　重量物
heavy fuel oil　重油
heavy load　重負荷
heavy metal　重合金, 重金属【比重5以上の金属】
heavy oil　重油, 重質油
heavy oil engine　重油機関
heavy sea　荒天
heel　ヒール, 横傾斜, 外端部, 後端
heel angle　ヒール角, 横傾斜角
heeling　傾斜
height　高さ
height gauge　高さゲージ
height of thread　ねじ山の高さ
height ratio of blading　翼高比
Heinrich's law　ハインリッヒの法則
Heisler chart　ハイスラー線図
held　動「hold」の過去・過去分詞
Hele Shaw pump　ヘルショウポンプ
helical　形傾斜のある, らせん状の

helical angle　つる巻角
helical gear　ヘリカルギア，はすば歯車【円筒上に歯がらせん状に切られている歯車】
helical potentiometer　ヘリカルポテンショメータ【可変抵抗器の一種】
helical spring　コイルばね，つる巻ばね
helical spur gear　はす歯平歯車
helical teeth　はす歯
helically　副らせん状に
helicoid　名らせん面　形らせん形[状]の
helicoid surface　ねじ面
helicoid[al] form　らせん型
helium　ヘリウム　記He
helix　らせん，つる巻線
helix angle　らせん角，ねじれ角
helix angle of thread　ねじれ角
helix tube　らせん管
helm　かじ，舵の柄，操舵装置
helm indicator　舵角指示器
helmet　ヘルメット，安全帽
help　援助，支援　動助ける，役立つ　関of great help in：〜に非常に役立つ
hemispherical　形半球の
hemp　麻
hence　副それゆえ
Henry　ヘンリー【インダクタンスの単位：H】
Henry's law　ヘンリーの法則
here　副ここで(へ，に)
heptane　ヘプタン【パラフィン系炭化水素の一つ】
herculoy　ハーキュロイ【鍛錬用銅ケイ素合金】
hermaphrodite caliper　片パス
herringbone gear　ヘリングボーン歯車，やまば歯車
Hertz　ヘルツ【周波数や振動数の単位：Hz】
heterogeneity　異質性，異種，不均質，不均質性，不均一，不均一性
heterogeneous combustion　不均一燃焼
hexadecimal　形16進法の
hexadecimal number　16進数
hexagon　六角形
hexagon head bolt　六角[頭]ボルト
hexagon nut　六角ナット
hexagon socket head cap bolt　六角穴付ボルト
hexagonal　形六角形の，六辺形の，六方晶系の
hexagonal close-packed lattice　最密六方格子
hexahedron　六面体
HFC (**Hydro-Fluoro-Carbon**)　ヒドロフルオロカーボン【冷媒の一種】
hidden outline　かくれ線
hide-out phenomena　ハイドアウト【ボイラ水中のリン酸イオン濃度が負荷によって変化する現象】
high　形高い，高位の，鋭い　副高く
high alloy steel　高合金鋼
high calorific value　高発熱量
high capacity　大容量
high carbon steel　高炭素鋼
high current　大(強)電流
high cycle fatigue　高サイクル疲労
high damping alloy　防振合金
high frequency　名高周波　形高周波の
high frequency generator　高周波発電機
high frequency hardening　高周波焼入れ
high frequency induction heating　高周波誘導加熱
high frequency seasoning　高周波乾燥
high frequency wave　高周波
high grade cast iron　高級鋳鉄
high lift safety valve　高揚程安全弁
high manganese steel　高マンガン鋼
high molecular compound　高分子化合物
high output　高出力
high pass filter　高域ろ波器
high power[ed]　形高馬力の，高

出力の，高性能の，強力な
high powered engine　強力エンジン
high precise　形高精度の
high pressure　形高圧の　名高圧
high pressure compressor　高圧圧縮機
high pressure cylinder　高圧シリンダ
high pressure gas turbine　高圧ガスタービン
high pressure injection　高圧注入
high pressure nozzle　高圧ノズル
high pressure stage　高圧段
high pressure turbine　高圧タービン
high purity　高純度
high resistance　高抵抗
high speed　形高速の　名高速
high speed circuit breaker　高速度遮断器
high speed cutting　高速切削
high speed engine　高速機関
high speed steel　高速度鋼
high speed vessel　高速船
high strength brass　高力黄銅
high strength cast iron　強じん鋳鉄
high strength malleable cast iron　高力可鍛鋳鉄
high strength steel　高力鋼，高強度鋼，高張力鋼，強じん鋼
high suction valve　上部吸込弁
high technology　ハイテク，先端技術，高度科学技術
high temperature　高温
high temperature corrosion　高温腐食
high temperature creep　高温クリープ
high tensile brass　高力黄銅
high tensile steel　高張力鋼
high tensile strength steel　高張力鋼
high tension　形高圧の　名高電圧，高張力
high tension circuit　高圧回路
high tension ignition　火花点火
high tension steel　高張力鋼
high velocity　高速，高速度
high voltage　高電圧
higher　形「high」の比較級
higher alcohol　高級アルコール
higher calorific value　高位発熱量
higher harmonic　高調波
higher heating value　高位発熱量
higher order　高次
higher order lag　高次遅れ
highest　形「high」の最上級，最高の
highest point　最高点，最上部
highlight　動強調する，目立たせる
highly　副高く，非常に
highly skewed propeller　ハイスキュードプロペラ
hinder　動妨げる，じゃまをする
hinge　ヒンジ，ちょうつがい
hinged door　開き戸
hinged joint　ピン継手
histogram　ヒストグラム，柱状図，柱状グラフ
hit　動打つ，点火する，はねる，当たる，当てる　名跳躍
hodo meter　走行距離計
hoist　ホイスト【小形の巻上げ機】
hoist drum　巻上げドラム
hold　名ホールド，保持，船倉
動保つ，持つ，収容する，保持する，保留する，固定する，開催する
hold-down stud　固定ボルト，押さえボルト
holder　ホルダ，保持器，取付部
holdfast　はどめ
holding　支持，保持
holding coil　保持コイル
holding current　保持電流
holding down bolt　すえ付けボルト
holding nut　取付けナット
holding tank　貯蔵タンク，汚水集合タンク
hole　ホール，穴，孔，正孔
hole basis　穴基準
hollow　形中空の，空洞の
名中空，穴，くぼみ，底，谷
動えぐる，くぼませる，中空にする
hollow blade　中空羽根
hollow cylinder　中空円筒
hollow key　くらキー
hollow propeller　中空プロペラ

hollow screw　袋ねじ
hollow shaft　中空軸
hollow tube　中空管
hollowization　空洞化
holographic memory　ホログラフィックメモリ【光メモリの一種】
home position　ホームポジション，原点，定位置
homogeneity　均一[性]，均質
homogeneous　形均質の，同種の，同族の，同質の，同次の
homogeneous combustion　均質燃焼
homogeneous series　同族列【分子式が一定の法則のもとであらわされる化合物】
homogeneous strain　一様ひずみ
homogeneous system　均質系，均一系，単相系
homologue　同族体【同族列中の化合物】
hone　動磨く，砥石で研ぐ
honeycomb radiator　蜂の巣形放熱器
honeycomb structure　ハニカム構造，蜂の巣構造
honing　ホーニング仕上げ【穴などの内面を精密に仕上げる加工】
hood　フード，ひさし
hook　名フック，留め金【かぎ状の機械部品】動かぎ形に曲げる，かぎで引っ掛ける
hook bolt　フックボルト
hook joint　フック継手
hook spanner wrench　引掛けスパナ
Hooke's law　フックの法則【弾性の法則】
hoop　名フープ，輪，帯鋼
動取り巻く，囲む，巻きつける
hoop iron　帯鋼
hoop stress　フープ応力【円筒に生じる円周方向の引張応力】
hoop tension　フープ応力，周応力，周方向応力
hopeless　形見込みのない，絶望して
Hopkinson's torsion meter　ホプキンソン式ねじり動力計
hopper　ホッパ【底のあいたラッパ状容器】
horizontal　形水平の，平面の，横向きの
horizontal axis　横軸，水平軸
horizontal beam　水平はり
horizontal boiler　横ボイラ
horizontal coupling　水平継手
horizontal engine　横型機関
horizontal line　横線，行，水平線
horizontal plane　水平面
horizontal scale　横目盛り
horizontally　副水平に，水平方向に
horizontally opposed engine　水平対向エンジン
HorsePower　HP，馬力【仕事率の単位】
horsepower of transmission　伝動馬力
horseshoe magnet　U字型磁石，馬てい形磁石
horseshoe type thrust bearing　馬てい型スラスト軸受
hose　ホース
hose connection　ホース継手
hose coupling　ホース継手
hose test　注水テスト
hostile　形適さない，厳しい条件の
hot　形熱い，高温の
hot bending test　熱間曲げ試験
hot blast　熱風
hot corrosion　熱腐食，高温腐食
hot dipping　溶融めっき
hot drawn tube　熱間引抜管
hot gas　高温ガス
hot hardness　高温硬さ
hot junction　熱接点
hot liquid　熱水，温液
hot metal　高温金属，溶けた金属
hot plate　ホットプレート
hot rolling　熱間圧延
hot set　熱間硬化
hot setting adhesive　加熱硬化接着剤，高温硬化接着剤
hot shortness　高温もろさ，赤熱脆性，赤熱もろさ
hot spot　ホットスポット
hot start　ホットスタート，温態

始動，高温始動
hot water 温水
hot water boiler 温水ボイラ
hot water heating 温水暖房
hot water radiator 温水放熱器
hot water supply 給湯
hot water tank 温水タンク
hot well ホットウェル，温水溜め
hot wire ammeter 熱線電流計
hot working 熱間加工

hour 1時間，時間
house 動保管する，収納する
housing ハウジング，ケース，ケーシング，容器，外被，覆い
hover 動空を舞う
hovercraft ホバークラフト
however 副しかしながら，けれども
HP（HorsePower） 馬力
h-s (enthalpy-entropy) diagram h-s（エンタルピ-エントロピ）線図
hub ハブ，中心，中枢
hue 色相，色合い，色調
hull 船体，船殻，胴体
hull strength 船体強度
hull structure 船体構造

human 形人の，人間の 名人，人間
human being 人類
human engineering 人間工学
human error ヒューマンエラー，人為的ミス

hume [concrete] pipe ヒューム管
humid 形湿気のある
humid air 湿り空気
humid volume 湿り比容積

humidification 加湿
humidifier 給湿機，加湿器
humidify 動湿らす，湿気を与える
humidity 湿度，湿気
humidity chart 湿度線図
humidity control 湿度調節，湿度制御

hundredfold 形百倍の
hunt 動追跡する，不規則に動く，揺れる
hunting ハンチング，追従，乱調
hunting gear 追従装置

hurry 動急ぐ
hybrid 名ハイブリッド，混成のもの 形ハイブリッドの，混成の
hybrid engine ハイブリッドエンジン

hydrant 消火栓，給水栓
hydrate 含水化合物
hydration 水和，水和作用
hydraulic 形油圧の，水圧の，水力の，流体の
hydraulic accumulator 蓄圧器，水力溜，油圧アキュムレータ
hydraulic actuator 油圧アクチュエータ，流体操作部
hydraulic amplifier 油圧増幅器
hydraulic brake 水圧ブレーキ，油圧ブレーキ，液圧ブレーキ
hydraulic cement 水硬セメント
hydraulic chuck 油圧チャック【油圧で工作物を取り付ける工具】
hydraulic circuit 油圧回路，流体回路
hydraulic control 油圧制御
hydraulic control valve 油圧制御弁
hydraulic coupling 流体継手
hydraulic crane 油圧クレーン
hydraulic cylinder 油圧シリンダ
hydraulic damper 油圧ダンパ【衝撃などを緩和する装置】
hydraulic diameter 水力直径
hydraulic direction control valve 油圧方向制御弁
hydraulic drive 油圧駆動
hydraulic dynamometer 水動力計
hydraulic efficiency 水力効率
hydraulic engineering 水力工学
hydraulic equipment 油圧機器
hydraulic fluid 作動油，作動液
hydraulic governor 油圧式ガバナ
hydraulic horsepower 流体動力
hydraulic jack 油圧ジャッキ
hydraulic lock 流体固着
hydraulic loss 水力損失
hydraulic machine 油圧機械
hydraulic mechanism 油圧機構
hydraulic motor 油圧モータ，水

力原動機
hydraulic nozzle　流体ノズル
hydraulic oil　作動油, 油圧油
hydraulic power　水力, 流体動力
hydraulic power transmission　流体伝達, 流体伝動装置, 水力伝達, 水力変速機
hydraulic power unit　油圧ユニット
hydraulic press　液圧プレス
hydraulic pressure　油圧, 水圧
hydraulic pressure test　水圧試験
hydraulic propeller　ジェットプロペラ
hydraulic pump　油圧ポンプ, 水圧ポンプ
hydraulic ram　水圧ラム
hydraulic seal　油圧シール
hydraulic servomechanism　油圧サーボ機構
hydraulic starter　油圧始動装置
hydraulic steering gear　油圧式舵取り装置
hydraulic system　油圧系統, 油圧装置, 油圧系
hydraulic system pump　油圧ポンプ
hydraulic test　水圧試験
hydraulic torque converter　流体トルクコンバータ, 流体変速機
hydraulic transformer　水圧減速装置
hydraulic transmission　油圧駆動装置, 流体変速機, 流体伝動装置, 水力伝達
hydraulic transmitter　水力継手
hydraulic turbine　水車
hydraulic unit　油圧ユニット
hydraulic valve　油圧弁
hydraulic winch　油圧ウインチ
hydraulically　副流体力学的に, 油圧で, 水圧で, 水力で, 流体で
　hydraulically operated valve　油圧駆動弁
hydraulics　水力学
hydrazine　ヒドラジン【還元剤, ロケット燃料】
hydrocarbon　炭化水素
hydrochloric acid　塩酸
HydroChloroFluoroCarbon　HCFC, ヒドロクロロフルオロカーボン【冷媒の一種】
hydrodynamic　形流体力学の, 水力学の, 水力の, 水圧の
　hydrodynamic lubrication　流体潤滑, 動圧潤滑
hydrodynamics　流体力学, 水力学
hydroelectric steering gear　電動油圧操舵装置
hydro-extractor　脱水機
HydroFluoroCarbon　HFC, ヒドロフルオロカーボン【冷媒の一種】
hydrofoil　水中翼船, 翼形
hydrogen　水素　記H
　hydrogen bond　水素結合
　hydrogen brittleness　水素脆(もろ)さ
　hydrogen chloride　塩化水素
　hydrogen electrode　水素電極
　hydrogen embrittlement　水素脆化, 水素脆性
　hydrogen engine　水素機関
　hydrogen fluoride　フッ化水素
　hydrogen ion　水素イオン
　hydrogen ion concentration　水素イオン濃度
　hydrogen ion exponent　pH, 水素イオン指数
　hydrogen peroxide solution　過酸化水素
　hydrogen plating　水素被膜
　hydrogen sulfide　硫化水素
hydrogenation　水素添加
hydrograph　ハイドログラフ, 水位図, 流量図
hydrolysis　加水分解
hydromechanics　流体力学
hydrometer　比重計, うきばかり
hydrometry　流量測定, 比重測定
hydrophilic group　親水基
hydrophobic group　疎水基
hydropower　水力, 水力電力
hydrostatic　形静水の, 流体静力学の
　hydrostatic bearing　静圧軸受
　hydrostatic curve　排水量等曲線図
　hydrostatic head　静水頭

hydrostatic lubrication　静圧潤滑
hydrostatic pressure　静圧，静水圧
hydrostatic pressure test　水圧試験
hydrostatic stress　静水応力
hydroxide　水酸化物
hydroxyl group　水酸基
hygrometer　湿度計
hygroscopic　形 吸湿性の
hygroscopic degree　吸湿度
hygroscopic moisture　湿分
hygroscopic property　吸湿性
hygrothermograph　温湿度計
hyperbola　双曲線
hyperbolic（**hyperbolical**）　形 双曲線の
hyperbolic equation　双曲形方程式
hyperboloidal gear　食違い歯車
hypereutectoid steel　過共析鋼
hypersonic　形 極超音速の【音速の5倍以上の速さ】
hypersonic flow　極超音速流
hypo-　接頭「亜」「下に」「以下の」の意
hypocycloid　ハイポサイクロイド，内サイクロイド，内転サイクロイド

hypoeutectoid steel　亜共析鋼
hypoid gear　ハイポイド歯車【食違い軸間に運動を伝える円すい形の歯車】
hypoid gearing　ハイポイドかみあい
hypothermia　ハイポサーミア，低体温，低体温症
hypothesis　仮定，仮説，憶測
hypothetical accident　仮想事故
hypotrochoid　内[転]トロコイド
hysteresis　ヒステリシス，履歴現象
hysteresis curve　ヒステリシス曲線，履歴曲線
hysteresis damping　履歴減衰
hysteresis error　ヒステリシス差，ヒステリシス誤差
hysteresis loop　ヒステリシス環線
hysteresis loss　ヒステリシス損[失]【鉄心にヒステリシス現象が生じたときのエネルギ損失】
hysteresis motor　ヒステリシスモータ

I-i

I bar I形材
I section I形断面
I steel I形鋼
IC (Integrated Circuit) 集積回路
ice 名氷
　動氷で冷やす，凍る，凍らす
　ice box　冷蔵箱
　ice chamber　冷蔵庫，冷凍室
　ice engine　製氷機
　ice machine oil　冷凍機油
　ice making capacity　製氷能力
　ice point　氷点
idea　アイデア，考え，発想
ideal　形理想の，理想的な　名理想
　ideal cycle　理想サイクル
　ideal efficiency　理想効率
　ideal fluid　理想流体
　ideal gas　理想ガス，理想気体
identical　形同一の，一致する，等しい
identification　確認
identify　動確認する，識別する，同一視する，気づく
identity　アイデンティティ，身元，識別，一致性，同一性，恒等式
　identity card　身分証明書，IDカード
　identity elastic alloy　恒弾性合金【熱弾性係数が一定の合金】
idle　形遊んでいる，使用されていない　動空回りする
　idle gear　遊び歯車，中間歯車
　idle pulley (wheel)　遊び車
idler　中間車，遊び車
idling　アイドリング，無負荷運転，空運転
i.e. (id est)　すなわち
if　接もし〜ならば，〜の場合，〜かどうか
　関 even if：例え〜でも（としても）
　if applicable　該当する場合は
　if appropriate　適切であれば
　if necessary　副必要ならば
　if possible　副できるならば
　if required　もし必要ならば
ignitability　着火性，発火性
ignite　動点火する，発火する
igniter　イグナイタ，点火器
　igniter cam　点火カム
ignition　イグニッション，点火，着火，発火，点弧
　ignition advance　点火進角
　ignition analyzer　イグニッションアナライザ
　ignition armature　点火コイル
　ignition burner　点火バーナ
　ignition coil　点火コイル
　ignition delay (lag)　着（発）火遅れ
　ignition device　点火装置
　ignition index　発火指数
　ignition miss　着火不良
　ignition order　点火順序
　ignition plug　点火栓，点火プラグ
　ignition point　着火点，発火点，着火温度
　ignition quality　発火性，着火性
　ignition source　発火源，点火源，引火源
　ignition temperature　発火温度，着火点，着火温度，点火温度
　ignition time　点火時間
　ignition timing　点火時期
ignitionability　着火性
ignore　動無視する，無知な
IHP (Indicated HorsePower)　図示馬力
illuminate　動照らす，電飾を施す
illumination　照明，照度
　illumination photometer　照度計
illuminator　照明器，反射鏡
illuminometer　照度計
illustrate　動説明する，さし絵を入れる，図解する
illustration　イラストレーション，説明図，さし絵，図解
image　イメージ，像，画像
　image processing　画像処理

imaginary 形 想像上の, 仮想の, 虚数単位の 名 虚数
　imaginary number　虚数
imagine 動 想像する, 心に描く
imbalance 不均衡
imitate 動 模倣する, 模造する
immediate 形 即座の, 直接の, 差し迫った
　immediate effect　即効
immediately 副 すぐに, 直ちに
immersion 浸漬, 沈入【浸すこと】
immunity 不活性態【腐食が起こらない状態】
impact 衝撃, 衝突, 打撃
　impact fatigue　衝撃疲労
　impact force　衝撃力
　impact hardness　衝撃硬さ
　impact load　衝撃荷重
　impact producing flexure　衝撃曲げ
　impact resistance test　耐衝撃試験
　impact strength　衝撃強さ
　impact stress　衝撃応力
　impact test　衝撃試験
　impact value　衝撃値
　impact wrench　インパクトレンチ【圧縮空気などを用いたナット締付け工具】
impair 動 減じる, 弱める, 害する, 損なう
impart 動 [分け]与える, 伝える
impedance インピーダンス【交流回路における電気抵抗】
　impedance angle　インピーダンス角
　impedance drop　インピーダンス降下
　impedance voltage　インピーダンス電圧
impede 動 妨げる, 遅らせる
impediment 障害, 妨害
impel 動 推し進める
impeller インペラ, 羽根車
　impeller blade　羽根, 羽根車の羽根
　impeller casing　羽根車ケーシング
　impeller pump　インペラポンプ
impend 動 差し迫る
imperfect 形 不完全な, 不十分な
　imperfect combustion　不完全燃焼
　imperfect lubrication　不完全潤滑
impinge 動 作用する, 衝突する
impingement 衝突, 衝撃, 侵害
　impingement attack　衝撃腐食
implantation 注入, 移植
implement 動 遂行する
imply 動 [暗に]意味する, ほのめかす
import 動 輸入する
importance 重要, 重要性
important 形 重要な, 重大な
impose 動 課す, 負わせる, 〜のせいにする, 押し付ける
impossibility 不可能[性]
impossible 形 不可能な
impregnant 浸透剤, 含浸剤
impregnated 形 浸透した
impregnating oil 透浸油
impregnation 含浸, 浸透, 注入【しみこませること】
impress 動 押す, 印加する, 印象を与える
impressed voltage 印加電圧
improper 形 異常な, 誤った, 不適当な, 不十分な
improve 動 改良する, 改善する, 向上させる
improvement 改善, 改良
improver 向上剤
impulse インパルス, 力積, 推進力, 衝撃, 衝動
　impulse blade　衝動羽根
　impulse circuit　インパルス回路
　impulse coupling　インパルスカップリング【エンジンの始動装置】
　impulse current　衝撃電流
　impulse force　衝撃力
　impulse function　インパルス関数
　impulse input　インパルス入力
　impulse response　インパルス応答
　impulse stage　衝動段
　impulse starter　インパルススタータ
　impulse stroke　運動(衝撃)行程
　impulse turbine　衝動タービン
　impulse type　衝撃型

impulse voltage　衝撃電圧
impulse wave　衝撃波
impulsive　形衝撃的な，衝動的な
　impulsive force　衝撃力
　impulsive load　衝撃荷重
　impulsive pressure　衝撃圧力
　impulsive sound　衝撃音
impurity　不純物，きょう雑物
in ～　前～の中に
　in ～ ing　～する際に
　in accordance with ～　前～に一致して，～に従って
　in addition　副その上
　in addition to　前～に加えて，～の他に
　in case of ～　前～の場合には，もし～したら
　in front of ～　前～の前に
　in order that ～　接～するように，～する目的で
　in order to ～　～するために，～するように
　in other words　副言い換えれば，つまり，換言すれば
　in place of ～　前～の代わりに
　in proportion to ～　前～に比例して
　in short　副手短に，つまり
　in so doing　そのようにしながら
　in spite of ～　前～にもかかわらず
　in the event of ～　前[万一]～の場合には
　in turn　副順番に，交替で
inaccurate　形不正確な，誤った
inactivation　不活性化
inadvertent　形不注意な，うっかりした
inboard　形船内の　副船内に
incandescence　白熱
incandescent　形白熱の
incapability　不能，不適当，無資格
inch　インチ【1inch = 25.4mm】
inching　インチング，微調整，寸動
incidence　入射，入射角，投射，結合，頻度
　incidence angle　入射角
　incidence rate　発生率

incident　形入射する，起こりやすい，付随する　名出来事，付随事件・条件・事項・義務
　incident angle　入射角
　incident ray　入射光線
incineration　焼却
incinerator　焼却炉
inclination　傾斜，勾配
　関 angle of inclination：傾斜角
incline　名斜面，勾配　動傾く
　incline precision level　傾斜測定用精密水準器
inclined　形傾斜した
　inclined plane　斜面
　inclined stress　傾斜応力
　inclined tube manometer　傾斜管式圧力計，傾斜マノメータ
inclinometer　傾斜計
include　動含む
included angle　挟角，開先角度
　included angle of thread　ねじ山の角度
inclusion　含有物，介在物
incombustible　形不燃性の
　incombustible material　不燃材料
incoming　形入ってくる
　incoming line　引込み線
　incoming panel　受電盤
incomplete　形不完全な，不十分な
　incomplete combustion　不完全燃焼
　関 complete combustion：完全燃焼
　incomplete lubrication　不完全潤滑
　incomplete thread　不完全ねじ部
incompletely　副不完全に，不十分に
incompressibility　非圧縮性
incompressible　形圧縮できない
　incompressible fluid　非圧縮性流体
inconclusive　形決定的でない，結論の出ない
Inconel　インコネル【ニッケル合金の一種】
incorporate　動合併する，組み入れる，具体化する，混合する　形合併した，結合した，具体化した
incorrect　形間違った，不適切な
incorrectly　副間違って，不正確に

increase 動増加する 名増加
increasing pitch propeller 漸増ピッチプロペラ
incredibly 副非常に, すごく, 信じられないほどに
incremental 形増分の
　incremental system　増分方式
incrustation 湯あか 同 scale
indeed 副実に, 本当に, 実際に
indentation 押込み, くぼみ, へこみ, 圧入, ぎざぎざ
　indentation hardness　押込み硬さ, 圧入硬度
　indentation test　押込み試験
indenter 圧子
independence 独立
independent 形独立した, 独立の, 独自の, 単独の, 無関係の
　independent bilge pump　独立ビルジポンプ
　independent brake valve　単独ブレーキ弁
　independent control　独立制御, 単動
independently 副独立して, 単独で, 独力で
index インデックス, 指数, 指針, 指標, 率, 目盛, 索引, 見出し
　index error　目盛誤差, 指示誤差, 指標誤差, 器差
　index number　指数
　index of refraction　屈折率
indexing インデキシング, 割出し
indicate 動指示する, 図示する, 示す, 表示する
indicated 形指示された, 表示された, 図示された
　indicated cylinder output　図示シリンダ出力, 図示仕事
　Indicated HorsePower　IHP, 図示馬力
　indicated mean effective pressure　図示平均有効圧
　indicated micrometer　指示マイクロメータ
　indicated power　図示出力
　indicated thermal efficiency　図示熱効率
　indicated work　図示仕事
indicating 指示, 表示
　indicating controller　指示制御器, 調節計
　indicating instrument　指示計器
　indicating lamp　表示灯
　indicating range　指示範囲
　indicating wattmeter　指示電力計
indication 指示, 図示, 示度, 表示, 徴候, 検出
indicator インジケータ, 指示器, 表示器, 指圧器, 表示計, 指示薬
　indicator card　インジケータ用紙
　indicator cock　指圧器弁
　indicator diagram　インジケータ線図, 指圧線図
　indicator lamp　表示灯
　indicator valve　インジケータバルブ, 指圧器弁
indicial 形指示する, 兆候のある
　indicial response　インディシャル応答
indirect 形間接の
　indirect control　間接制御
　indirect expansion system　間接膨張式
　indirect load　間接荷重
　indirect measurement　間接測定
　indirect transmission　間接伝動
indirectly 副間接的に
indispensable 形不可欠な, 必須の
indium インジウム 記 In
individual 形単独の, 個々の, 個人の, 独特の, 特有の
　individual drive　単独運転
　individual system　並列式
indoor transformer 屋内変圧器
induce 動吸引する, 誘導する
induced 形誘導された, 誘発された
　induced current　誘導電流
　induced draft　吸引通風, 誘引通風
　induced draft fan　吸引通風機
　induced drag　誘導抗力
　induced electromotive force　誘導起電力

induced oxidation　誘導酸化
induced reaction　誘導反応
induced velocity　誘導速度
inducer　インデューサ，入口導翼
inductance　インダクタンス，誘導係数
inductance regulator　誘導電圧調整器
induction　吸入，吸引，帰納，誘導
induction annealing　高周波焼なまし
induction coil　誘導コイル
induction current　誘導電流
induction field　誘導磁界
induction furnace　誘導電気炉
induction generator　誘導発電機
induction hardening　高周波焼入れ
induction heater　誘導加熱器【誘導電流によって発熱させる加熱器】
induction machine　誘導機
induction manifold　吸気マニホルド
induction motor　誘導電動機
induction quenching　高周波焼入れ
induction regulator　誘導電圧調整器
induction tempering　高周波焼もどし
induction type relay　誘導型継電器
induction valve　吸込み弁，吸気弁
inductive circuit　誘導性回路
inductive load　誘導負荷
inductive reactance　誘導リアクタンス
inductor　インダクタ，誘導子
inductor type generator　誘導子形交流発電機
inductor type magnet　誘導形マグネット，誘導子回転形マグネット
industrial　形産業の，工業の
industrial alcohol　工業用アルコール
industrial gasoline　工業用ガソリン
industrial instrument　工業計器
industrial standards　工業規格
Industrial Tele Vision　工業用テレビ
industry　工業
inefficient　形能率の悪い，非効率な，効果のない
inelastic　形弾性のない，非弾性の

inelastic collision　非弾性衝突
inelastic region　非弾性域
inelasticity　非弾性
inequality　不等[式]，差
inert　形不活性な，化学作用を起こさない
inert gas　不活性ガス
inert gas arc welding　イナートガスアーク溶接
inert gas metal arc welding　MIG溶接，イナートガス金属アーク溶接
inert gas smothering system　イナートガス消火装置
inert gas tungsten arc welding　TIG溶接，イナートガス[タングステン]アーク溶接
inertia　イナーシア，慣性
関 the law of inertia：慣性の法則
inertia couple　慣性偶力
inertia effect　慣性効果
inertia force　慣性力
inertia governor　慣性調速機
inertia guidance system　慣性誘導装置
inertia supercharging　慣性過給
inexhaustible　形無尽蔵の
inconclusive　形決定的でない，結論の出ない
inextensional deformation　伸びなし変形
infiltration　溶浸，侵入，浸透
infinite　形無限の
infinite element　無限要素
infinite series　無限級数
infinitesimal　形微小の，無限小の　名極微量，無限小
infinity　無限，無限大
inflame　動燃え上がる
inflammability　可燃性，引火性
inflammability limit　可燃限界
inflammable　形可燃性の
inflammable gas　可燃性ガス
inflammable liquid　引火性液体
inflammable material　可燃材料
inflammation point　引火点
inflection　変曲，屈折，湾曲，屈曲

inflow 流入, 流入量
influence 图影響, 作用, 感化, 誘導 動影響を及ぼす
inform 動知らせる, 情報を提供する
information インフォメーション, 情報, 資料
　information interchange　情報交換
　information processing　情報処理
　information system　情報システム
　Information Technology　IT, 情報技術
infrared 形赤外[線]の 图赤外線
　infrared gas analyzer　赤外線ガス分析計
　infrared radiation　赤外線
　infrared radiation thermometer　赤外線ふく射温度計
　infrared ray　赤外線
　infrared spectrum　赤外線スペクトル
ingot インゴット, 鋳塊
　Ingot Metal　IM 合金
　　㊟ powder metal：PM 合金
　ingot pipe　収縮管【インゴットのパイプ状の空洞】
　ingot steel　インゴットスチール, 溶製鋼
ingress 進入, 入口
inhale 動吸い込む
inherent 形固有の, 特有の
inhibition 抑制, 禁止
inhibitor インヒビタ, 抑制剤, 防せい剤, 防止剤
initial 形最初の, 初期の
　initial charge　初充電
　initial condition　初期条件
　initial cost　イニシャルコスト, 初期経費, 原価
　initial design　初期設計
　initial phase　初位相
　initial pressure　初圧, 初期圧力
　initial running in period　初期慣らし運転
　initial strain　元ひずみ
　initial velocity　初速度
initialization 初期化, 初期設定

initially 副初めに, 最初に
initiate 動始める
initiator イニシエータ, 起爆剤
inject 動噴射する, 注入する
injecting atomizer 噴霧器
injection 噴射
　injection air　噴射空気
　injection beginning　噴射始め
　injection delay (lag)　噴射遅れ
　injection end　噴射終わり
　injection lag　噴射遅れ
　injection loss　噴射損失
　injection molding　射出成形
　injection nozzle　噴射ノズル
　injection period　噴射期間
　injection pipe　噴射管
　injection pressure　噴射圧
　injection pump　噴射ポンプ
　injection quantity　噴射量
　injection rating　噴射率
　injection temperature　射出温度
　injection timing　噴射時期
　injection valve　噴射弁, 燃料弁
　injection valve opening pressure　開弁圧, 噴射開始圧
　injection velocity　噴射速度
injector インゼクタ, インジェクタ, 燃料噴射装置, 噴射器, 注入装置
inlet 入口　⊗ outlet：出口
　inlet and outlet　出入口
　inlet angle　入口角, 流入角
　inlet blade angle　羽根入口角
　inlet bucket　前翼
　inlet header　入口管寄せ
　inlet length　助走距離
　inlet manifold　吸込みマニホールド
　inlet pipe　入口管, 吸込み管
　inlet port　入口孔, 吸気口
　inlet pressure　入口圧力
　inlet region　助走区間
　inlet temperature　入口温度
　inlet valve　入口弁, 吸入弁
　inlet velocity　入口速度
INMARSAT (**International Maritime Satellite Organization**)

インマルサット，国際海事衛星機構
inner 形内側の，内部の
　inner bottom plate　内底板
　inner crystal crack　結晶粒内割れ
　inner cylinder　内筒
　inner diameter　内径
　inner face　内面
　inner flame　内部フレーム
　inner friction　内部摩擦
　inner gearing　内かみあい
　inner layer　内層
　inner liner　インナライナ
　inner micrometer　内側マイクロメータ
　inner potential　内部ポテンシャル，内部電位
　inner product　スカラ積，内積
　inner race　内レース【玉軸受の内輪】
　inner ring　内輪
　inner rotor　内部回転体
　inner shell electron　内殻電子
　inner side　内側
　inner structure　内部構造
　inner surface　内部表面，内面
　inner tube　内管
innovation　イノベーション，革新
inoperative　形作用しない
inorganic　形無機質の，無機の
　inorganic chemical　無機化合物
　inorganic material　無機質材料
in-phase　形同相の
　in-phase component　同相分
inplane shear　面内せん断
inplane stress　面内応力
input　名インプット，入力，入力信号　動インプットする，入力する，投入する
　input circuit　入力回路
　input device　入力装置
　input information　入力情報
　input offset voltage　入力オフセット電圧
　input-output　入力-出力
　input output port　I/O ポート【コンピュータと外部装置との接続装置】

　input point　入力点
　input resistance　入力抵抗
　input signal　入力信号
　input voltage　入力電圧
inquire　動問う，尋ねる
inrush　突入，侵入，流入
　inrush current　突入電流
inscribe　動記入する，刻みつける
inscribed circle　内接円
insensitive　感知(反応)しにくい
insert　名インサート，挿入，埋め金，入れ子　動挿入する
　insert bearing　インサート軸受，交換軸受
　insert/delete　挿入／削除
in-service inspection　供用期間中検査
inside　名内部，内側　形内部の　前～の内部に，以内に
　inside calipers　内パス，穴パス
　inside diameter　内径
　inside micrometer　内マイクロメータ
　inside wall　内壁
insist　動主張する，要求する
insofar as ～　接～する限りでは
insolubility　不溶性
insoluble　名不溶分　形不溶解性の
　insoluble matter　不溶分(物)
　insoluble residue　不溶残分
inspect　動検査する，調べる
inspection　検査，点検
　inspection door　点検窓，検査ふた
　inspection gauge　検査ゲージ
　inspection hole　のぞき穴，検査穴
　inspection panel　点検窓，点検パネル
　inspection tank　検査タンク
instability　不安定，不安定性
install　動インストールする，据付ける，設置する，設備する，取付ける
installation　インストール，設置，装置，据付け，取付け，設定
instance　例，場合
　関 for instance：例えば
instant　形即時の，緊急の，インス

タントの 名瞬間 閏 for an instant：ほんのちょっとの間
instantaneous 形瞬間の
  instantaneous governing　瞬時調速
  instantaneous power　瞬時電力
  instantaneous value　瞬時値
instantly 副即座に
instead 副その代わりに
  instead of ～　前～の代わりに
institute 名協会，学会，研究所 動設ける，制定する，設立する，実施する
  Institute of Electrical Engineers　電気学会
  Institute of Mechanical Engineers　機械学会
instruct 動教える，知らせる，説明する，指示する
instruction 取扱説明書，指示，命令
instrument 機器，計器，器具，装置
  instrument panel　計器盤
  instrument transformer　計器用変成器【変圧器の一種】
instrumental analysis 機器分析
instrumental error 器差，計器誤差
instrumentation 計装，計測
  instrumentation drawing　計装図
  instrumentation symbol　計装用記号
insufficient 形不十分な
insulate 動絶縁する，断熱する，保温する
insulating 絶縁[性]
  insulating bolt　絶縁ボルト
  insulating material　絶縁材，断熱材
  insulating oil　絶縁油
  insulating paper　絶縁紙
  insulating quality　絶縁性
  insulating tape　絶縁テープ
  insulating varnish　絶縁ワニス
insulation 絶縁，防熱材
  insulation breakdown　絶縁破壊
  insulation of vibration　振動防止
  insulation oil　絶縁油
  insulation resistance　絶縁抵抗
  insulation test　絶縁試験
  insulation transformer　絶縁トランス
insulator インシュレータ，絶縁体，碍子，断熱材，保温材
insurance 保険
insure 保証する，確実にする
intake 入口，取入口
  intake air temperature　吸気温度
  intake manifold　吸気マニホルド
  intake pipe　吸気管
  intake port　吸気ポート
  intake pressure　吸気圧力
  intake stroke　吸入行程，吸気行程
  intake valve　吸気弁
integer 整数，完全体
integral 形積分の，整数の，不可欠な 名積分，全体
  integral action　積分動作
  integral construction　一体構造
  integral control　積分制御
  integral control action　I動作，積分動作
  integral element　積分要素
  integral equation　積分方程式
  integral method　積分法
  integral network　積分回路
  integral time　積分時間
integrally 副一体的に，完全に
integrate 動まとめる，集約する，積分する，統合する
Integrated Circuit IC，集積回路
integrating 積分
  integrating circuit　積分回路
  integrating coefficient　積分係数
  integrating meter　積算計，積算計器
  integrating watt-meter　積算電力計
integration constant 積分定数
integrator 積分器，求積器
intelligence 知能
  intelligence robot　知能ロボット
intelligent 形賢い，知的な，聡明な
intend 動～するつもりである，～しようとする
intended 形意図された，予定された，計画された
intense 形激しい，強度の
intensifier 増圧器
intensify 動強める

intensifying accumulator 増圧アキュムレータ
intensity 強度, 強さ, 明暗度, 拡大
　intensity control　強度制御
　intensity of electric field　電界の強さ
　intensity of illumination　照度
　intensity of magnetic field　磁界の強さ
　intensity of pressure　圧力の強さ
　intensity of stress　応力度
intensive　形示強の, 集中的な
　intensive property　示強性, 示強変数, 示強的性質, 強度的性質
　intensive quantity　示強量
intention　意図
intentionally　副意図的に, 故意に
inter-　接頭「中」「間」「相互」の意
interact　動相互に作用する, 互いに影響しあう, 情報を交換する
interaction　相互作用
interchange　交替, 交換
　関 information interchange：情報交換
interchangeability　互換性
interchangeable　形互換性のある
intercondenser　中間コンデンサ, 中間復水器, 内部復水器
interconnect　動相互に接続する
intercooler　中間冷却器, 給気冷却器
intercrystalline corrosion　[結晶]粒界腐食
intercrystalline crack　[結晶]粒界割れ
interest　名関心, 興味　動興味を起こさせる, 関心を持たせる
interface　インタフェース, 界面, 境界面, 共有領域
interfacial tension　界面張力
interfere　動干渉する, 妨害する
interference　干渉, 締めしろ, 妨害
　interference dilatometer　光干渉式熱膨張計
　interference drag　干渉抗力
　interference fit　締まりばめ
　interference in exhaust process　排気干渉
　interference in intake process　吸気干渉
　interference of equal thickness　等厚干渉
　interference of light wave　光波干渉
　interference of pulsation wave　脈動干渉
interferometer　干渉計
intergranular　形粒間の, 粒子間の
　intergranular corrosion　粒界腐食
　intergranular crack[ing]　粒界割れ
　intergranular failure　粒界破断
　intergranular fracture　粒界破壊
　intergranular stress corrosion cracking　粒界応力腐食割れ
interior　形内部の, 内側の
　名内部, 内側, 屋内
　interior angle　内角
　interior wiring　屋内配線
interlinkage　鎖交
interlock　動連動する, 連結する, 組み合う
　名インタロック, 連動装置
　interlock arrangement　連動装置
　interlock circuit　インタロック回路
interlocking device　連続装置
interlocking gear　連動装置
intermediate　形中間の, 中級の
　intermediate connector　媒介節
　intermediate gear　中間歯車
　intermediate inspection　中間検査
　intermediate pressure turbine　中圧タービン
　intermediate product　中間生成物
　intermediate shaft　中間軸
　intermediate shaft bearing　中間軸受
　intermediate valve　中間弁
intermetallic compound　金属間化合物
intermit　動断続する, 間欠する
intermittent　形断続的な, 間欠的な
　intermittent action　間欠的動作
　intermittent current　断続電流, 間欠電流, 脈動電流, 間欠流
　intermittent duty rating　断続使用定格
　intermittent firing　間欠燃焼

intermittent gear　間欠歯車
intermittent injection　間欠噴射
intermittent motion mechanism　間欠機構
intermittent movement　間欠運動
intermittent recorder　間欠記録, 打点式記録計
intermittent ringing　断続信号
intermittent working　間欠作動, 間断作業

intermixing　混合
intermolecular condensation　分子内縮合
intermolecular force　分子間力
internal　形 内部の
　internal combustion engine　内燃機関
　internal diameter　内径
　internal efficiency　内部効率
　internal energy　内部エネルギ
　internal force　内力
　internal friction　内部摩擦
　internal gear pump　内歯歯車ポンプ
　internal impedance　内部インピーダンス
　internal latent heat　内部潜熱
　internal layer　内層
　internal loss　内部損失
　internal passage　内部通路
　internal pressure　内圧
　internal resistance　内部抵抗
　internal stress　内部応力
　internal surface　内壁面, 内側面
　internal thermal efficiency　内部熱効率
　internal thread　めねじ
　internal type desuperheater　内置式過熱もどし器
　internal work　内部仕事
internally fired boiler　内だきボイラ
internally reversible process　内部可逆過程
international　形 国際的な
　International Organization for Standardization　ISO, 国際標準化機構
　international standard meter　国際メートル原器
　International System of Units　SI（フランス語でSystème International d'Unités）, 国際単位系
　関 gravitational unit：重力単位
　international unit　国際単位
internet　インターネット
interpole　補極
interpose　動 挿入する
interpret　動 解釈する, 通訳する
interpreter　インタプリタ, 解釈プログラム, 翻訳機, 通訳者
interrupt　動 さえぎる, 中断する
interruption　中断, 不通, 妨害
intersect　動 交差する
intersecting line　交線
intersection　交差, 交点, 論理積, 横断, 共通部分
　intersection angle　交角
　intersection axis　交差軸
　intersection point　交点
interstitial solid solution　侵入型固溶体
interval　インターバル, 間隔
into　前 ～の中へ【「in」はすでに中にある状態,「into」は外から中への移動を伴う場合に用いる】
intricate　形 複雑な, 精巧な
intrinsic energy　固有エネルギ
intrinsic semiconductor　真性半導体
intrinsic viscosity　固有粘度
introduce　動 導入する, 挿入する, 紹介する
introduction　紹介, 挿入, 導入, 概要
inturn　内曲がり, 内転
invar　インバール, アンバ, 不変鋼
　invar alloys　インバール合金【熱膨張係数が極めて小さい合金】
invariable　形 一定の, 不変の
　invariable steel　不変鋼【温度変化による性質変化の小さい鋼】
invariance　不変, 不変性
invariant　形 不変の, 一定の, 不変式の　名 不変量, 不変式
invasion　侵入
invent　動 発明する

invention 発明
inventory インベントリ，在庫品，備品目録
inverse 形逆の，反対の
 名逆，逆関数，反対
　inverse cam 逆カム
　inverse Carnot cycle 逆カルノーサイクル
　inverse Laplace transform 逆ラプラス変換
　inverse proportion 反比例
　inverse time relay 反限時リレー
inversely 副反対に
inversion インバージョン，反像，逆変換，逆[転]，反転
 関 phase inversion：転相
inverted 形逆さの，反転した
　inverted cam 逆カム，従動カム
inverter インバータ，変換器
investigate 動調査する，研究する
investigation 研究，調査
inviscid 形無粘性の，粘性のない
　inviscid flow 非粘性流
　inviscid fluid 非粘性流体
invite 動依頼する，勧誘する，招く，もたらす，招待する，勧める，引き起こす
involute 名インボリュート，伸開線，閉旋回，内巻き
 形内巻きの，複雑な
　involute curve インボリュート曲線
　involute gear インボリュート歯車
　involute tooth profile インボリュート歯形
involve 動含む，必要とする，巻く，巻き込む，巻き込まれる，参加する，携わる，伴う
inward 副中へ，内側へ，内部へ
 形中の，内側にある，内部の
 関 inward turning：内回り
iodide ヨウ化物
ion イオン【帯電した原子または原子団のこと】
　ion beam machining イオンビーム加工
　ion exchange イオン交換

　ion exchange resin イオン交換樹脂
　ion pair イオン対
　ion plating イオンめっき
ionic bond イオン結合
ionization 電離，イオン化
　ionization potential イオン化ポテンシャル，イオン化電位
iridium イリジウム 記 Ir
iron 鉄 記 Fe 関 α-iron：α鉄
　iron carbon equilibrium diagram 鉄-炭素平衡状態図
　iron core 鉄心
　iron hydroxide 水酸化鉄
　iron loss 鉄損
　iron ore 鉄鉱石
　iron oxide 酸化鉄
　iron plate 鉄板
　iron sand 砂鉄
irrational number 無理数
irregular 形不規則な，不ぞろいな
　irregular error 不規則誤差
　irregular injection 不整噴射
irregularly 副不規則に
irreversibility 不可逆性
irreversible 形逆にできない，不可逆の
　irreversible change 不可逆変化
　irreversible cycle 不可逆サイクル
　irreversible engine 非可逆機関
irrotational flow 渦なし流れ，非回転流
isenthalpic 形等エンタルピの
 名等エンタルピ
　isenthalpic change 等エンタルピ変化
isentrope 等エントロピ線
isentropic 形等エントロピの
 名等エントロピ
　isentropic change 等エントロピ変化
　isentropic curve 等エントロピ線
　isentropic efficiency 等エントロピ効率
　isentropic expansion 等エントロピ膨張
　isentropic flow 等エントロピ流れ
　isentropic line 等エントロピ線

iso- [接頭]「等しい」「同じ」の意
ISO（**International Organization for Standardization**） 国際標準化機構
  ISO screw thread　ISO ねじ
isobar 等圧線
isobaric [形]等圧の
  isobaric change　等圧変化
  isobaric expansion　等圧膨張
  isobaric line　等圧線
  isobaric process　等圧過程
  isobaric specific heat capacity　定圧比熱
  isobaric surface　等圧面
isobutane イソブタン【無色の引火性気体】
isochoric [形]等容の
  isochoric change　等積変化, 定容変化, 等容変化
  isochoric specific heat capacity　定容比熱, 定積比熱
isochronous governor 等速調速機
isoconcentration 等濃度
isodynamic [形]等強度の, 等磁力線の
isolate [動]切り離す, 孤立させる, 絶縁する
isolated [形]孤立した, 絶縁した〈電機〉
  isolated system　孤立系
isolation 絶縁, 分離, 遊離, 孤立, 隔離, 遮断
  isolation valve　隔離弁, 遮断弁
isolator アイソレータ, 防音装置, 断路器, 絶縁装置
isomer 異性体
isometric [形]等角の, 等大の, 等軸の, 等方の
  isometric change　等容変化, 定容変化
  isometric drawing　等角投影図, 等測図, 等角図
  isometric projection　等角投影[法]
isooctane イソオクタン【炭化水素の一種】
isopiestic [形]等圧の
  isopiestic change　等圧変化
  isopiestic line　等圧線
  isopiestic surface　等圧面
isosceles triangle 二等辺三角形
isostere 等比容線
isotherm 等温線
isothermal [形]等温の
  isothermal change　等温変化
  isothermal compression　等温圧縮
  isothermal curve　等温線
  isothermal efficiency　等温効率
  isothermal elastic modulus　等温弾性係数
  isothermal expansion　等温膨張
  isothermal line　等温線
  isothermal solidification　等温凝固
  isothermal transformation　恒温変態, 等温変態
isotope アイソトープ, 同位元素
isotropy 等方性
isovolumetric [形]等積の
  isovolumetric change　等容変化, 等積変化, 定容変化
issue [動]発行する, 出版する, 出る, 支給する
  [名]問題, 発刊, 流出, 刊行物, 号
IT（**Information Technology**） 情報技術
item 項目, 細目
iterative method 反復法
ITV（**Industrial Tele Vision**） 工業用テレビ
Izod impact test アイゾット衝撃試験

J-j

J ジュール【エネルギの単位】
jack ジャッキ
  jack bolt　ジャッキボルト
  jack screw　止めねじ
  jack shaft　副軸
jacket ジャケット,外筒
  jacket and piston cooling fresh water pump　ジャケット兼ピストン冷却清水ポンプ
  jacket cooling fresh water cooler　ジャケット清水冷却器
  jacket cooling fresh water pump　ジャケット冷却清水ポンプ
  jacket water　ジャケット水
jacketed gasket ジャケット形ガスケット
jacketed pipe 二重管
jacking oil pump ジャッキングオイルポンプ【起動時用の潤滑油ポンプ】
jacking screw 高さ調整ねじ
Jacob's ladder なわばしご
Jakob number ヤコブ数
jam 動動かなくなる
  jam nut　緩み止めナット
jamming 妨害,電波妨害
Janney pump ジャンネーポンプ
Japanese Industrial Standards JIS,日本工業規格
Japanese syllabary かな文字
jar 名ジャー,びん,つぼ,衝撃,振動 動振動させる,衝撃を与える
jaw chuck ジョーチャック
jerky 形ぎくしゃくした,引っかかる
jet 名ジェット,噴流,噴射 動噴出する
  jet blower　噴射送風機
  jet carburetor　気化器
  jet condenser　噴射復水器
  jet deflector　噴射そらせ板
  jet engine　ジェットエンジン
  jet flow　噴流
  jet fuel　ジェット燃料
  jet lubrication　ジェット潤滑
  jet pipe　噴射管
  jet propeller　ジェットプロペラ
  jet propulsion　ジェット推進
  jet propulsion gas turbine　ジェット推進ガスタービン
  jet pump　ジェットポンプ
  jet velocity　噴射速度
jetty 突堤,桟橋
jewel bearing 宝石軸受
jib ジブ〈荷役装置〉
  jib crane　ジブクレーン
jig ジグ【機械加工などで工具を制御・案内する装置】
jigsaw 糸のこ
jim crow ジンクロ【曲がりを修正する工具】
JIS (Japanese Industrial Standards) 日本工業規格
joggled lap joint 段付継手
joggling 段付け
Johnson valve ジョンソン弁
join 動接合する
joint ジョイント,継手,節点,接合,継ぎ目
  joint box　接続箱
  joint efficiency　継手効率
  joint pin　継手ピン
  joint plate　継ぎ板
  joint sheet　シートガスケット
Jominy curve ジョミニ曲線
Josephson effect ジョセフソン効果
joule ジュール【エネルギの単位】
  joule heat　ジュール熱
  Joule effect　ジュール効果
  Joule's equivalent　熱の仕事当量
  Joule's law　ジュールの法則
journal ジャーナル,軸,軸端,軸頸【回転軸の軸受にはまり合う部分】
  journal bearing　ジャーナル軸受
  journal bearing wedge　受金押え
  journal box　軸箱

journal pin　ジャーナルピン
joystick　ジョイスティック
judge　動判断する
judg[e]ment　判断
judging from ～　副～から判断すれば
jump　名ジャンプ，跳躍　動ジャンプする，飛ぶ，急に高くなる
　jump scavenging　ジャンプ式掃気
　jump spark ignition　火花点火
jumping　ジャンピング

junction　接合，結合
　junction box　接続箱
junk ring　押え輪
Junker's calorimeter　ユンカース熱量計
Junker's dynamometer　ユンカース動力計
jute　ジュート【コーヒ，綿花などの袋の材料】

K-k

K ケルビン【温度の単位】
Kadenacy effect カデナシ効果
Kalium カリウム 記K
Kaplan turbine カプラン水車
Karman カルマン
 Karman vortex カルマン渦
 Karman vortex street カルマン渦列
 Karman's constant カルマン定数
keel キール, 竜骨
 keel batten キール定規
keep 動保つ, 維持する, ～させ続ける
 名キープ, 押え, 保持すること
 keep clean 動整頓する
 keep in close contact with ～ 動～と連絡を密に取る
 keep one's eye on ～ 動～に注意する
 keep plate 押え板
 keep relay キープリレー, 保持リレー
kelmet ケルメット【軸受用の鉛銅合金】
kelp 海草
Kelvin ケルビン【温度の単位：K】
kerf 切り口, ひき目, 切り溝
kerosene ケロシン, 灯油
 kerosene engine 石油機関
 kerosene lamp 石油ランプ
 kerosene tank 軽油タンク
Kevlar ケブラー【高強力繊維の一種】
key 名キー, 鍵 形重要な
 key-board キーボード
 key-boss キーボス
 key-groove キー溝
 key-hole 鍵穴
 key-less propeller キーレスプロペラ【プロペラ軸とプロペラボスをキーを使わずに摩擦力で固定するプロペラ】
 key-seat rule キー溝定規
 key-slot キー溝
 key-stone ring キーストンリング
 key-way キー溝
 key-wrench 箱スパナ
keying circuit 開閉回路
kickback キックバック【逆方向の推力】
kill 動(動きを)止める
killed steel キルド鋼【十分に脱酸した鋼】 関rimmed steel：リムド鋼
kiln キルン【窯の一種】
kilocalorie kcal, キロカロリ
kilowatt kW, キロワット
 kilowatt hour kWh, キロワット時
kind 種類
kinematic 形運動学的な, 運動学上の
 kinematic chain 連鎖
 kinematic hardening 移動硬化
 kinematic pair 対偶
 kinematic viscosity 動粘度
kinematics 運動学
 kinematics of machinery 機構学
kinetic 形運動の, 動的な
 kinetic energy 運動エネルギ
 kinetic friction 動摩擦, 運動摩擦
 kinetic potential 運動ポテンシャル, 運動電位
 kinetic pressure 動圧
 kinetic pressure bearing 動圧軸受
 kinetic pump カイネチックポンプ
 kinetic viscosity 動粘度, 動粘性
kinetics 動力学
king pin キングピン, 中心ピン
Kinghorn's metallic valve キングホーン弁
kingston valve キングストン弁【海水の主吸込み弁】
kink [コード, ロープなどの]よじれ, もつれ
Kirchhoff's law (principle) キルヒホッフの法則
kit キット【組立用部品一式】
kitchen キッチン, 台所
Klinger's water level gauge クリン

ガ式水面計

klystron クライストロン，速度変調管

knee 曲がり，ひざ
　knee bend　エルボ　⊜elbow
　knee plate　ニープレート

knew 動「know」の過去形

knife ナイフ，小刀，刃
　knife edge　ナイフエッジ，刃形台【荷重の支え方の一形態】
　knife switch　ナイフスイッチ，刃形開閉器【開閉器の一種】

knob 取っ手，握り，ノブ

knock 動打つ，たたく　名ノック，ノッキング
　knock-bolt　ノックボルト
　knock-down　ノックダウン，分解，組立式のもの
　knock-limited compression ratio　臨界圧縮比
　knock off　動［仕事を］終える
　knock-phenomenon　ノック現象
　knock-pin　ノックピン
　knock-property　ノック性

knocking ノッキング，たたき音

knot 名ノット【船速を表す単位：1ノット = 1,852m/h】
　動結ぶ，結び目を作る

know 動知る，分かる
　know how　ノウハウ【[製造などの]技術やこつ】

knowledge 知識，情報
　knowledge engineering　知識工学

known 動「know」の過去分詞
　形既知の，周知の

knuckle 関節，ひじがね
　knuckle joint　ナックル継手【軸継手の一種】
　knuckle pin　ナックルピン，リストピン【V型機関の副連接棒と主連接棒をつなぐピン】
　knuckle screw thread　丸ねじ

knurl ローレット

knurled nut ローレットナット，刻み付ナット

knurling ローレット切り【握り部にギザギザの滑り止めをつけること】
　knurling tool　ローレット工具【滑り止めをつける工具】

Kort nozzle コルトノズル【推進器の周りに取り付けられたノズル】

Kovar コバール【Fe-Ni-Co合金】

Kramer system クレーマ方式【巻線型誘導電動機の速度制御方法の一つ】

KS steel KS鋼

L-l

La Mont boiler ラモントボイラ
labor saving 省力化
laboratory 実験室, 研究所
laborious 形面倒な, 困難な
labyrinth 迷路
  labyrinth packing ラビリンスパッキン【蒸気タービンの軸封装置】
lace 名レース, ひも
  動締める, 通す
lacing wire レーシングワイヤ, 押え金
lack 名不足, 欠乏
  動欠く, 足りない
ladder はしご
  ladder diagram ラダー図【制御回路図の一種】
laden 形〜を帯びた
  関moisture-laden：湿気を帯びた
lag 名遅れ, ずれ 動遅れる
  lag element 遅れ要素
lagging ラギング【保温材で被覆すること】, 保温材, [断熱]被覆材
  lagging current 遅れ電流
  lagging material 保温材
lamella ラメラ, 層板, 薄板, 薄膜
lamellar 形薄板状の, 薄膜の, 薄層の, 層板の
  lamellar spring 重ね板ばね
  lamellar structure 層状組織
laminar 形層流の, 層状の
  laminar boundary layer 層流境界層
  laminar convection 層流対流
  laminar flame 層流火炎
  laminar flow 層流
  laminar sublayer 層流底層
laminate 形薄層(板)状の
  名ラミネート, 層状材料
  動薄板(層)にする, 積層する
  laminate molding 積層形成
laminated core 成層鉄心, 積層鉄心
laminated spring 重ね板ばね
lamination ラミネーション【鋼板に生ずる欠陥の一つ】

Lami's theorem ラミーの定理
lamp ランプ, 明かり
  lamp ball 電球
  lamp ballast 安定器
  lamp oil 灯油
LAN (Local Area Network) ラン, ローカルエリアネットワーク, 構内情報通信網
Lancashire boiler ランカシャボイラ【円筒形ボイラの一種】
land 陸 動着地する
  land trial 陸上試運転
  関sea trial：海上試運転
landing edge 縦継手
language 言語
lantern ring ランタンリング【パッキンリングの一種】
lap 名重なり 動摺り合わせる, 包む, 巻く, 重ねる
  lap joint 重ね継手
  lap seam welding 重ねシーム溶接
  lap welding 重ね溶接
  lap winding 重ね巻
Laplace transformation ラプラス変換
Laplacian operator ラプラシアン演算子
lapping ラッピング, ラップ仕上げ, 摺合わせ, 研磨【精密仕上げの一種】
  lapping machine ラップ盤
  lapping tool ラッピング用工具
lapse 経過, 消滅, 推移
  lapse rate 減率
  lapse time 経過時間
lard oil ラード油, 豚油
large 形大きな, 大型の
  Large Scale Integration LSI, 大規模集積回路
  large diameter 形大口径の
  large size diesel engine 大型ディーゼル機関
largely 副主として, 大部分は
laser レーザ

(同) Light Amplification by Stimulated Emission of Radiation
laser beam machining　レーザ加工
laser scanner　レーザスキャナ【光学的自動読みとり装置の一種】
lash　(動)縛る，結ぶ　(名)ラッシュ，可動部品間のすきま，遊び
lashing　ラッシング，なわ，鎖，固縛
last　(形)最後の，最近の　(動)続く，長持ちする
laster plotter　ラスタプロッタ【作図器の一種】
latch　ラッチ，掛け金，外れ止め，保持
　latch bolt　ラッチボルト【自動式ばねボルトの一種】
　latch circuit　ラッチ回路
　latch rod　掛金棒
latching relay　ラッチリレー【保持リレーの一種】
late　(形)遅い，最近の　(副)遅く，遅れて，最近
　late combustion　後燃え
latent heat　潜熱
　(関) sensible heat：顕熱
　latent heat of fusion　融解潜熱
　latent heat of vaporization　蒸発[潜]熱，気化熱
later　(形)「late」の比較級，その後，より最近の　(副)後で，その後，遅れて
lateral　(形)横の，横向きの，側面の　(名)側面
　lateral bending　横曲げ
　lateral construction　横縮み
　lateral dimension　横方向寸法
　lateral direction　横方向
　lateral extensometer　横伸び計
　lateral force　横力
　lateral load　横荷重
　lateral pressure　側圧
　lateral resistance　横抵抗
　lateral strain　横ひずみ
　lateral vibration　横振動
　lateral view　側面図
latest　(形)最近の

lathe　旋盤
latitude　緯度　(関) longitude：経度
latter　(形)後者の，後期の，後の
　(関) the latter：後者
　(反) the former：前者
lattice　格子
　lattice constant　格子定数
　lattice defect　格子欠陥
launch　(動)進水させる，［ボートを］降ろす　(名)ランチ，艇，進水
law　法則，定理
　law of action and reaction　作用・反作用の法則
　law of conservation of energy　エネルギ保存の法則
　law of conservation of mass　質量保存の法則
　law of conservation of momentum　運動量保存の法則
　law of dynamical similarity　力学的相似則
　law of inertia　慣性の法則
　law of similarity　相似則
　law of succession　連続の法則
laws of motion　運動の三法則
lay　(動)置く，横たえる，配置する
layer　(動)層状にする　(名)層
　(関) boundary layer：境界層
layout　配置，レイアウト
LCD（Liquid Crystal Display）　液晶ディスプレイ装置
lead　(動)導く，誘う　(名)リード，進み，導線，鉛　(記) Pb
　lead-acid battery　鉛蓄電池
　lead angle　リード角，進み角
　lead covered wire　被鉛線
　lead equivalent　鉛当量
　lead hammer　鉛ハンマ
　lead pipe　鉛管
　lead screw　親ねじ
　lead storage battery　鉛蓄電池
　lead tester　リードテスタ【ねじのピッチ誤差測定器】
　lead wire　リード線，導線
leading current　進み電流
leading edge　前縁〈プロペラ〉

leading screw 親ねじ
leaf spring 板ばね
leak 動漏れる，漏らす 名漏れ，漏電 関 gas leak：ガス漏れ
 leak detector 漏洩検知器
 leak resistance 漏れ抵抗
 leak test 漏れ試験
leakage 漏れ
 関 air leakage：空気漏れ／water leakage：漏水
 leakage coefficient 漏れ係数
 leakage current 漏れ電流
 leakage detector 漏洩検知器
 leakage flux 漏れ磁束
 leakage inductance 漏れインダクタンス
 leakage loss 漏れ損失
 leakage reactance 漏れリアクタンス
 leakage test 漏れ試験
leak-off pipe line 燃料戻し管
leaky 形漏れやすい，漏れがある
lean 名傾き，傾斜 動傾く，もたれる 形貧弱な
 lean burn engine リーンバーンエンジン，希薄燃焼エンジン
 lean rich combustion 濃淡燃焼
learning control 学習制御
least 形最も少ない，最も小さい
 関 at least：少なくとも
 least energy 最小エネルギ
 least square method 最小二乗法
 least work 最小仕事
leather 革，なめし革
leave 動放置する，離れる，〜のままにしておく
 leave through 動を通して出る
leaving loss 排出損失
led 動「lead」の過去・過去分詞
LED（Luminescent Electric Diode） 発光ダイオード
lee side 風下
left 名左 形左の 動「leave」の過去・過去分詞
 left hand rule ［フレミングの］左手の法則

 left hand screw 左ねじ
 left handed propeller 左回りプロペラ
 left open 動開いたまま
leg 脚（あし），支柱
 leg length 脚長
length 長さ，距離
 関 wave length：波長
 length breadth ratio 長さ幅比
 length overall 全長
lengthen 動長くする，延ばす
lens レンズ
 lens ring gasket レンズリングガスケット【ガスケットの一種】
Lenz's law レンツの法則
less 形「little」の比較級，より少ない，より小さい 代より少ないもの
 関 much less than：〜よりもっと少ない
 less noble metal 卑金属
 less than 〜より少ない
 less than or equal to 〜以下
 less A than B AというよりB
lessen 動少なくする，小さくする，減らす
let 動〜させる，〜するのを許す
letter 文字，手紙
level 名レベル，水位，液面，水平，水準，水準器，準位 動水平にする，平らにする
 level controller 液面調節器
 level gauge レベルゲージ，液位計
 level indicator レベル表示器，液面計，水準計
 level meter レベル計
 level switch レベルスイッチ
 level vial 水準器
leveler ひずみ取り機
leveling レベリング，水平出し，表面平滑化
lever てこ，レバー
 関 hand lever：手動レバー
 lever and spring loaded safety valve ばね平衡式安全弁
 lever handle 取っ手
 lever relation てこの関係

lever rule　てこの原理
lever safety valve　てこ安全弁
lever spring　てこばね
leverage　てこの作用，てこ比
levitation　[空中]浮揚
liability　責任
liberate　動 遊離させる，放出する，解放する，作用させる
lisence　回 license
license　名 ライセンス，認可，免許　動 許可を与える，認める
license pressure　認可圧力
lid　蓋
lie　動 横たわる，置く，展開する，位置する，停泊する
life　生命，寿命
life boat　救命艇
life buoy　救命ブイ
life jacket　救命胴衣
life preserver　救命用具
life raft　救命いかだ
life rope　命綱
life saving appliance　救命設備
lift　名 エレベータ，昇降機，揚力，揚程　動 持ち上げる，引き上げる
lift check valve　リフト逆止め弁
lift coefficient　揚力係数
lift pump　吸上げポンプ
lifting　リフティング，巻上げ
lifting bolt　引上げボルト
light　名 ライト，照明，光　形 軽い，明るい，薄い，楽な　動 明かりをつける，火をつける
light alloy　軽合金
light dark　明暗
light emitting diode　発光ダイオード
light equation　光差
light flux　光束
light gauge steel　軽量形鋼
light load　軽負荷
light metal　軽金属【比重4以下の金属】
light oil　軽油
light oil tank　軽油タンク
light quantum　光量子
light source　光源
light wave　光波
light-weight　形 軽量の
lighten　動 明るくする，軽くする
lightening　軽量化
lighting coil　点灯コイル
lightly　副 軽く，少し
lightness　明度
lightning conductor　避雷針
lignumvitae　リグナムバイタ【硬い木材で船尾管軸受材として用いる】
like　動 好む　前 ～のような
likely　形 起こりえる，～しそうである　副 多分，おそらく
likewise　副 同様に，さらに
limber　リンバ【船底の汚水路】
limber hole　リンバホール【あか水を流すために肋材などに開けた一連の穴】
lime　ライム，石灰
limestone　石灰石
limit　名 限度，限界，境界，制限，極限　動 制限する
limit design　極限設計
limit gauge　限界ゲージ
limit load　極限荷重
limit of elasticity　弾性限度
limit of inflammability　可燃限度
limit of proportionality　比例限度
limit ring　制限リング
limit size　限界寸法
limit switch　リミットスイッチ
limitative　形 限界の
limited　形 限られた
limiter　リミッタ，制限器
limiter function　制限関数
limiting　形 制限する，限定的な
limiting creep stress　クリープ制限応力
limiting speed　限界速度，制限速度
limiting stress　限界応力
limiting value　極限値
line　動 並べる，～の内側を覆う，覆う，裏打ちする，線を引く　名 ライン，パイプライン，線，行，ひも，系統，管系，管路，回線，電線路

㊀ straight line：直線／curved line：曲線／parallel lines：平行線／broken line：破線／dotted line：点線／horizontal line：水平線／vertical line：垂直線
line chart　折れ線グラフ
line contact　線接触
line current　線電流
line of action　作用線
line of contact　接触線
line of discontinuity　不連続線
line of electric force　電気力線
line of flow　流線
line of force　力線，磁力線
line of magnetic force　磁力線
line of shearing stress　せん断応力線
line shaft　線軸
line spectrum　線スペクトル
line start motor　直入れ電動機
line starting　直入れ始動
line-to-line voltage　線間電圧
line-to-neutral voltage　相電圧
line up　㊁合わせる，並べる
line voltage　線間電圧
linear　㊂リニア，直線的な，線の，直線の，線形の，一次の
　linear actuator　リニアアクチュエータ【油圧などの動力を直線運動に変換する装置】
　linear equation　一次式，一次方程式
　linear expansion coefficient　線膨張率
　linear induction motor　リニア誘導モータ
　linear load　線圧
　linear momentum　線形運動量
　linear motion　直線運動，線運動
　linear motor　リニアモータ【可動部分が直線運動をする電動機】
　linear motor ship　リニアモータ船
　linear power density　線出力密度
　linear rolling bearing　リニア軸受
　linear servomechanism　線形サーボ機構
　linear strain　線ひずみ
　linear system　線形システム
　linear transformation　一次変換
　linear velocity　線速度
linearity　直線性
linearization　線化
liner　ライナ，かぶせ金，はさみ金
　㊀ cylinder liner：シリンダライナ
　liner ring　ライナリング
　liner wear　ライナの摩耗
linger　㊁残存する
lining　ライニング，内面，内層，内面被覆材，内張り，裏張り，裏打ち
　lining brick　内張りれんが
link　㊃リンク，結合，連係
　㊁つなぐ，連結する
　link mechanism　リンク機構
　link switch　タンブラスイッチ
　link work　リンク装置
linkage　リンク機構，リンク装置，結合，連携，連鎖，つながり
linoleum　リノリウム【樹脂の一種】
linotape　リノテープ【絶縁テープの一種】
lip　リップ，縁，切れ刃，唇
　lip packing　リップパッキン【Uパッキン，Vパッキンなどがある】
　lip seal　リップシール，縁シール
lipophilic group　親油基
Lipowitz metal　リポウィッツ合金【低温度可溶合金の一種】
liquefaction　液化
Liquefied Natural Gas　LNG，液化天然ガス
Liquefied Petroleum Gas　LPG，液化石油ガス
liquefier　液化機，液化装置
liquid　㊂液体(状)の，流動性の
　㊃液体
　㊀ fluid：流体／solid：固体
　liquid ammonia　液体アンモニア
　liquid carburizing　液体浸炭法
　liquid cooled engine　液冷機関
　liquid crystal　液晶
　Liquid Crystal Display　LCD，液晶ディスプレイ装置
　liquid damping　液体制動
　liquid flow　流体の流れ
　liquid friction　流体摩擦

liquid fuel 液体燃料
関 gaseous fuel：気体燃料
liquid hammer リキッドハンマ
liquid helium 液体ヘリウム
liquid honing 液体ホーニング【仕上加工の一種】
liquid hydrogen 液体水素
liquid impregnated capacitor 液体含浸コンデンサ
liquid level meter 液面計
liquid manometer 液柱型圧力計，液体圧力計
liquid metal 液体金属
liquid metal fuel 液体金属燃料
liquid metal polish 金属磨きペースト
liquid nitriding 液体窒化法
liquid nitrogen 液体窒素
liquid oxygen 液体酸素
liquid packing 液体パッキン
liquid penetrant inspection 浸透探傷
liquid phase 液相
liquid refrigerent 液冷媒
liquid sealing 液体シーリング
liquid separator 液分離器
liquid temperature 液温
liquidity 流動性
liquidus curve 液相線，液化曲線
list 名リスト，表，目録，一覧表 動表にする，列挙する
listener 聞く人，受信者
listening rod 聴音棒
liter リットル【容量の単位】
literal 定数
literally 副文字通りに
lithium リチウム 記Li
lithium cell リチウム電池
lithium chloride 塩化リチウム
lithium complex grease リチウムコンプレックスグリース【耐熱性を向上させたグリース】
lithium grease リチウムグリース【多目的グリースの一種】
litmus paper リトマス紙
little 形小さい，わずかな，ほとんどない 副少しは，やや，多少は，まったく…ない，少しも…ない
little by little 副少しずつ，徐々に
live 形作動中の，通電状態の
live load 活荷重，動荷重
live steam なま（生）蒸気
living quarter 居住区域
LNG（Liquefied Natural Gas） 液化天然ガス
LO（Lubricating Oil） 潤滑油
load 名ロード，負荷，荷重，積み荷 動荷重をかける，負荷する
load brake 荷重ブレーキ
load capacity curve 負荷性能曲線
load cell ロードセル【荷重測定器】
load characteristic curve 負荷特性曲線
load condition 負荷条件，負荷状態
load current 負荷電流
load curve 負荷曲線
load deformation curve 荷重-たわみ曲線，荷重-ひずみ曲線，荷重-変形曲線
load draft 満載喫水線
load elongation diagram 荷重-伸び線図
load factor 負荷率
load in tension 引張荷重
load indicator 負荷指示計
load line 満載喫水線
load operation 負荷運転（動作）
load rate 負荷率
load rating 動定格荷重
load resistance 負荷抵抗
load shedding 負荷遮断
load test 荷重試験
loaded 形荷を積んだ，負荷された
loaded condition 満載状態，負荷状態
loaded governor おもり調速機
loaded standard condition 基準負荷状態
loading ロード，負荷，荷重，荷役，積荷，充填
Lobster adjustable wrench モンキーレンチ
local 形局部的な，局所的な

local acceleration 局所加速度
local action 局部作用
Local Area Network LAN，ラン，ローカルエリアネットワーク，構内通信網
local boiling 局所沸騰
local buckling 局部座屈
local cell 局部電池
local circuit 局部回路
local contraction 局部縮み
local corrosion 局部腐食
local current 局部電流
local elongation 局部伸び
local hardening 局部焼入れ
local heating 局部加熱
local illumination 局部照明
local oscillation 局部振動
local panel 現場計器盤
local strain 局部ひずみ
local stress 局部応力
local thermal equilibrium 局所熱平衡
local variometer 磁気偏差計
local vibration 局部振動
local view 局部投影図
localize 動限定する，局部集中する
localized 形局部的な
  localized corrosion 局部腐食
locally 副局部的に
locate 動位置する，設置する，決める，示す
locating 位置決め
  locating cone 位置決め円すい
  locating pin 位置決めピン
location 位置，場所，配置
  location tolerance 位置公差
locator ロケータ，探知機
lock 名ロック，錠 動固定する
  lock nut 止めナット
  lock test 拘束試験
  lock washer 止め座金
locked 形固定の
  locked chain 拘束連鎖
locker ロッカ，戸棚
lock-in effect ロックイン効果
locking ロッキング，施錠，弛み止め，固定，締付け
  locking bib cock 錠前カラン
  locking bolt 締付けボルト
  locking circuit 保持回路
  locking coil 保持コイル
  locking device 固定装置，制限装置，割ピン
  locking nut ロックナット，止めナット
  locking pin 止めピン
  locking valve 位置固定弁
locomotive 名機関車 形運動する，移動力のある，機関車の
  locomotive crane ロコクレーン
  locomotive engine 機関車
locus 軌跡
log 航海日誌，運転記録，対数
  log mean temperature difference 対数平均温度差
  log normal distribution 対数正規分布
logarithm 対数
logarithmic 形対数の
  logarithmic function 対数関数
  logarithmic law 対数法則
  logarithmic mean temperature difference 対数平均温度差
  logarithmic scale 対数目盛
  logarithmic strain 対数ひずみ
logbook ログブック，航海日誌，業務日誌
logger ロガー【記録装置の一種】
logic ロジック，論理
  logic circuit 論理回路
  logic control 論理制御，シーケンス制御
  logic product 論理積
  logic symbol 論理記号
  logic valve ロジック弁
logical 形論理的な，論理回路上の，必然の
  logical circuit 論理回路
  logical product 論理積
  logical sum 論理和
  logical symbol 論理記号
logistics ロジスティクス【物流管

long

long 形長い 副長く，長い間
  long column　長柱
  long nose & side cutting pliers　ラジオペンチ　関 pliers：ペンチ
  long shunt　外分巻
  long stroke engine　ロングストローク機関
  long term (period)　長期
  long ton　ロングトン【重量の単位】
  long wave　長波
longer　「long」の比較級
  関 no longer：もはや～ではない
longitude　経度，縦
  関 latitude：緯度
longitudinal 形縦の，経度の
  longitudinal bending　縦曲げ
  longitudinal bulkhead　縦隔壁
  longitudinal force　縦力
  longitudinal load　縦荷重
  longitudinal metacenter　縦メタセンタ
  longitudinal oscillation　縦振動
  longitudinal section　縦断面
  longitudinal shear　縦せん断
  longitudinal stay　長手控え
  longitudinal strain　縦ひずみ
  longitudinal strength　縦強度
  longitudinal stress　縦応力
  longitudinal vibration　縦振動
longitudinally 副縦方向に
longwearing 形長持ちのする
look 動見る，調べてみる，～に見える，～らしい
  look for　動探す
  look up　動調べる，上向く
loop　ループ，輪，環，閉回路
  loop control　ループ制御
  loop gain　ループゲイン【入力と出力の振幅比】
  loop loss　ループ損，絞り損失
  loop ratio　ループ比
  loop scavenging　ループ掃気
  loop seal　ループ封止【空気除去装置の一種】
  loop transfer function　ループ伝達関数
loose 形ゆるんだ，ゆるい，自由な，開放された
  loose coupling　遊動継手
  loose pulley　空回り車
loosen 動緩める
looseness　ゆるいこと，がた
loran　ロラン【自分の位置を割り出す装置】
Lorentz force　ローレンツ力
lose 動失う，浪費する，なくす
loss　ロス，損，損失，欠損，損耗，減量，減少
  loss evaluation　損失評価
  loss factor (coefficient)　損失係数
  loss head　損失ヘッド，損失水頭
    同 loss of head
  loss of power　電源喪失
lost motion　ロストモーション，空転，空動き
lost work　無効仕事
loud 動(音量が)が大きい，派手な，不快な
  loud speaker　スピーカ
louder 形「loud」の比較級
louver　ルーバ，鎧戸，鎧窓
low 形低い
  low alloy steel　低合金鋼
  low calorific value　低発熱量
    同 low calorific power
  low carbon steel　低炭素鋼
  low cycle fatigue　低サイクル疲労
  low excess air burning　低空気比燃焼
  low expansion alloy　低膨張合金
  low frequency　低周波
  low friction　低(滑らかな)摩擦
  low fuel consumption　低燃費
  low grade 形低品質の，軽度の 名ローグレード，下級品，低級
  low lift safety valve　低揚程安全弁
  low load　低負荷
  low melting metal　低温溶融金属
  low oxygen combustion　低酸素燃焼
  low pass filter　低周波フィルタ
  low pressure 名低圧 形低圧の

low pressure compressor　低圧圧縮機
low pressure cylinder　低圧シリンダ
low pressure nozzle　低圧ノズル
low pressure stage　低圧段
low pressure steam generator　低圧蒸気発生器
low pressure steam turbine　低圧蒸気タービン　関 high pressure turbine：高圧タービン
low resistance　低抵抗
low sea suction valve　低位海水吸入弁
low selector relay　ローセレクタリレー【空気圧を利用したリレーの一種】
low speed　低速
low speed engine　低速機関　同 slow speed engine
low speed nozzle　低速ノズル
low stress fracture　低応力破壊
low suction valve　低位吸込弁
low temperature　名低温，形低温の
low temperature annealing　低温焼なまし
low temperature brittleness (embrittlement)　低温脆性
low temperature carbonization　低温乾留
low temperature corrosion　低温腐食，硫酸腐食
low temperature creep　低温クリープ
low temperature grease　低温グリース
low temperature resistance　耐寒性
low temperature strength　低温強度
low temperature welding　低温溶接
low tension　名低電圧，低張力　形低電圧の
low tension coil　低圧コイル
low tension electric ignition　低圧電気点火
low tension side　低圧側
low tension terminal　一次端子
low voltage　低電圧
low water alarm　低水位警報［器］
low water level　低水位

lower　動下げる，低くする
　形下部の，下位の，低い
　lower calorific value　低位発熱量，真発熱量
　lower dead center　下方死点
　lower explosion limit　爆発下限
　lower half　下半分
　lower heating value　低位発熱量，真発熱量
　lower limit　最低限界，下限，下限値，最小寸法
　lower pair　面対偶，低次対偶
　lower part　下部
　lower yield point　下降伏点
LPG（Liquefied Petroleum Gas）　液化石油ガス
LSI（Large Scale Integration）　大規模集積回路
lubricant　潤滑剤，減摩剤，潤滑油
lubricate　動潤滑する，油を塗る，滑らかにする
lubricated　形潤滑された
lubricating　潤滑
　lubricating device (gear)　潤滑装置
　lubricating film　油膜
　Lubricating Oil　LO，潤滑油
　lubricating oil consumption　潤滑油消費量
　lubricating oil cooler　LOクーラ，潤滑油冷却器
　lubricating oil drain tank　潤滑油ドレンタンク
　lubricating oil gravity tank　潤滑油重力タンク
　lubricating oil pressure　潤滑油圧力
　lubricating oil pump　LOポンプ，潤滑油ポンプ
　lubricating oil purifier　潤滑油清浄機
　lubricating oil settling tank　潤滑油澄ましタンク
　lubricating oil sludge tank　潤滑油スラッジタンク
　lubricating oil storage tank　潤滑油貯蔵タンク
　lubricating oil strainer　潤滑油こし器
　lubricating oil sump tank　潤滑油サ

ンプタンク
lubricating oil temperature　潤滑油温度
lubricating oil transfer pump　潤滑油移送ポンプ
lubricating system　潤滑方式, 潤滑系統

lubrication　潤滑
(関) bath lubrication：油浴潤滑／boundary lubricatio：境界潤滑／built-in lubrication：無給油潤滑
Lubrication Oil　LO, 潤滑油

lubricator　ルブリケータ, 潤滑装置, 給油器, 注油器, 油差し

lug　出張り, 突出部
lug bolt　ラグボルト, 耳付きボルト

lumen　ルーメン【光束の単位：lm】

luminescence　ルミネセンス, 冷光【温度放射によらない発光】
luminescence plastics　発光プラスチック

Luminescent Electric Diode
LED, 発光ダイオード

luminous　形光を発する, 輝く, 明るい
luminous and spectrometric analysis　発光分光分析
luminous body　発光体
luminous flame　輝炎
luminous flux density　光束密度
luminous intensity　光度
(関) candela：カンデラ
luminous paint　発光塗料, 夜光塗料
luminous watch　夜光時計

lurgi metal　ルルギメタル【鉛-アルカリ土合金の一種】

luster　光沢

lux　ルクス【照度の単位：lx】

luxmeter　照度計

M-m

M0 check list　M0 チェックリスト
M alkalinity　M アルカリ度
Mach　マッハ【音速に対する速度の比】
　Mach angle　マッハ角
　Mach number　マッハ数
　Mach wave　マッハ波
machinability　機械加工性, 切削性, 被削性, 可削性
machine　名機械
　動機械加工する, 仕上げる
　machine casting　機械鋳造
　machine design　機械設計
　machine drawing　機械図
　machine element　機械要素
　machine finish　機械仕上げ
　machine layout drawing　機械配置図
　machine oil　マシン油, 機械油
　machine parts　機械部品
　machine riveting　機械締め
　machine room　機械室
　machine shop　工作室, 機械工場
　machine tool　工作機械
　machine work　機械加工
machinery　機関, 機械, 機械類
　machinery arrangement　機関配置, 機械配置
　machinery arrangement in engine room　機関室全体配置図
　machinery department　機関部
　machinery efficiency　機関効率
　machinery end bulkhead　機関室端囲壁
　machinery fitting　機関ぎ装
　machinery room (space)　機関室, 機械室
　machinery specification　機関仕様書
　machinery steel　機械部品用鋼
　machinery weight　機関重量
machining　加工, 機械加工
　machining accuracy　加工精度
　machining allowance　仕上げ代(しろ)
　machining of metals　切削加工
　machining time　加工時間
macro-　接頭「大きい」「巨大な」の意
　macrocell　マクロ電池
　macroetching　マクロエッチング, マクロ腐食
　macromolecular　形高分子の
　macromolecule　巨大分子
　macroscopic cross section　巨視的断面図
　macroscopic examination　マクロ組織検査
　macroscopic test　肉眼試験
　macrosegregation　マクロ偏析
　macrostructure　マクロ組織
made　動「make」の過去・過去分詞
　made of　動～で作られている
　made up of　動～で構成されている, ～から成る
magnaflux inspection　磁気探傷検査
magnaflux method　マグナフラックス法, 磁気探傷法
magnaflux test　電磁試験
magnesium　マグネシウム　記Mg
　magnesium carbonate　炭酸マグネシウム
　magnesium sulfate　硫酸マグネシウム
magnet　磁石
　magnet brake　電磁ブレーキ
　magnet coil　電磁コイル
　magnet core　鉄心, 磁石鉄心
　magnet stripe　磁気ストライプ
magnetic　形磁気の
　magnetic action　磁気作用
　magnetic amplifier　磁気増幅器
　magnetic attraction　磁力
　magnetic axis　磁軸
　magnetic bearing　磁気軸受
　magnetic blow-out　磁気遮断器
　magnetic brake　電磁ブレーキ
　magnetic charge　磁荷, 磁気量
　magnetic circuit　磁気回路
　magnetic clutch　電磁クラッチ

magnetic coil 磁気コイル
magnetic compass 磁気コンパス
magnetic cone brake 電磁円すい型ブレーキ
magnetic contactor 電磁接触器
magnetic core memory 磁心記憶装置, 磁気コアメモリ
magnetic coupling 電磁継手
magnetic damping 磁気制動
magnetic defect inspection 磁気探傷法
magnetic drum 磁気ドラム
magnetic field 磁界, 磁場
magnetic field examination 磁粉探傷検査
magnetic field intensity (strength) 磁界の強さ
magnetic figure 磁力線図
magnetic fluid 磁性流体
magnetic flux 磁束
magnetic flux density 磁束密度
magnetic force 磁力, 磁気力
magnetic head 磁気ヘッド
magnetic hysteresis 磁気ヒステリシス
magnetic induction 磁気誘導
magnetic inspection 磁気探傷検査
magnetic line of force 磁力線
magnetic material 磁気材料, 磁性体
magnetic moment 磁気モーメント
magnetic oil strainer 磁力油こし器
magnetic particle test 磁粉探傷試験, 磁気探傷検査
magnetic path 磁路
magnetic permeability 透磁率
magnetic pole 磁極
magnetic potential 磁気ポテンシャル, 磁位
magnetic powder 磁気粉末, 磁粉
Magnetic Resonance Imaging MRI, 磁気共鳴映像法
magnetic saturation 磁気飽和
magnetic separator 磁気分離器
magnetic shield[ing] 磁気遮へい
magnetic steel 磁石鋼, マグネット鋼
magnetic strain 磁気ひずみ
magnetic strainer 磁力こし器
magnetic substance (material) 磁性体
magnetic survey 磁気探査
magnetic susceptibility 磁化率
magnetic transformation 磁気変態
magnetic variation 磁気偏差, 磁気変動, 磁気変化
magnetic wave 磁波
magnetism 磁気, 磁性
magnetite 磁鉄鉱
magnetization 磁化, 励磁
　magnetization current 磁化電流
　magnetization curve 磁化曲線
magnetize 動磁化する
magnetizing 磁化
　magnetizing current 磁化電流
　magnetizing force 磁化力
magneto 磁石発電機, マグネット発電機
magnetoelectric ignition マグネト点火【高圧電気点火の一方法】
magnetogenerator マグネト発電機
MagnetoHydroDynamic power generation MHD 発電, 電磁流体力学発電
magnetometer 磁気計, 磁力計
Magneto Motive Force MMF, 起磁力
magneto-optical disk 光磁気ディスク
magnetoresistor マグネトレジスタ
magnetotype ball bearing マグネト形玉軸受
magnetron マグネトロン, 磁電管
magnification 倍率, 拡大, 拡大図
　同 magnification factor：倍率
magnify 動拡大する
magnifying glass 拡大鏡
magnifying lens ルーペ, 拡大レンズ
magnitude マグニチュード, 大きさ, 等級, 絶対値, 量, 規模, 振幅
　magnitude of complex number 複素数の絶対値
　magnitude of electric field 電界強度
main 名主要部, 幹線
　形主な, 主要な

main air reservoir　主空気槽
main bearing　主軸受
main boiler　主ボイラ
main circuit　主回路
main condenser　主復水器
main contact　主接点
main (electric) power source　主電源
main engine　主機, 主機関
main engine trial　主機試運転
main exhaust　主機排気
main feed water pump　主給水ポンプ
main flow　主流
main gear wheel　主大歯車
main journal　[主]ジャーナル, 主軸受
main magnetic flux　主磁束
main pipe　主管
main pole　主磁極
main power supply　主電源
main propulsion turbine　主推進[用]タービン
main shaft　主軸
main spindle　主軸, スピンドル軸
main steam　主蒸気
main steam control valve　主蒸気加減弁
main steam pipe　主蒸気管
main [steam] stop valve　主塞止弁, 主止め弁, 主蒸気止め弁
main stream　主流
main switch　主開閉器, 主スイッチ
main switch board　主配電盤
main turbine　主タービン
main valve　主弁
main wheel　大歯車, 主車輪
mainly　副主に
maintain　動維持する
maintainability　保全度, 保全性
maintenance　メインテナンス, 保守, 整備, 維持, 保全
maintenance manual　整備マニュアル
maintenance record　保守記録
maintenance work　保守(整備)作業
major　形主要な, 重大な, 大規模な, 副すごく　関 of major importance：非常に重要である
major accident　重大事故
major axis　長径
major critical speed　主危険速度
majority　[大]多数, 大部分, 過半数
majority carrier　多数キャリア
make　動作る, ～させる, ～にする, ～になる, [回路]を閉じる　名製作, ～製, 接続
make A from B　AをBでつくる【Bの質が変化してAになる場合】
make A of B　AをBでつくる【AとBの質が変わらない場合】
make a round　動見回りをする
make and break contact　開閉接点
make contact　メーク接点, a接点
make for　動用意する
make it possible to　動～することを可能にする
make possible　動可能にする
make ready for ～　動～を用意する
make sure　動必ず～する, 確認する, 注意する
make up of ～　動構成する
make use of ～　動～を利用する
maker　メーカ, 製造元
make-up　名補給, 補充, 構成, 組立
make-up pump　補給水ポンプ
make-up valve　補給弁
make-up water　補給水
同 make-up feed
mal-　接頭「不良」,「不完全な」,「不」の意
mal-alignment　アライメント不良
malachite green　マラカイトグリーン, 青竹【けがきの際に工作面に塗る塗料】
male screw　おねじ
malfunction　誤動作, 機能異常
malleability　展性, 展延性, 可鍛性
malleability and ductility　展延性
malleable　形可鍛性の, 展性のある
malleable cast iron　可鍛鋳鉄
mallet　木づち
man machine interface　マン・マシ

ンインタフェース【人と機械が情報の授受を行う境界部分】
man machine system 人間-機械系
man-made 形人工の, 合成の
manage 動管理する, 操縦する
management 管理
mandrel マンドレル, 主軸, 心棒
maneuver 動操作する, 動かす 名操作 ◎ manoeuvre《英》
maneuverability 運動性, 操縦性
maneuverable 形操作できる, 扱いやすい
maneuvering 操縦
　maneuvering air　操縦用空気
　maneuvering gear　操縦装置
　maneuvering lever　操縦レバー
　maneuvering platform　操縦台
　maneuvering valve　操縦弁
manganese マンガン 記Mn
　manganese bronze　マンガン青銅
manhole マンホール, 人孔
manifold 形多くの 名マニホールド, 集合管, 多岐管
　manifold air pressure　吸気圧力
manipulate 動手で動かす, 巧みに操る
manipulated variable 操作量, 操作変数
manipulating force 操作力
manipulation 操作, 加工, 操縦, 取扱い, 処置
manipulator マニピュレータ, マジックハンド, 操縦者
manned 形有人の
manner マナー, 方法, 習慣, 態度, 様式, 礼儀
manoeuvrability 操縦性, 運動性, 機動性
manoeuvring 操縦
manoeuvre 動操縦する, 操作する 名運動
manometer マノメータ, 液柱圧力計, 圧力計, 検圧器
manometric efficiency マノメータ効率, 圧力計効率
manometry マノメトリ, 検圧

manual 名マニュアル, 便覧, 手引き書 形手動の
　manual control　手動制御
　manual operation　手動
　manual valve　手動弁
manually 副手動で
manufactory 製造所, 工場
manufacture 名製造, 生産, 製作, 製品 動生産する, 製造する
manufacturer メーカ, 製造業者
manufacturing 製造
many 形多くの, 多数の
margin マージン, 限界, 境界, 余白, 余裕, 縁, へり
　margin line　限界線
　margin of safety　安全限界, 安全率, 安全範囲
marine 形海の, 舶用の, 船舶の
　marine auxiliary machinery　船用補機
　marine boiler　舶用ボイラ
　marine engine　舶用機関
　marine growth　海洋生成物
　marine growth preventing equipment　海洋生成物付着防止装置
　marine [steam] turbine　舶用蒸気タービン
　marine water tube boiler　船用水管ボイラ
Marisat (Maritime satellite) マリサット, 海事通信衛星
Maritime Disaster Inquiry Agency 海難審判庁
maritime satellite communication system 海事衛星通信システム
mark 名マーク, しるし, 記号 動印をつける, 目盛る
marked degree 目標値
marker マーカ, 目印, 目盛, 標識
marking-off けがき
marking-off pin けがき針
marline 細索
MARPOL convention MARPOL条約【海洋汚染防止を目的とした条約】
marquench マルクエンチ【鋼の熱処理の一種】
martempering マルテンパ【鋼の熱

処理の一種】
Martens extensometer マルテンス伸び計
martensite マルテンサイト【焼入れ鋼の組織の一種】
MASER(**Microwave Amplification by Stimulated Emission of Radiation**) メーザ, マイクロウェーブ発信器
masking マスキング, 遮蔽
mass 質量, 塊り, 大量
 (関) the mass of ～：～の大部分／a mass of ～：多数の～, 多量の～
 mass defect 質量欠損
 mass effect 質量効果
 mass flow controller マスフローコントローラ【気体の質量を計測, 制御する機器】
 mass flow [rate] 質量流量
  (関) volume flow rate：体積流量
 mass number 質量数
 mass point 質点
 mass-produce 動大量生産する
 mass production 大量生産
 mass ratio 質量比
 mass spectrograph 質量分析器, 質量分光器
 mass spectrometer 質量分析計
 mass spectrometry 質量分析[法]
 mass spectrum 質量スペクトル
 mass velocity 質量速度
massive 形巨大な, 大量の, 大規模な
master 船長, 熟練職人
 master cam 親カム, 基本カム
 master controller 主制御器
 master gauge 親ゲージ, 主ゲージ
 master gear 主歯車, 親歯車
 master screw 親ねじ
 master valve 主弁
 master wheel 大歯車
match 名マッチ, 突合わせ, 整合 動突き合わせる, 調和させる, 匹敵する
 match-mark 合いマーク, 合印
matching マッチング, 整合
material 材料, 原料, 物質, 物体
 material analysis 材料分析
 material marks 材料記号
 material particle 質点
 material test 材料試験
materially 副著しく, 大いに
mathematical 形数学的な
 mathematical expression 数式
mating surface 合わせ面, 接着面
matrix マトリックス, 行列, 基質, 基盤
 matrix mechanics マトリックス力学
matter 物質, 事柄, 問題, 質
 (関) particle matter：粒子状物質／solid matter：固体
mature 形円熟した, 成長した, 分別のある
 動円熟する, 成熟する・させる
maul hammer 大ハンマ
maximum 名最大[値], 最大級, 最高, 極大 形最高の, 最大の
 maximum allowable working pressure 最大許容使用圧力
 maximum and minimum thermometer 最高最低温度計
 maximum bending moment 最大曲げモーメント
 maximum blade width ratio 最大翼幅比
 maximum clearance 最大すきま
 Maximum Continuous Output MCO, 連続最大出力
 Maximum Continuous Rating MCR, 連続最大定格, 連続最大出力
 maximum effciency 最大(高)効率
 maximum effective head 最高有効落差
 maximum elastic strain energy 最大弾性エネルギ
 maximum elevation 最大仰角
 maximum flexural strength 最大曲げ強さ
 maximum interference 最大締めしろ
 maximum load 最大荷重, 最大負

荷,破壊荷重
maximum main stress　最大主応力
maximum output　最大出力
Maximum Permissible Concentration　MPC,最大許容濃度
maximum pressure　最高圧力
maximum principal strain　最大主ひずみ
maximum principal stress　最大主応力
maximum rating　最大定格
maximum section coefficient　最大横断面係数
maximum shear stress　最大せん断応力
maximum speed　最大速力
maximum thermometer　最高温度計
maximum thickness　最大翼厚
maximum torque　最大トルク
maximum value　最大値
maximum vapor pressure　最大蒸気圧
maximum work　最大仕事
maximum working pressure　最大使用圧力
may 助〜かもしれない,〜する可能性がある【約50％の確実性を示す】
may be　よくある
may well　〜するのも最もだ,多分〜だろう
MCCB (**Moulded Case Circuit Breaker**)　配線用遮断器
MCO (**Maximum Continuous Output**)　連続最大出力
MCR (**Maximum Continuous Rating**)　連続最大定格,連続最大出力
mean 名平均,平均値,中間　形中間の,平均の　動意味する
mean blade width ratio　平均翼幅比
mean deviation　平均偏差
mean diameter　平均径,平均直径
Mean Effective Pressure　MEP,平均有効圧
mean error　平均誤差
mean free path　平均自由行程
mean life　平均寿命
mean piston speed　平均ピストン速度
mean pitch　平均ピッチ
mean pressure　平均圧力
mean rate of vaporization　平均蒸発率
mean specific heat　平均比熱
mean speed　平均速度
mean stress　平均応力
mean temperature　平均温度
mean temperature difference　平均温度差
Mean Time Between Failure　MTBF,平均故障間隔【故障しやすさを示す指標の一つ】
Mean Time To Failure　MTTF,平均故障寿命【故障するまでの平均時間】
mean value　平均値
meaning 意味
means 手段,方法
　関 by means of：〜によって,〜を用いて,〜の方法により
means and end　手段と目的
means by which　〜する手段
measure 名メジャー,測定,計測,尺度,物差し,測量,計量,手段,基準　動測定する
measured value 測定値
measurement 測定,測定値,計測,尺度,寸法,容積
measurement deviation　測定誤差
measures 対策,処置
measuring 形測定の,測量の　名測定,計測
measuring cylinder　メスシリンダ
measuring equipment　測定器
measuring instrument　計器,計測器,測定器
measuring machine　計測器
measuring microscope　測定顕微鏡,読みとり顕微鏡
measuring range　測定範囲
measuring tank　計量タンク
mechanical 形機械の,機械的な,

力学的な, 力学の
mechanical automation　メカニカルオートメーション
mechanical brake　機械ブレーキ
mechanical comparator　機械的コンパレータ
mechanical damping　機械制動
mechanical design　機械設計
mechanical draft　強制通風, 人工通風
mechanical drawing　機械製図
mechanical efficiency　機械効率
mechanical efficiency of power transmission　動力伝達効率
mechanical energy　機械的エネルギ, 力学的エネルギ
mechanical engineering　機械工学
mechanical equivalent of heat　熱の仕事当量
mechanical equivalent of light　光の仕事当量
mechanical friction　機械摩擦
mechanical governor　機械式調速機
mechanical impedance　機械インピーダンス
mechanical indicator　機械的インジケータ
mechanical injection engine　機械的噴射機関, 無気噴射機関
mechanical load　機械負荷
mechanical loss　機械損失
mechanical part　機械部品
mechanical power　機械[的]動力
mechanical property　機械的性質
mechanical reduction gear　歯車減速装置
mechanical refrigeration　機械冷凍
mechanical resistance　機械抵抗
mechanical seal　メカニカルシール【軸封装置の一種】
mechanical strength　機械的強度
mechanical stress　機械的応力
mechanical test　機械的試験
mechanical unit　力学単位
mechanical vibration　機械振動
mechanical vibrometer　機械的振動計
mechanical work　機械仕事
mechanically　副機械的に
　mechanically operated fuel valve　機械的作動燃料弁
mechanics　力学, 機械学
　mechanics of plasticity　塑性力学
mechanism　機構, 仕組み
mechanization　機械化
mechanize　動機械化する
mechanochemical　形機械化学の
mechatronics　メカトロニクス, 機械電子工学
media　媒体
median　名メジアン, 中点
　形中央の
medium　形中位の, 中間の
　名中位, 中間, 媒体, 媒質, 溶液
　例heating medium：加熱媒体
　medium carbon steel　中炭素鋼
　medium speed engine　中速機関
　medium wave　中波
Meehanite cast iron　ミーハナイト鋳鉄
meet　動接触する・させる, 交わる, 適合する, 接合する, 集まる
mega-　接頭「大きい」「百万」の意
　mega-hertz　メガヘルツ【MHz】
　mega-watt　メガワット【MW】
megger (megohmmeter)　メガー【メグオーム計の商品名：絶縁抵抗計】
megohm　メグオーム【電気抵抗の単位】
melt　動溶ける
melted state　溶融状態
melting　融解, 溶融
　melting furnace　融解炉
　melting loss　溶解損失, 溶解減耗
　melting point　融点
　melting temperature　融(解)点, 溶解温度
member　メンバ, 部材, 部, 要素, 項
membrane　膜, 薄膜, 皮膜
　membrane force　膜力
　membrane separation　膜分離
　membrane wall　メンブレンウォー

ル，板状溶接壁
memorial circuit 記憶回路
memorize 動記憶する
memory メモリ，記憶，記憶装置
　memory capacity 記憶容量
　memory card メモリカード
mend 動修繕する，改良する
meniscus 三日月状のもの
mental 形知的な
mention 動述べる，言及する
mephitic 形悪臭のある，有害な
MEP (Mean Effective Pressure) 平均有効圧
merchant ship 商船
mercuric nitrate 硝酸第二水銀
mercury 水銀　記Hg
　mercury arc lamp 水銀灯
　同 mercury vapor lamp
　mercury manometer 水銀マノメータ，水銀圧力計【水銀を用いた液柱圧力計】
meridian 子午線，経線
merit メリット，利点，価値，長所
mesa transistor メサ型トランジスタ
mesh 名メッシュ，網目，かみ合い　動かみ合わせる，調和させる
　mesh connection 環状接続，環状結線
meshing angle 嚙み合い角
met 動「meet」の過去・過去分詞
metacenter メタセンタ，傾心
metacentric diagram GM曲線，メタセンタ曲線
metacentric height GM，メタセンタ高さ
metal メタル，金属
　関 bearing metal：軸受メタル
　metal arc cutting 金属アーク切断
　metal arc welding 金属アーク溶接
　metal halide lamp メタルハライドランプ【超高圧放電灯】
　metal ion 金属イオン
　metal[lic] mold 金型
　metal[lic] wear 金属摩耗
　metal material 金属材料
　Metal Oxide Semiconductor diode MOSダイオード，金属酸化膜半導体ダイオード
　metal particles 金属片，金属微片
　metal plating 金属めっき
　metal pyrometer 金属高温計
　metal salt 金属塩
　metal soap base grease 金属石けん基グリース
　metal spraying 金属溶射
　metal surface 金属表面
　metal touch メタルタッチ，金属接触
　metal vapor 金属蒸気
　metal wire 金属線
　metalwork 金属加工
metalized carbon 金属化炭素
metallic 形金属の
　metallic bond 金属結合
　metallic coating 金属被膜
　metallic contact 金属接触
　metallic element 金属元素
　Metallic Inert Gas welding MIG溶接，ミグ溶接
　metallic joint 金属継手
　metallic material 金属材料
　metallic packing 金属パッキン
　metallic sodium 金属ナトリウム
metallographic structure 金属組織
metallographical microscope 金属顕微鏡
metalloid 半金属，非金属
metallurgy 冶金，冶金学
meter 名メータ，計器，メートル【長さの単位】動計量する
　meter bar メートル原器
　meter constant 計器定数
　meter-in circuit メータイン回路【速度制御する油圧回路の一種】
　meter-out circuit メータアウト回路【速度制御する油圧回路の一種】
　meter reading メータ[の]示度
　meter-run 計測管，計測流路
　meter transformer 計器用変成器
　meter wave メートル波
metering 計量，調量
　metering device 計量装置

metering pump 定量ポンプ,流量調節ポンプ
metering tank 計量タンク
metering valve 絞り弁
methane メタン【天然ガスの主成分】
methanol メタノール
method 方法,方式,手段
method of double weighting 二重ひょう量法
method of least square 最小二乗法
method of projection 投影法
methyl alcohol メチルアルコール
methyl group メチル基
metric メートルの,メートル法の
metric coarse screw thread メートル並目ねじ
metric fine screw thread メートル細目ねじ
metric horsepower メートル馬力
metric screw thread メートルねじ
metric system メートル法
metrical 形測定の,測量の
metrical instrument 計量機械
MHD(Magneto Hydro Dynamic) 電磁流体力学
mho モー【電気伝導率の単位:℧】
mica マイカ,雲母
関 mica condenser:マイカコンデンサ
micell[e] ミセル
Michell thrust bearing ミッチェルスラスト軸受
micro 形微小の
micro- 接頭「小」「微」「百万分の一」の意
microanalysis 微量分析
microbarometer 微圧計
microbe 微生物,細菌
microcomputer マイコン,超小型電算機
microlux ミクロルックス【コンパレータの一種】
micromachine マイクロマシン,微小機械
micromanometer 微圧計
micrometer マイクロメータ,測微計
micromotor マイクロモータ,ミニアチュアモータ
micron ミクロン,百万分の一メートル【μ】
microphotograph マイクロ写真,微小写真,顕微鏡写真
microprocessor マイクロプロセッサ
microscope 顕微鏡
microscopic test 金属組織検査
microsecond マイクロセカンド,百万分の一秒【μs】
microstructure ミクロ組織,顕微鏡組織
microswitch マイクロスイッチ【高感度スイッチの一種】
microvibrograph 微振動計
microvision マイクロビジョン,超高精度画像処理
microwave マイクロ波,極超短波
microwave oven 電子レンジ
mid 形中間の,中央の
mid gear ミッドギア,中間歯車,中央位置
middle 形中央の,中間の 名中央
middle point 中間点,中点
midship engine 中央機関
MIG(Metal Inert Gas)welding ミグ溶接,金属不活性ガス溶接
might 助「may」の過去形,〜かもしれない.
migration 移行,移動,拡散
mil ミル【1 ミル=千分の一インチ】
mild steel 軟鋼 関 soft steel
mile マイル【距離の単位:land mile=1609m, nautical mile=1852m】
milkiness 乳化
milking 乳濁
mill 名ミル,粉砕機
動製粉する,粉にひく
mill finish 圧延仕上げ
mill scale ミルスケール,黒皮
miller cycle engine ミラーサイクルエンジン
milli- 接頭「千分の一」の意

millimeter 千分の一メートル，ミリメートル
milling cutter フライス【周囲に刃のついた回転切削工具】
milling machine フライス盤
millionth 形百万分の一の
millivolt meter ミリボルト計
mineral 名ミネラル，鉱物，鉱石，無機物 形鉱物の，無機の
　mineral oil　鉱油，鉱物油，石油
mini- 接頭「小」「小型の」の意
miniature bearing ミニチュア軸受
miniature coupling ミニチュアカップリング【軸継手の一種】
miniaturization 小型化
miniaturize 動小型化する
minimeter ミニメータ，微測計
minimize 動最小限にする，最小化する
minimum 名最小，最低[値]，極小 形最小の 関 a minimum of：少なくとも
　minimum clearance　最小すきま
　minimum interference　最小締めしろ
　minimum load operation　最低負荷運転
　minimum pressure　最低(小)圧力
　minimum radius of gyration　最小断面二次半径
　minimum running current　始動電流
　minimum second moment of area　最小断面二次モーメント
　minimum speed　最低回転数，最小速度
　minimum thickness　最小厚さ
minium 光明丹 同 red lead
minor axis 短径
minus 名マイナス 形マイナスの，負の
　minus thread　マイナスねじ
minute 形微小な，微細な，精密な 名分
　minute particles　微粒子
mirror finishing 鏡面仕上げ
misalignment ミスアライメント，軸心のずれ，不良位置合わせ，位置のずれ
misfire 名ミスファイア，失火，不点火 動点火しない
misoperation 誤操作
miss 名ミス，失敗，回避 動失敗する，点火しない
mission 任務
mist ミスト，霧【気体中の微小液体粒子】
　mist flow　噴霧流
　mist lubrication　ミスト潤滑法
mistake 名ミス，間違い，過失，誤解，失敗 動間違える，誤る
mistaken 形誤った，誤解した
misunderstand 動誤解する
miter gear マイタ歯車【かさ歯車の一種】
miter valve 円すい弁
mix 名ミックス，混合 動混ぜる，混合する
mixed 形混合した
　mixed based crude oil　混合基原油
　mixed combustion　混焼
　mixed convection　共存対流
　mixed cycle　混合サイクル，複合サイクル
　mixed firing　混焼
　mixed flow　混相流
　mixed flow compressor　混流圧縮機，斜流圧縮機
　mixed flow fan　混流送風機
　mixed flow pump　混流ポンプ
　mixed flow turbine　混流式タービン
　mixed gas　混成ガス
　mixed layer　混合層
　mixed pressure turbine　混圧タービン
mixer ミキサ，混合，攪拌機
mixing ミクシング，混合
　mixing chamber　混合室
　mixing-in　混合
　mixing layer　混合層
　mixing ratio　混合比
mixture 混合，混合気，混合物
　mixture ratio　混合比
MMF (**Magneto Motive Force**)

起磁力
mobile 形可動性の
　mobile crane　移動式(自走)クレーン
　mobile grease　モビルグリース
　mobile oil　モビル油, 自動車油【エンジン油の一種】
mobility　流動性, 移動度, 可動性
mock-up　モックアップ, 実物大模型
mode　モード, 型, 形式, 型式, 形態, 方式, 方法, 様式
　mode of vibration　振動モード
model　名モデル, 模型, 型　動模型を作る, 型どる
　model experiment　模型試験
　model test　模型試験
modeling　モデリング, モデル化
modem　モデム, 変復調装置【電話回線を通じてインターネットに接続する装置】
moderate　形適度な, 穏やかな　動和らげる, 加減する, 下げる, 減速する
moderating ratio　減速率
moderator　減速材
modern　形現代の, 最近の
modification　修正, 変更, 変形, 加減, 限定, 改修, 修飾, 改質
modified　形修正された
modifier　調節剤
　関 surface tension modifier：界面活性剤
modify　動加減する, 修正する, 緩和する
modular　モジュラ, モジュール, モジュール方式
modularization　モジュラ化
modulate　動調整(節)する, 変調する
modulation　変調, 調節, 調整, 転調
　modulation frequency　変調周波数
　modulation loss　変調損
　modulation wave　変調波
modulator　調節器, 変調器
　modulator and demodulator　変復調器
module　モジュール, 基準寸法, 測定基準
modulus　率, 係数, 絶対値, 母数, 引張応力
　modulus fiber　モジュラス繊維
　modulus of decay　減衰率
　modulus of direct elasticity　縦弾性係数, ヤング率, ヤング係数
　modulus of elasticity　弾性係数, 弾性率
　modulus of elasticity of volume　体積弾性係数
　modulus of longitudinal elasticity　縦弾性係数
　modulus of rigidity　せん断弾性係数, 剛性率, 横弾性係数
　modulus of section　断面係数
　modulus of shearing elasticity　横弾性係数, せん断弾性係数
　modulus of transverse elasticity　せん断弾性係数, 横弾性係数
　modulus of volume elasticity　体積弾性係数
moist　形湿った
　moist air　湿り空気
　moist steam　湿り蒸気
moisture　湿気, 水分, 水蒸気
　moisture content　湿分
　moisture loss　湿り損失
　moisture proof　防湿
　moisture resistance　耐湿性, 透湿抵抗
mol　略 mole
　mol concentration　モル濃度
molar　形 1 モル当たりの, モル濃度の, 質量の
　molar concentration　モル濃度
　molar fraction　モル分析
　molar heat　分子熱, モル熱, モル比熱
　molar mass　モル質量
　molar ratio　モル比
　molar specific heat　モル比熱
molarity　モル濃度
mold　名鋳型, 型板　動鋳る, 成形する, 形づくる
molded dimension　型寸法
molding　成形, 型込め, 面取り
　molding box　型枠

molding machine 型込機, 造型機, 面取機
- **mole** モル【物質量の単位】
   mole fraction モル分率
   mole ratio モル比
- **molecular** 形分子の
   molecular film 分子膜
   molecular formula 分子式
   molecular layer 分子層
   molecular motion 分子運動
   molecular weight 分子量
- **molecule** 分子, グラム分子
- **Mollier chart (diagram)** モリエル線図
- **molten** 形溶けた
   molten metal 溶融金属
   molten steel 溶鋼
- **molybdenum** モリブデン 記Mo
   molybdenum disulphide greese 硫化モリブデングリース
- **moment** モーメント, 回転偶力, 能率, 瞬間 関at the moment：今のところ, 現在は, ちょうど今
   moment of area 断面一次モーメント, 面積モーメント
   moment of couple 偶力モーメント
   moment of force 力のモーメント
   moment of impulse 力積モーメント
   moment of inertia 慣性モーメント
   moment of inertia of area 断面二次モーメント
   moment of momentum 運動量モーメント, 角運動量
   moment of resistance 抵抗モーメント
   moment of rotation 回転モーメント
   moment of torsion ねじれモーメント
- **momentary** 形瞬間的な
   momentary breakdown 瞬間破壊
   momentary load 瞬時負荷
   momentary power 瞬時電力
- **momentum** 運動量
   momentum equation 運動量方程式
- **monitor** モニタ, 監視器
- **monitoring** モニタリング, 監視

   monitoring panel 監視パネル
- **mono-** 接頭「1」「単」の意
   monoblock casting 一体鋳造, 単体鋳造
   monochromatic 形単色の
   monochrome モノクロ, 単色
   monochrometer 単色計
   monocrystal 名単結晶 形単結晶の
   monocyclic 形単周期の
   monolayer 単一層
   monomer 単量体
   monomial 単項式
   monotectic reaction 偏晶反応
   monotube manometer 単管式圧力計
- **Monte-Carlo method** モンテカルロ法
- **Moody diagram** ムーディ線図
- **moor** 動［船を］つなぐ, 停泊させる
- **mooring** 係留, 係船
   mooring equipment 係留装置
   mooring trial 係船運転
- **more** 形「many」「much」の比較級, より多い, より多くの, より大きい 代より多くのもの, 以上のもの 関no more than：たった〜, 〜しかない／not more than：多くても〜, せいぜい〜
   more and more 副ますます
   more or less 副多かれ少なかれ, 多少, およそ, 実際に
   more than 〜以上, 〜より多い
   more 〜 than… …より多くの〜
- **moreover** 副さらに, その上
- **mortar** モルタル, 漆喰(しっくい)
- **mortise wheel** はめば歯車
- **most** 形「many」「much」の最上級, 大多数の, たいていの 関at most：多くとも
   most economical rating 経済定格
   most modern 最新の, 最も近代的な
   most of 〜の多く, 〜の大部分
   most probable value 最確値
- **motion** 名運動, 動き, 作動, 機構, 動作, 移動, 動揺 動合図する 関in motion：運動中

motion analysis 動作分析
motive power 動力, 原動力
motor モータ, 電動機, 発動機
  motor circuit 動力回路
  motor driven 形電動の
  motor driven blower 電動送風機
  motor load 電動機負荷
  motor oil モータ油
  motor operated valve 電動弁
  motor starter 電動機始動装置
motoring test モータリング試験
motorized pulley モータプーリ【モータと減速装置を内蔵したプーリ】
mottled cast iron まだら鋳鉄
mould 同 mold
moulding 同 molding
mount 動取り付ける, 据え付ける, 備える 名取付台
mounted 形取り付けられた, 固定した
mounting 取付け, 配置, 据付け
  mounting hardware 取付金具
mouse マウス
mouth piece 口金
mouth ring マウスリング, 口環
movable 形可動性の, 移動する
  movable coil 可動コイル
  movable contact 可動接触子
  movable element 可動子
  movable end 可動端
  movable guide vane 可動案内羽根
  movable pulley 動滑車
  movable vane 可動羽根, 可動翼
move 動動く, 移動する, 動揺する
  move away from ～ 動～から離れる
  move up and down 上下(に移動)する
movement 運動, 動作, 移動, 変動
mover 発動機
moving 形可動式の, 運転中の, 動いている, 作動している
  moving average 移動平均
  moving blade 動翼, 回転羽根
  moving coil 可動コイル
  moving coil instrument 可動コイル型計器
  moving contact 可動接触子, 可動接点, しゅう動接点
  moving element 可動部分
  moving iron type instrument 可動鉄片型計器
  moving load 移動荷重
  moving magnet 可動磁石
  moving part 可動部
  moving vane 回転羽根
MPC (**Maximum Permissible Concentration**) 最大許容濃度
MTBF (**Mean Time Between Failure**) 平均故障間隔【故障しやすさを示す指標の一つ】
MTTF (**Mean Time To Failure**) 平均故障寿命【故障するまでの平均時間】
much 形多くの, 多量の
副大変に, はるかに, 大いに
名大部分, 多量
関 too much:副過剰(度)に
  much less than はるかに少ない
mud 泥
  mud box マッドボックス, 泥箱
muff coupling スリーブ継手, 筒形軸継手
muffle 名マッフル, 覆い
動[音を]消す, 小さくする, 包む
  muffle furnace マッフル炉【電気炉の一種】
muffler マフラ, 消音器
multi- 接頭「多くの」の意
  multi-blade fan シロッコファン, 多翼ファン
  multi-component 形多成分の
  multi-conductor cable 多心ケーブル
  multi-cylinder engine 多シリンダ機関
  multi-flow 多流
  multi-fuel burner 混合燃料バーナ
  multi-fuel engine 多種燃料機関
  multi-fuel fired boiler 混焼ボイラ
  multi-function 多機能
  multi-layer 多層
  multi-layer cylinder 多層円筒

multi-layer insulation 積層断熱, 多層断熱
multi-million 名数百万 形数百万の
multi-pass 多流
multi-phase flow 混相流
multi-pipe 多管
multi-port burner 多孔バーナ
multi-pressure condenser 複圧式復水器
multi-purpose 多目的
multi-purpose grease 万能形グリース
multi-shaft compressor 多軸圧縮機
multi-stage 形多段の
multi-stage centrifugal pump 多段渦巻ポンプ
multi-stage compressor 多段圧縮機
multi-stage expansion 多段膨張
multi-stage pump 多段ポンプ
multi-tubular boiler 多管ボイラ
multi-tubular heat exchanger 多管式熱交換器
multi-vane 多翼
multi-vibrator マルチバイブレータ【弛張(しちょう)発振器の一種】
multinomial 多項式
multiple 形多量の, 多数の, 多重の, 複式の
  multiple cylinder 多シリンダ
  multiple defense 多重防護
  multiple disc clutch 多板クラッチ
  multiple hole nozzle 多孔ノズル
  multiple screw ship 多軸船
  multiple [stage] centrifugal pump 多段渦巻きポンプ
  multiple system 多重システム, 並列式
  multiple thread screw 多重ねじ
multiplex winding 多重巻
multiplexing 多重化
multiplication 乗法, 掛け算, 増殖
multiplier 掛算器, 乗算器, 倍率器
multiply 動増加する, 掛け合わせる
multipolar 形多極の
multipole 多極, 多重極
  multipole moment 多重極モーメント, 多極[子]モーメント
  multipole radiation 多重極放射, 多極放射
Munsell scale マンセル表色尺度【塗装色の表示法】
muriatic acid 塩酸
mushroom valve きのこ形弁
music wire ピアノ線
must 助~しなければならない, ~する必要がる
  must be ~であるに違いない
mutual 形相互の
  mutual action 相互作用
  mutual impedance 相互インピーダンス
  mutual inductance 相互インダクタンス
  mutual induction 相互誘導
  mutual interference 相互干渉
mutually 副互いに, 相互に
MW (mega-watt) メガワット

N-n

N ニュートン【力のSI単位】
N type semiconductor N型半導体
nail 名くぎ 動くぎを打つ, 鋲を打つ
name plate 銘板
namely 副すなわち, つまり
NAND circuit ナンド回路, NAND回路
nanometer ナノメートル【長さの単位：10^{-9}m】
naphtha ナフサ【粗製ガソリン】
naphthalene ナフタリン
naphthenic hydrocarbon ナフテン系炭化水素
narrow 形狭い 動狭くする, 縮小する
Nash pump ナッシュポンプ, 液封式真空ポンプ
natrium ナトリウム【sodiumの旧称】記 Na
  natrium soap ナトリウム石けん
natural 形天然の, 自然の, 固有の
  natural air cooling 自然空冷
  natural angular frequency 固有角周波数
  natural aspiration 無過給
  natural circulation boiler 自然循環ボイラ
  natural convection 自然対流
  natural convector 自然対流放熱器
  natural draft 自然通風
  natural frequency 固有振動数, 自然周波数
  natural gas 天然ガス
  natural head 自然落差, 全落差
  natural logarithm 自然対数
  natural oscillation 固有振動, 自然振動
  natural period 固有周期
  natural rubber 天然ゴム
  natural science 自然科学
  natural seasoning 自然乾燥
  natural strain 自然ひずみ
  natural vibration 固有振動
naturally 副自然に, 当然, 本来
  naturally aspirated engine 無過給機関
nature ネーチャ, 性質, 特質, 内容, 本性, 本質, 自然［界］
nautical 形航海の, 海事の, 船舶の, 船員の
  nautical mile 海里【1海里＝1852m】
naval 形船の, 船舶の
  naval architecture 造船工学
  naval brass ネーバル黄銅
Navier-Stokes equation ナビエ-ストークスの式
navigate 動航海する, 操縦する, 運転する
navigation 航行, 航法
NC（Numerical Control） 数値制御
neap tide 小潮 関 spring tide：大潮
near 形近い, 密接な 副近く, 密接に 前〜の近くに
nearby 形近くの 副近くに（で）
nearly 副ほとんど, ほぼ
  nearly closed ほぼ閉じた
necessarily 副必ず
  関 not necessarily：必ずしも〜でない
necessary 形必要な
  necessary condition 必要条件
necessitate 動必要とする
neck bush ネックブッシュ【弁棒やインペラ軸の摩耗防止に用いる青銅製段付中空円筒】
neck pinion カムワルツ
necking くびれ
need 名ニーズ, 要求, 需要, 必要, 必要性 動必要とする, 要する
needle ニードル, 針, 指針, 可動子, 弁体
  needle bearing 針状軸受
  needle pliers ニードルプライア
  needle regulator ニードル調整弁
  needle roller ニードルころ

needle roller bearing　針状ころ軸受
needle valve　ニードル弁，針弁
negative　形負の，陰性の　名否定，負数，負号，陰電気　動否定する
negative characteristics　負特性
negative charge　陰電荷，負電荷
negative electricity　陰電気，負電気
negative electrode　陰極，負極
negative electron　電子，陰電子
negative feedback　負帰還
negative feedback control　負帰還制御
negative ion　アニオン，負イオン，陰イオン
negative phase　逆相
negative pole　陰極
negative pressure　負圧
negative resistance　負性抵抗，負抵抗
negative sign　負号，マイナス
negative slip　負失脚
negative stability　負復原力
negative strain　負ひずみ
negative terminal　陰端子，負端子
neglect　動無視する，軽視する，怠る　名無視，放置
negligible　形無視できるほどの
negotiate　動交渉する，取り決める
neither A nor B　AもBもない
neon　ネオン　記Ne
nephelometer　比濁計
nest　ネスト，群れ，一群，巣
nest of tube　管群
net　名網　形正味の
　関 gross：総体の，全体の
net calorific value　真発熱量
net efficiency　正味効率
net head　正味落差
Net HorsePower　NHP，正味馬力，制動馬力
Net Positive Suction Head　NPSH，有効吸込水頭，有効吸込ヘッド
net pump head　全揚程
net thermal efficiency　正味熱効率
net tonnage　純トン数
net weight　正味重量
net work　正味仕事
network　ネットワーク，回路網
network structure　網状組織
neuro-　「神経[組織][の]」「神経性系[の]」「神経系[の]」の意を表す連結形
neuro computer　ニューロコンピュータ
neuro control　ニューロ制御
neutral　形中立の，中性の，帯電しない，くすんだ，はっきりしない，中間色の　名中性，中性点
neutral axis　中立軸
neutral catalyst　中性触媒
neutral conductor　中性導体，中性線
neutral equilibrium　中性つり合い
neutral flame　中性炎
neutral line　中立線，中性線
neutral oil　中性油
neutral plane (surface)　中立面
neutral point　中性点
neutralization　中和
neutralization number　中和価
neutralization titration　中和滴定
neutralize　動中和する
neutralizer　中和剤
neutron　中性子　関 proton：陽子
neutron source　中性子源
never　副決して～しない，これまで一度も～したことがない
nevertheless　副それでも，しかも，しかしながら
Newtonian fluid　ニュートン流体
Newton's equation of motion　ニュートンの運動方程式
Newton's law of cooling　ニュートンの冷却則
Newton's law of viscosity　ニュートンの粘性法則
next　形次の，来～，翌～，隣の　副次に，今度は
next [to]　前～の隣に，～の次に
NFB (No Fuse Braker)　ノーヒューズ遮断器
NHP (Net HorsePower)　正味馬力，制動馬力

NHP（**Normal HorsePower**）　最大連続馬力，正規馬力
nichrome wire　ニクロム線
nick　［切り］傷，刻み目
　nick bend test　切欠き曲げ試験【疲労試験の一種】
　nick break test　切欠き破断試験
nickel　ニッケル　㊜ Ni
　nickel alloy　ニッケル合金
　nickel bronze　ニッケル青銅
　nickel cast iron　ニッケル鋳鉄
　nickel chrome cast iron　ニッケルクロム鋳鉄
　nickel chrome molybdenum steel　ニッケルクロムモリブデン鋼
　nickel chrome steel　ニッケルクロム鋼
　nickel plating　ニッケルめっき
　nickel silver　洋銀【銅合金の一種】
　nickel steel　ニッケル鋼
nicking　ニッキング，筋目立て
niobium　ニオブ　㊜ Nb
nip　動つかむ，挟む
nipper　ニッパ【切断工具の一種】
nipple joint　ニップル継手【管継手の一種】
nitrate　硝酸［塩］
nitration　ニトロ化，硝化
nitric　形窒素の，硝石の
　nitric acid　硝酸
　nitric monoxide　NO，一酸化窒素
nitridation　窒化硬化法，窒化作用
nitride　窒化［物］
nitriding　窒化，窒化処理
　㊝ gas nitriding：ガス窒化
　nitriding steel　窒化用鋼
nitrile rubber　ニトリルゴム
nitrocarburizing　浸炭窒化
nitrogen　窒素　㊜ N
　nitrogen dioxide　二酸化窒素
　nitrogen hardening　窒化
　nitrogen oxides　NOx，窒素酸化物
nitroglycerin　ニトログリセリン【ダイナマイトの原料】
nitrous　形窒素の，亜硝酸の
NO（**nitric monoxide**）　一酸化窒素

nodal point　節点
no-go side　止まり側
no load　無負荷
　no load characteristics　無負荷特性
　no load current　無負荷電流
　no load loss　無負荷損
　no load running　無負荷運転
　no load saturation curve　無負荷飽和曲線
　no load speed　無負荷速度
　no load test　無負荷試験
　no load voltage　無負荷電圧
no longer　副もはや〜しない
no voltage relay　無電圧リレー
noble　形不活性の，貴(希)の
　㊝ less noble metal：卑金属
　noble gas　希ガス
　noble metal　貴金属
nodal analysis　節点解析法
node　交点，接続点，節，節点【ロープの結び目／振動波において変位がゼロのところ】
　node average　節点平均
nodular [**graphite**] **cast iron**　ノジュラ鋳鉄，球状黒鉛鋳鉄
noise　ノイズ，雑音，騒音
　noise control　防音
　noise controller　騒音防止装置
noisiness　騒音
noisy　形騒々しい
nominal　形公称の，名ばかりの
　nominal area　見かけ接触面積
　nominal diameter　呼び径
　nominal diameter of screw　ねじの呼び径
　nominal dimension　呼び寸法
　Nominal HorsePower　NHP，公称馬力
　nominal induced electromotive force　公称誘導起電力
　nominal rating　公称定格
　nominal size　呼び寸法
　nominal speed　通常速力，公称速力
　nominal stress-strain diagram　公称応力ひずみ図
　nominal voltage　公称電圧

nomogram ノモグラム, 計算図表
non- 接頭「非」「不」「無」の意
  non-absorbent material 不浸材料
  non-circular gear 半円形歯車
  non-combustible material 不燃材料
  non-condensable gas 不凝結ガス
  non-conducting material 絶縁物
  non-conductor 不導体, 絶縁体
  non-destructive inspection (test) 非破壊検査
  non-drying oil 不乾性油
  non-electrolyte 非電解質
  non-ferrous metal 非鉄金属
    関 ferrous metal：鉄金属
  non-flammable 形不燃性の
  non-flammable paint 耐火ペイント
  non-freezing solution 不凍液
  non-homing ノンホーミング〈制御〉【スタート位置またはホーム位置に戻らないこと】
  non-homogeneity 不均質性
  non-inductive load 無誘導負荷
  non-linear 形非線形の
  non-linear control 非線形制御
  non-luminous flame 不輝炎
  non-magnetic material (substance) 非磁性体
  non-magnetic steel 非磁性鋼
  non-metal 非金属
  non-metallic constituent 非金属成分
  non-metallic material 非金属材料
  non-return valve 逆止め弁
  non-reversible engine 非可逆機関
  non-stationary 非定常
  non-stick oil 固着防止剤
  non-steady heat conduction 非定常熱伝導
  non-steady state 非定常状態
  non-step variable speed gear 無段変速機
  non-supercharged engine 無過給機関
  non-uniform flow 非一様流
    関 uniform flow：一様流
  non-uniform pitch propeller 変動ピッチプロペラ
  non-volatile matter 不揮発性物質
NOR circuit ノア回路
normal 形正常の, 正規の, 標準の, 通常の, 直角の, 垂直の 名標準, 平均, 正常, 垂直線(面), 法線
  normal axis 垂直軸
  normal blade N羽根
  normal condition 基準状態, 通常状態
  normal cubic meter 標準立方メートル【標準状態における $1m^3$ の体積】
  normal discharge 正規流量
  normal distribution 正規分布
  normal force 垂直力
  normal fracture 分離破壊
  normal function 正規関数
  Normal HorsePower NHP, 連続最大馬力
  normal line 法線
  normal load 垂直荷重, 常用負荷
  normal operation 正常(通常)運転(操作)
  normal output 常用出力, 正規出力
  normal pitch 法線ピッチ, 垂直ピッチ, 正ピッチ
  normal position 平常位置, 定位
  normal pressure 標準圧力, 常用圧力
  normal slide valve 正すべり弁
  normal solution 規定液
  normal speed 常用速度
  normal state 正常状態, 基底状態, 通常状態, 標準状態
  normal strain 垂直ひずみ
  normal start 正常(通常)始動
  normal stress 垂直応力, 法線応力, 公称応力
  normal stress-strain diagram 公称応力ひずみ図
  normal structure 標準組織
  normal temperature 常温
  normal thrust 常用スラスト
  normal valve 正すべり弁
normality 規定度
normalize 動焼ならしする, 基準化する, 正規化する, 規格化する

normalized coordinate 正規座標
normalized function 正規関数
normalized mode 正規モード
normalizing 焼ならし【内部応力除去などの目的で行う熱処理】
normally 副正常に，いつもは，標準的に，普通に，普通は，直角に
  normally aspirated engine 無過給機関
  normally closed 常時閉
  normally open 常時開
  normally opened connective point NO接点，a接点
north pole N極
nose circle 先端円
NOT circuit ノット回路
not only A but also B Aのみならず Bでも
notation 記号，表記法
notch ノッチ，切欠き
  notch bar test 切欠き試験
  notch brittleness 切欠きもろさ，切欠き脆性
  notch effect ノッチ効果，切欠き効果
  notch geometry 切欠き形状
  notch sensitivity factor 切欠き感度係数
  notch toughness 切欠き靱性，切欠きねばさ
notched specimen 切欠き試験片
note 名ノート，注意，注記，注釈 動注意する，気づく，注記する
notice 名通知，警告，注意，掲示 動気がつく，注意する，注目する
  notice board 掲示板
noticeable 形きわだった，顕著な
notify 動通知する，知らせる
notion 概念，意志
now 副今，現在，ところで，さて 関 for now：今のところ／right now：まさに今
NOx 窒素酸化物
  関 SOx：硫黄酸化物
  NOx removal equipment 脱硝装置
noxious 形有害な，危険な

noxious gas 有毒ガス
nozzle ノズル，噴射孔，火口
  nozzle angle ノズル角
  nozzle blade ノズル羽根
  nozzle box ノズル箱
  nozzle control governing ノズル加減調速
  nozzle control valve ノズル加減弁，ノズル制御弁
  nozzle cutout governing ノズル締切り調速
  nozzle efficiency ノズル効率
  nozzle flapper ノズルフラッパ
  nozzle flow coefficient ノズル流量係数
  nozzle hole ノズル噴口
  nozzle loss ノズル損失
  nozzle throat ノズルのど
  nozzle throat area ノズルのど面積
  nozzle tip ノズルチップ
  nozzle valve ノズル弁
  nozzle velocity coefficient ノズル速度係数
npn transistor npnトランジスタ
NPSH（Net Positive Suction Head） 有効吸込みヘッド
nuclear 形核の
  nuclear fission 核分裂
  nuclear force 核力
  nuclear fusion 核融合
  nuclear power 原子力
  nuclear power propulsion 原子力推進
  nuclear reaction 核反応
  nuclear resonance energy 原子核共鳴エネルギ
  nuclear ship 原子力船
nucleate boiling 核沸騰
nuclei 「nucleus」の複数形
nucleonics [原子]核工学
nucleus 原子核，核，中心，起点，核心
nugget かたまり，ナゲット
null ヌル，中立，ゼロ点
  null method 零位法
number 名数，数字，指数，係数

numberless

　(動)数える，総計～になる
　(関) a small (large) number of ～：小(多)数の～／a number of ～：いくらかの～，多数の～／the number of ～：～の数
　number of active coils　有効巻数
　number of critical revolution　危険回転数
　number of moles　モル数
　number of revolution　回転数
　number of stroke　行程数
　number of turns　巻数
　number of vibration　振動数
numberless　(形)無数の
numeral　数字
numeration　計算，計算法
numerator　分子
　(関) denominator：分母

numerical　(形)数値の，数値的な，数字で表す
　numerical analysis　数値解析
　Numerical Control　NC，数値制御
　numerical expression　数式
　numerical statement　統計
numerically　(副)数値的に，数字上で
　Numerically Controlled machine tool　NC工作機械
numerous　(形)多数の
Nusselt number　ヌッセルト数
nut　ナット　(関) bolt：ボルト
nylon　ナイロン
　nylon resin　ナイロン樹脂
Nyquist diagram　ナイキスト線図
Nyquist stability criterion　ナイキストの安定条件

O-o

O ring Oリング
obey 動従う
object 名オブジェクト，物体，対象，目的
　object glass　対物レンズ
objectionable　許容できない
objective 名対物レンズ，目標，目的 形客観的な
oblate 形扁円の
oblige 動強いる，余儀なくさせる
oblique 形斜めの 名傾斜
　oblique angle　斜角，傾斜角
　oblique circular cone　斜円すい
　oblique circular cylinder　斜円筒
　oblique drawing　斜め投影図
　oblique line　斜線
　oblique projection　斜投影法
　oblique section　斜め断面
　oblique stress　斜め応力
　oblique stroke　スラッシュ，斜線
obliquity　傾斜，傾斜角，傾度
　obliquity rod　斜向棒
obliterate 動消す，取り除く
observance　遵守，慣例，観察，習慣
observation　観察，監視，観測，天測
　observation equipment　観測装置
　observation tank　検油タンク
observe 動観察する，観測する，遵守する，監視する，注意する
observer　オブザーバ，観察者，監視者
obstruct 動遮る，妨害する
obstruction　妨害，じゃま物，閉塞
obtain 動得る，もたらす，確保する
obtainable 形入手できる
obtuse 形鈍い，鈍角の
　obtuse angle　鈍角
obvious 形明白な
obviously 副明白に
occasion　場合，時，機会
occasional survey　臨時検査
occasionally 副時折，時々，たまに
occupied area　居住区
occupy 動占める，居住する
occur 動生ずる，起こる
occurrence　発生，出来事
oceanography　海洋学
OCR（**Optical Character Reader**）光学文字読取装置
octagon　八角形
octahedron　八面体
octane number（**value**）オクタン価【ガソリンのアンチノック性を示す指数】
odd 形半端な，奇数の
　odd number　奇数
odometer　オドメータ，走行距離計
odor（**odour**）臭い，臭気
of ～ 前～の，～から，～という，～に関して(の)，～について(の)
　of aluminium　アルミニウム製
off 前から離れて(外れて)
　副離れて，外れて，切れて，停止して，狂って
　off center arrangement　偏心装置
offer 動提供する，提案する，表す
official trial　公試運転
offset　オフセット，偏差，心違い，片寄り，偏り 動相殺する，ずらす
　offset angle　オフセット角
　offset bearing　片寄り軸受
　offset cam　片寄りカム
　offset cylinder　片寄りシリンダ
　offset die　かみあわせダイ
　offset electrode holder　オフセット電極ホルダ
　offset link　オフセットリンク
　offset wrench　めがねレンチ
off-site control　オフサイト制御
often 副しばしば
ogival section　弓形断面
ohm　オーム【電気抵抗の単位：Ω】
ohmic drop　抵抗損，銅損
ohmic resistance　オーム抵抗
ohmmeter　オームメータ，オーム計，電気抵抗計

Ohm's law オームの法則

oil 名オイル，油
動油を注ぐ，油をさす
oil barge 油タンク船
oil bath 油浴【熱処理の方法】
oil bath type stern tube bearing 油潤滑式船尾管軸受
oil brake 油ブレーキ
oil burner 油バーナ
oil burning boiler 油焚きボイラ
oil can 油差し
oil carrier 油送船，油タンカ
oil cleaner 油清浄器
oil condenser 油コンデンサ
oil consumption 油消費量
oil content meter 油分計
oil control ring 油かきリング
oil cooled transformer 油冷変圧器
oil cooler オイルクーラ，油冷却器
oil damping 油制動
oil drain tank 油ドレンタンク
oil droplet 油滴
oil extractor 油分離器
oil feeder 油差し，給油装置
oil feeding tank 給油タンク
oil fence オイルフェンス
oil filling pipe 油取入れ管
oil film 油膜
oil film pressure 油膜圧力
oil film thickness 油膜厚さ
oil filter オイルフィルタ，油ろ過器
oil free 形油の入ってない
oil free bearing 無潤滑軸受
oil gas オイルガス
oil gauge 油量(面)計
oil groove 油みぞ
oil gun オイルガン【潤滑油注入器の一種】
oil hardening 油焼入れ
oil head tank 油重力タンク
oil heater 油加熱器
oil hole 油穴(孔)
oil hydraulic motor 油圧モータ
oil hydraulic pump 油圧ポンプ
oil immersed transformer 油入変圧器

oil immersion 油浸
oil injection system （分離)給油装置
oil insulation 油絶縁
oil leak[age] 油漏れ
oil level gauge 油面計，検油棒
oil lift 静圧揚力軸受
oil lubricated stern tube bearing 油潤滑式船尾管軸受
oil lubricated type 油潤滑式
oil mist オイルミスト【霧状になったオイル】
oil mist detector オイルミストディテクタ，オイルミスト検知器
oil mist lubricator 噴霧給油装置
oil paint 油性ペイント，油性塗料
oil pan 油受
oil pipe 油管
oil pollution 油汚染
oil pressure 油圧
oil pressure burner 圧力噴射式重油バーナ
oil pressure circuit diagram 油圧回路図
oil pressure gauge 油圧計
oil pressure regulating valve 油圧調整弁
oil proof 形耐油性の
oil pump 油ポンプ
oil purifier 油清浄機
oil quenching 油焼入れ
oil record book 油記録簿
oil refining 石油精製
oil reservoir 油溜め，油タンク
oil resistance 耐油性
oil retaining bearing 含油軸受
oil return type oil burner 戻り油形油[圧式]バーナ
oil ring オイルリング
oil ring bearing オイルリング軸受
oil scraping ring 油かきリング
oil seal オイルシール，油密
oil separator 油分離器
oil sight 検油器
oil sludge オイルスラッジ
oil solubility 油溶性
oil spill 油流出

oil sprayer 噴油器
oil stone 油砥石
oil strainer 油こし器
oil sump tank サンプタンク, 油だめタンク
oil syringe 注油器
oil tank オイルタンク, 油タンク
oil temperature 油温
oil transfer pump 油移送ポンプ
oil way 油みぞ
oil whip オイルホイップ【軸受内での油の振れ回り現象】
oil whirl オイルホワール
oiliness 油性【油の分子が金属面に吸着する性質】
oiliness agent (improver) 油性向上剤
oiling オイリング, 給油, 注油, 潤滑
oiling system 給油方式, 給油系統
oilless bearing 無給油軸受
oil-tight 油密
oil-tight joint 油密継手
oil-tight test 油密試験
oily 形油性の, 油質の
oily water 油水
older engine 以前のエンジン
olefin オレフィン
olefin hydrocarbon オレフィン系炭化水素
oleo damper 油圧ダンパ
oleometer オレオメータ, 油比重計
oleophilic group 親油基
oleophobic group 疎油基
omission 省略
omission error 省略エラー
omit 動省略する
omnibus bar 母線
omnimeter オムニメータ
on ～ ing 前～するや否や, ～したときに
on account of ～ 前～のために
on-line オンライン
on load 負荷時
on-off control オンオフ制御
on-off control action オンオフ動作
on-site control オンサイト制御

once 接一旦～すると 副一度, かつて 関 at once：直ちに, 一度に, 同時に
once through boiler 貫流ボイラ
oncoming 迫りくる
ondometer 波長計
one 形1つの, 単一の, ある～, 一方の 名1, 1つ
one another 代互いに
one dimensional flow 一次元流れ
one end 一端
one man control system ワンマンコントロール方式
one or more 1つ以上
one piece molding 一体成型
one side 片側, 一面
one touch joint 急速継手
one way valve 一方向弁
only 形唯一の 副ただ～だけ
関 not only A but also B：AだけでなくBも
only if ～ 接～という[条件の]場合にだけ
onto ～ 前～の上へ
ooze 動しみ出る
open 形開いた 動開く, 開放する 名開路【電流が流れない状態】
open air condenser 大気復水器
open bucket type trap うきトラップ
open chamber type combustion chamber 単室式燃焼室
open circuit test 無負荷試験, 閉回路試験
open cycle 開放サイクル
open hearth steel 平炉鋼
open loop control 開ループ制御
open loop transfer function 一巡伝達関数, 開ループ伝達関数
open propeller efficiency 単独プロペラ効率
open system オープンシステム, 開いた系, 開放系
open type 開放型
open water test プロペラ単独試験
opening 口, 開き, 開口
opening pressure 開口圧力

operability 操作性
operable 形操作(利用)可能な
operate 動運転する,作動する
operated 形〜式の,駆動の,操作の
　関 air-operated:形空気駆動の／steam-operated:形蒸気駆動の
operating 形作動している,操作上の,運転中の,運転上の
　operating condition 運転(操作,動作)条件(状態)
　operating current 動作電流
　operating handle 操作ハンドル
　operating life 動作(運転)寿命
　operating point 運転点,動作点
　operating pressure 動作圧力,運転圧力,作動圧
　operating signal 動作信号
　operating temperature 動作温度,運転温度
　operating time 動作時間
　operating voltage 動作電圧
　operating water 作動水
operation オペレーション,操作,運転,動作,働き,作用,加工
　関 in operation:運転(動作)中
　operation and maintenance 運転管理
　operation expense 運転費
　operation factor 運転率
　operation panel 操作盤
　operation time 運転時間
operational 形操作の
　operational amplifier 演算増幅器
　operational circuit 演算回路
Operations Research OR, オペレーションズリサーチ
operator オペレータ,運転者
opportunity 機会
oppose 動〜に反対する,妨害する,〜を向かい合わせにする
opposed piston engine 対向ピストン機関
opposite 反対 形反対側の,逆の 前〜の反対(向かい)側に
　opposite angle 対角
　opposite direction 反対方向,逆方向
　opposite effect 逆効果
　opposite end 反対端
　opposite phase 逆相,逆位相
opposition 抵抗,反対,対立
　関 in opposition to:〜に抵抗して,〜に反対して
optical 形光学の,光学式
　optical communication 光通信
　optical fiber cable 光ファイバケーブル
　optical flat オプチカルフラット【平面度を測る測定器】
　optical glass 光学ガラス
　optical indicator 光[学式]インジケータ
　optical instrument 光学機器
　optical isolator 光アイソレータ【光を一方向だけに通す装置】
　optical marking フォトマーキング【光学的けがき法】
　optical memory 光メモリ
　optical property 光学的性質
　optical pulse scale 光学的パルススケール
　optical pyrometer 光高温計
　optical sensor 光センサ
optics 光学
optimal (optimum) 形最適の
　optimal control 最適制御
　optimal feedback control 最適フィードバック制御
　optimal setting 最適調整
　optimal value 最適値
optimality 最適性
optimeter オプチメータ【光学的比較測定器の一種】
optimization 最適化
optimum 形最適の,最高の
　optimum condition 最適条件
　optimum control 最適制御
option オプション,選択
or 接…かまたは〜,…でも〜でも,すなわち,[否定語の後で]…もまた〜でない,言いかえれば
OR circuit OR回路,論理和回路
orbit 名軌道 動軌道に乗せる
　orbit circular 円軌道

orbital 形軌道の
orbiting 形軌道を回る
　orbiting electron　軌道電子
　orbiting motion　軌道運動
order 名オーダ，命令，状態，指示，次数，順序，度，等級
　動指示する，命令する，注文する
　関 in order：[機械などが] 調子よく／in order to ～：～のために
　order drawing　注文図
ordinal number　序数
ordinary 形通常の，普通の，
　ordinary flow　常流
　ordinary temperature　常温
ordinate　縦座標
organic 形有機の
　organic chemistry　有機化学
　organic compound　有機化合物
　organic matter　有機質
organization　組織，機構，協会，構成
organize 動組織する，創立する
orientation　オリエンテーション，位置決め，方向付け，配向
　orientation vector　方位ベクトル
orifice　オリフィス，開口部，穴口
　orifice meter　オリフィス流量計
origin　原点
original 形最初の，初期の，もとの　名オリジナル，原型，原点，原本
　original dimension　原寸
originally 副元来は
originate 動起こる，始まる，考案される，発明する
Orsat analyzing apparatus　オルザットガス分析計　同 Orsat [gas] analyzer／Orsat apparatus
orthogonal coordinates　直交座標
orthogonality　直交性
orthographic 形直角の
orthophosphate　燐酸
OS (Operating System)　オペレーティングシステム
oscillate 動振動する
oscillating combustion　振動燃焼
oscillating cylinder　揺動シリンダ
oscillating load　揺動荷重

oscillation　振動，動揺，発振，揺動
　oscillation circuit　発振回路
　oscillation constant　振動定数
　oscillation frequency　振動周波数
oscillator　発振器，振動子
oscillatory combustion　振動燃焼
oscillograph　オシログラフ【振動記録器】
oscilloscope　オシロスコープ[信号電圧の波形を観測する装置]
osmosis　浸透
osmotic pressure　浸透圧
other 形他の，別の，もう一方の，他方の，反対の，裏の
　関 the other　もう一方の，もう片方の
　other than 前／接～以外の(に)，～とは異なって，～とは別の
otherwise 副さもなければ，別のやり方で，その他の点では
　形他の，異なった
Otto cycle　オットーサイクル【電気着火機関の基本サイクル】
out-of-date 形旧式の，時代遅れの
out-of-order 形故障した
out-of-phase 形位相外れの
out-of-plane loading　面外荷重
out-of-round 形真円でない
out-of-roundness　真円度
out-of-step　同期外れ
outage　機能停止，停止，事故
outboard 形船外の　副船外に
outcome　結果，成果
outer 形外側の，外部の，外界の
　outer edge　外縁
　outer layer　外層
　outer product　外積
　outer ring　外輪
　outer surface　外表面
outfit　艤装，艤装品
outflow　流出　動流出する
outgassing　ガス放出
outgoing line　引出線
outlet　出口，ソケット
　outlet angle　出口角，流出角
　outlet blade angle　羽根出口角

outlet joint アウトレット継手【枝管を接続する管継手】
outlet valve 出口弁
outlet velocity 出口速度, 流出速度
outline アウトライン, 輪郭, 概略
output 名アウトプット, 出力, 排出量, 生産量 動出力する
output admittance 出力アドミタンス
output axis 出力軸
output capacitance 出力キャパシタンス
output circuit 出力回路
output density 出力密度
output factor 出力率
output horsepower 出力馬力
output impedance 出力インピーダンス
output meter 出力計
output resistance 出力抵抗
output transformer 出力トランス
output voltage 出力電圧
output winding 出力巻線
outside 名外側, 外部, 外面, 外界 形外側の, 外部の 前〜の外(側)に
outside air 外気
outside calipers 外パス
outside diameter 外径
outside diameter of thread ねじの外径
outside micrometer 外マイクロメータ
outstanding 形目立つ, 顕著な
outward 形外部の 副外へ, 外側へ, 外部へ
outward flow turbine 外向き流れタービン
oval 形卵形の, 楕円の
oval boiler 楕円形ボイラ
oval flow meter オーバル流量計
oven オーブン, 窯
over 〜 前〜の上に, 〜を覆って, 〜を越えて, 〜にわたって
over time 時間とともに, 徐々に
over- 接頭「過度に」「あまりに多くの」の意

overall 形全体の, 全部の 副全体に, 端から端まで
overall adiabatic efficiency 全断熱効率
overall characteristic 総合特性
overall coefficient of heat transfer 総[括]伝熱係数
overall coefficient of heat transmission 全熱通過率, 全熱貫流率
overall dimension 全寸法
overall efficiency 全効率, 総合効率
overall error 総合誤差
overall heat transfer 熱貫流, 熱通過
overall heat transfer coefficient 熱通過率, 熱貫流率
overall heat transmission 熱貫流, 熱通過
overall heat transmission coefficient 熱貫流率, 熱通過率
overall length 全長
overall reaction rate 全反応率
overall thermal efficiency 全熱効率
overassessment 過大評価
overboard 副船外(に)へ
overcame 「overcome」の過去形
overcharge オーバチャージ, 過充電, 超過挿入
overcome 動打ち勝つ, 打開する
overcurrent relay 過電流継電器
overexcitation 過励磁
overexpansion 過膨張
over-charging 過放電
overflow 名オーバフロー 動溢れ出る
overflow pipe あふれ管
overgrind 削りすぎ
overhaul 名オーバホール, 分解点検, 開放点検, 分解整備 動分解点検する, 分解修理する
overhaul inspection 開放検査
overhead 形頭上の, 高架の
overhead traveling crane 天井クレーン 同 overhead traveler
overhead valve engine 頭弁式機関
overheat 動過熱させる

名オーバヒート，過熱
overheating 過熱
overlap 動…に重なる
　名オーバラップ，重なり
　overlap of valves　弁重なり
overlay 名オーバレイ【金属製の被覆物】動覆う，めっきをする
overload 名オーバロード，過負荷 動過負荷を掛ける，過電流を流す
　overload capacity　過負荷容量
　overload condition　過負荷状態
　overload current　過負荷電流
　overload cut-out device　過負荷防止装置
　overload firing　過負荷燃焼
　overload operation　過負荷運転
　overload power　過負荷出力
　overload protection　過負荷防止
　overload relay　過負荷継電器
　overload test　過負荷試験
　overload trial　過負荷試験
overloading prevent device 過負荷防止装置
overlook 動見落とす
override 動無効にする，優先する
overshoot オーバシュート，行過ぎ量
overspeed オーバスピード，過速度，過回転数
　overspeed emergency trip device 過速度危急遮断器
　overspeed governor　オーバスピードガバナ，過速度調速機
　overspeed limit　超過速度限界
　overspeed relay　オーバスピードリレー，過速度継電器
　overspeed test　超過速度試験

oversquare engine オーバスクウェアエンジン【行程長さより大きなシリンダ内径のエンジン】
overstatement 過大表示
overstress 過大応力
overturn 動ひっくり返す，転覆する
overvoltage relay 過電圧継電器
owe 動［義務を］負う，［恩恵を］こうむる
owing to ～ 前～のために，～によって
own 動所有する　形特有の
oxidation 酸化
　oxidation catalyst　酸化触媒
　oxidation film　酸化皮膜
　oxidation inhibitor　酸化防止剤
　oxidation polymerization　酸化重合
　oxidation stability　酸化安定性
　oxidation treatment　酸化処理
oxide 酸化物
oxidize 動酸化する
oxidizing agent 酸化剤
oxidizing flame 酸化炎
oxyacetylene cutting 酸素アセチレン切断
oxyacetylene welding 酸素アセチレン溶接
oxygen 酸素　記O
　oxygen deficiency　酸素欠乏
　oxygen-free　無酸素
　oxygen sensor　O_2センサ
ozone オゾン
　ozone generator　オゾン発生器
　ozone (layer) depletion　オゾン層破壊

P-p

P action P動作, 比例動作
  圓 proportional control action
P alkalinity Pアルカリ度
P type semiconductor P形半導体
pack 動詰め込む, 包む, 固める
packet パケット, 束
packing パッキン, 詰物【運動部分の密封装置, 静止部分はgasketを使用】
  packing box パッキン箱
  packing coil コイルパッキン
  packing gland パッキン押え
  packing material 充てん材, パッキン材料
  packing piece 植え金
  packing ring パッキン輪
  packing steam パッキン蒸気
packless valve パックレス弁
pad 名パッド, まくら, 受台, 当て物
  動詰め物をする, 当て物をする
  pad bearing パッド軸受
  pad lubrication パッド注油, パッド潤滑【パッドの毛管作用を利用した潤滑法】
padding 肉盛り
paddle fan パドルファン【ラジアルファンの一種】
paint 名ペイント, 塗料
  動塗装する, 塗る, 塗布する
painting 塗装
pair 一対, ペア, 対偶
  圓 in pairs 2つ1組(ペア)で
  pair of element 対偶
palladium パラジウム 記Pd
pallet パレット, つめ
pan 受け皿
  pan head rivet なべ頭リベット
panel パネル, 板
  panel board 分電盤
  panel point 節点
  panel room 配電盤室
parabola パラボラ, 放物線
paraboloid 放物面
paraffin パラフィン, 直鎖族炭化水素
  paraffin base crude oil パラフィン基原油
  paraffin hydrocarbon パラフィン系炭化水素
  paraffin wax パラフィンワックス
paragraph パラグラフ, 項, 節, 段落
parallax パララックス, 視差
parallel 名平行, 並行, 並列
  形並列の, 平行な, 並行の
  動並列につなぐ, 平行する
  圓 in parallel：並列に
  parallel blade 平行羽根
  parallel branch 並列分岐
  parallel circuit 並列回路
  parallel connection 並列接続
  parallel crank mechanism 平行クランク機構
  parallel current 並流
  parallel drive 並列駆動
  parallel flow 平行流れ, 並流
  parallel flow turbine 並流タービン, 軸流タービン
  parallel flow type heat exchanger 並流式熱交換器
  parallel key 平行キー
  parallel line 平行線
  parallel motion 平行運動
  parallel nozzle 平行ノズル
  parallel operation 並列運転
  parallel plate capacitor 平行板コンデンサ
  parallel processing パラレル処理
  parallel resonance 並列共振
  parallel rod 平行連接棒
  parallel roller 円筒ころ
  parallel running 並行運転
parallelism 平行度, 平行性
parallelogram 平行四辺形
paramagnetic 形常磁性の
  paramagnetic material (substance)

常磁性体
paramagnetism 常磁性
parameter パラメータ，媒介変数，助変数，係数
parametric excitation パラメータ励振
parent metal 母材
parenthesis 括弧，丸括弧
Parsons number パーソンス数
Parsons turbine パーソンスタービン【軸流タービンの一種】
part 部品，部分　関 for the most part：大部分は，たいていは
 part assembly drawing　部分組立図
 part drawing　部品図
 part load　部分負荷
partial 形部分的な，不完全の
 partial admission turbine　部分流入タービン
 partial annealing　不完全焼なまし
 partial assembly drawing　部分組立図
 partial bond　部分結合
 partial bulkhead　部分隔壁
 partial combustion　部分燃焼，不完全燃焼
 partial differential equation　偏微分方程式
 partial load　部分負荷
 partial loss　分損
 partial pressure　分圧
 partial view　部分図，部分投影図
partially 副部分的に，不十分に
 partially fabricated item　半製品
 partially pulsating stress　部分片振り応力
participate 動加わる，参加する
particle [微]粒子，質点
 関 metal particles：金属[微]片
 particle velocity　粒子速度
particular 形特定の，特別な，特殊な，詳細な　名項目，事項，詳細
 particular sheet　要目表
particularly 副特に，特別に
particulate 名微粒子，ばいじん，粒子状物質　形粒状の
 particulate concentration　ばいじん濃度
 Particulate Matter　PM，粒子状物質
partition 名パーティション，仕切り，隔壁，分配
 動分配する，分割する，仕切る
partly 副一部分は，部分的に，ある程度は，いくぶん
parts 部品
 parts list　部品表
 parts per billion　ppb，十億分の一
 parts per million　ppm，百万分の一
Pascal パスカル【圧力の単位：Pa】
 Pascal's law　パスカルの法則
 Pascal's principle　パスカルの原理
pass 名通過，通行，通路
 動過ぎる，通過する，送る
 pass along　動伝える，知らせる
 pass over　動通り過ぎる，横切る
 pass through ～　動～を通りすぎる，～を貫く
 pass to　動～に移動する
passage パッセージ，通路，流路，通過，経過，通行，出入り口
 passage of turbine　流路〈タービン〉
 passage vortex　流路渦
passageway 通路，廊下
passameter パッサメータ【工作物の外径を精密に測定する計器】
passenger 乗客，通行人
passimeter パッシメータ【工作物の内径を精密に測定する計器】
passivation パシベーション，不動態化，不活性化，安定化
passive 形不活性の，受動的な
passivity 受動性，不動態
 同 passive state
past 形過ぎ去った，元の
 名過去　前～を通り過ぎて
paste 名ペースト【糊状物質】
 動のりで貼る，貼り付ける
pasteurizer 殺菌装置
patch 名当て金，補修剤
 動継ぎを当てる，当て金を当てる
 patch board　配線盤
patent パテント，特許
path 通路，進路，経路

path line 流跡線
path of flow 流路
pattern パターン，様式，模型，模様，紋，原型，木型〈鋳造〉
pattern control パターン制御
pause 動休止する
pawl つめ，歯止め
pawl washer つめ付ワッシャ
pay 動［注意などを］払う
peacock ピーコック【小さなコック】
peak ピーク，頂
peak contact 歯先接触
peak load ピーク負荷
peak pressure 最高圧力
peak stress ピーク応力
peak tank ピークタンク
peak to peak value ピークピーク値
pearlite パーライト【共析鋼の結晶組織】
pearlite cast iron パーライト鋳鉄
Peclet number ペクレ数
pedal ペダル
pedestal 台
peel 名ピール，剥離
動むける，はがれる
peen 打ち伸ばす，曲げる
peening ピーニング【金属をハンマなどでたたいて表面の硬化と疲れ限度を増す加工法】
peening effect ピーニング効果
peening hammer 打撃ハンマ
peening wear ピーニング摩耗
peep 名のぞき見 動のぞく
peep hole のぞき穴
pellet ペレット【小さな球や円柱等に造粒した材料のこと】
Peltier effect ペルチェ効果
Peltier element ペルチェ素子
pendant lamp ペンダントランプ
pendulum 振り子
pendulum governor 振り子調速機
pendulum impact tester 振り子衝撃試験機【シャルピーやアイゾット衝撃試験機がある】
penetrameter 透過度計
penetrant 形浸透性の 名浸透剤

penetrant inspection 浸透探傷検査
penetrant test 浸透探傷試験
penetrate 動貫通する，浸透する
penetrating 形貫通する，浸透する
penetrating agent 浸透剤
penetrating oil 浸透油
penetrating power 透過力
penetration 浸透［力］，透過［力］，貫通［力］，針入度，溶込み〈溶接〉
penetration degree 針入度
penetration leakage 浸透漏れ
penetration rate 浸透率
penetration test 針入度試験
penetrator 圧子
penetrometer X線透過度計，［半固体物質の］硬度計，針入度計
pent[a] 「5」の意の連結形
pentagon 五角(辺)形
pentahedron 五面体
per 前〜あたり，〜ごとに
per mill 千分率
percent パーセント，百分率
percent shear せん断破面率
percentage パーセンテージ，百分率，割合
percentage decrease 減少率
percentage increase 増加率
percentage of contraction 縮み率
percentage of elongation 伸び率
percolation ろ過，浸透，貫流
percussion 衝撃，打撃
percussion test 衝撃試験，打撃試験
percussion welding 衝撃溶接
perfect 形完全な
動仕上げる，完成する
perfect black body 完全黒体
perfect combustion 完全燃焼
perfect fluid 理想流体，完全流体
perfect gas 完全ガス，完全気体
perfect lubrication 完全潤滑
perfect scavenging 完全掃気
perfectly 副完全に
perfectly diffused scavenging 完全拡散掃気
perfectly stratified scavenging 完全層状掃気

perforate 動穴をあける
perforated 形穴のあいた
  perforated pipe　多孔管
  perforated plate　多孔板
perform 動行う，実施する
performance 性能，成績，動作
  performance characteristics　動作特性
  performance chart　動作線図
  performance curve　性能曲線
  performance function　評価関数
  performance index　評価指数，性能指数
  performance number　出力価
  performance test　性能試験
perhaps 副多分
perimeter 全周
period 期間，周期
  period of after burning　後燃え期間
  period of controllable combustion　制御燃焼期間
  period of ignition lag　発火遅れ期間
periodic 形周期的な，定期的な
  periodic blow down　間欠ブロー
  periodic damping　振動減衰
  periodic function　周期関数
  periodic law　周期律
  periodic maintenance　定期修理
  periodic motion　周期運動
  periodic table　周期律表
periodical survey 定期検査
periodically 副定期的に
periodicity 周期性
peripheral 形周辺の，周囲の
  peripheral (periphery) cam　周辺カム
  peripheral velocity　周速度
periphery 周囲，外側，円周
peritectic reaction 包晶反応
peritectoid 包析[晶]
perlite パーライト
permalloy パーマロイ【高ニッケル鋼の一種】
permanence 耐久度
permanent 形永久的な，耐久の
  permanent deformation　永久変形
  permanent gas　永久ガス
  permanent hardness　永久硬度
  permanent load　不変荷重
  permanent magnet　永久磁石
  permanent magnetism　残留磁気
  permanent set (strain)　永久ひずみ
permanently 副永久に，不変に
permeability 透磁率，透磁性
permissible 形許容できる，許される
  permissible concentration　許容濃度
  permissible creep rate　許容クリープ率
  permissible current　許容電流
  permissible dimensional deviation　寸法許容差
  permissible error　許容誤差
  permissible level　許容基準
  permissible limit　許容限度
  permissible load　許容荷重
  permissible maximum temperature　許容最高温度
  permissible stress　許容応力
  permissible temperature　許容温度
  permissible value　許容値
permission 許可，許容
permit 動許可する，可能にする，容認する　名許可証
  permit A to do　Aに〜することを認める
permittivity 誘電率
peroxide 過酸化物
perpendicular 名垂線，垂直　形垂直の，直立した
  perpendicular axis　垂直軸
  perpendicular line　垂線
perpendicularity 直立，垂直，直角度
perpendicularly 副垂直に
perpetual 形永久に続く
  perpetual engine of the first kind　第一種永久機関
  perpetual motion machine　永久機関
persist 動固執する，持続する，主張する，し続ける
personal 形一個人の，個人的な
personnel 職員，人員，人々，人事部
perspective 遠近法，透視図法
  perspective drawing　透視図
pertain 動関係する，付属する，適

perturbation 摂動
petcock 豆コック
petrol ガソリン 同 gasoline
  petrol engine　ガソリン機関
petrolatum ワセリン
petroleum 石油
  petroleum gas　石油ガス
  関 Liquefied Petroleum Gas：LPG, 液化石油ガス
  petroleum processing　石油精製
  petroleum product　石油製品
petvalve 豆弁
pH (hydrogen ion exponent) 水素イオン指数
  pH value　pH値
phase 相, 位相, 段階, 局面, 力率
  関 gas phase：気相／liquid phase：液相／solid phase：固相／multi phase flow：多相流／in phase with：〜と同相で
  phase adjuster　位相調整器
  phase angle　位相角
  phase boundary　相[境]界, 界面
  phase change　相変化, 相転移
  phase contrast　形位相差の 名位相差
  phase control　位相制御
  phase current　相電流
  phase delay　位相遅れ
  phase diagram　状態図, 平衡図
  phase difference　位相差
  phase distortion　位相ひずみ
  phase-down　漸減, 段階的減少
  phase equilibrium　相平衡
  phase indicator　位相計
  phase lag　位相遅れ
  phase lead (advance)　位相進み
  phase lock loop　位相同期回路
  phase lock oscillator　位相同期発振器
  phase locking　位相同期
  phase margin　位相余裕
  phase meter (indicator)　位相計
  phase modulation　位相変調
  phase-out　段階的停止, 段階的廃止
  phase rotation　相回転
  phase rule　相律
  phase sequence　位相順序, 相順
  phase shift　位相ずれ
  phase shifter　位相調整器
  phase transition　相転位
  phase velocity　位相速度
  phase voltage　相電圧
  phase winding　相巻線
phenol resin フェノール樹脂, ベークライト【熱硬化性樹脂の一種】
phenomena　「phenomenon」の複数形
phenomenon 現象
philosophy 考え方
pH-meter pHメータ
phon フォン, ホン【音の強さの単位】
phone meter 音量計
phosgene ホスゲン【毒性が強く刺激臭のある無色のガス】
phosphate りん酸[塩]
  phosphate of lime　りん酸カルシウム
phosphide りん化物
phosphor 蛍光体, 蛍光物質, りん
  phosphor bronze　りん青銅
phosphoric acid りん酸
phosphorus りん, 燐　記 P
photo 写真
photo- 接頭「光」「写真」の意
  photocell　光電池, 光セル
  photochemical smog　光化学スモッグ
  photoconductive cell　光導電セル
  photocoupler　光結合器
  photocurrent　光電流
  photodiode　フォトダイオード
  photoelasticity　光弾性
  photoelectric　形光電子の, 光電式
  photoelectric cell　光電池, 光電セル
  photoelectric conversion method　光電変換方式
  photoelectric device　光電素子
  photoelectric dew point hygrometer　光電管露点計
  photoelectric effect　光電効果
  photoelectric pyrometer　光電高温計

photoelectric switch　光電スイッチ
photoelectric tube　光電管
photoelectric tube pyrometer　光電管高温計
photoelectricity microscope　光電顕微鏡
photoelectrochemical cell　光電気化学電池
photoelectron　光電子
photoemission　光電子放出
photoetching　フォトエッチング
photoforming　フォトフォーミング
photograph　写真
photoluminescence　フォトルミネッセンス【蛍光塗料の原理】
photolysis　光分解
photomarking　フォトマーキング，投影けがき
photometer　光度計
photomicrograph　顕微鏡写真
photon　光子，光量子
photoresist　フォトレジスト，光硬化性樹脂
phototelegraphy　写真電送
phototransistor　フォトトランジスタ
phototube　光電管
photo-type smoke meter　光透過式排気濃度計
photovoltaic cell　光起電力セル，光電池
photovoltaic power generation　太陽光発電
PHP (Propeller HorsePower)　プロペラ馬力
PHP (Pump HorsePower)　ポンプ馬力
pH-value　pH 値，pH 価
physical　形物理的な，物理学上の，自然の，天然の，肉体的な，身体の
physical change　物理的変化
physical delay　物理遅れ
physical development　物理現象
physical property (characteristic)　物理的性質，物性値
physical quantity　物理量
physical vapor deposition　物理蒸着

pi　円周率【記号：π】
PI (Proportional plus Integral) action　PI 動作，比例積分動作
piano wire　ピアノ線
pick up　動取り上げる，かき集める
pick-up voltage　引上電圧
pickling　酸洗い
pickling test　酸洗い検査
pickle　動［金属製品を］酸洗いする
pictorial　形絵で表した，絵入りの
picture　絵，写真，状況，事態
picture element　画素
PID (Proportional plus Integral plus Derivative) action　PID 動作，比例積分微分動作
pie chart (graph)　円グラフ
piece　断片，破片，一個，一部分，要素　関 a piece of 〜：ひとつの〜
piece work　単位仕事
pier　桟橋
piercing　穴あけ
piezoelectric　形圧電気の
piezoelectric actuator　圧電アクチュエータ
piezoelectric effect　圧電効果
piezoelectric element　圧電素子
piezoelectric materials　圧電材料
piezoelectric vibrometer　圧電形振動計
piezoelectricity　圧電気，ピエゾ電気
piezometer　ピエゾメータ，圧縮計，圧度計
pig　生子(なまこ)鉄
pig iron　銑鉄
pigment　顔料【塗料やインクの着色に用いる】
pile　動重ねる，集積する，盛る
pillar　ピラー，柱
pillow block　ピロー形軸受
pilot burner　パイロットバーナ，口火，種火
pilot cell　表示電池
pilot current　監視電流
pilot exciter　パイロット励磁器
pilot flame　パイロット火炎，口火
pilot injection　先立ち噴射

pilot lamp　パイロットランプ，表示灯
pilot valve　パイロット弁，案内弁
pin　ピン【小径の丸棒】
　pin gear　ピン歯車
　pin hole　ピンホール
　pin joint　ピン継手
　pin key　ピンキー
　pin wrench　ピン付きレンチ
pinboard　ピンボード，配電盤
pinch　名ひねり，つまみ，はさみ　動つまむ，はさむ，かむ
　pinch cock　ピンチコック
　pinch valve　ピンチ弁
pinion　ピニオン，小歯車【一対の歯車のうち歯数の少ない歯車の呼び名，多い方はギアあるいはホイールと呼ばれる】
pintle　ピントル，舵針，分配軸
　pintle nozzle　ピントルノズル
pipe　名パイプ，管　動配管する，導く，供給する
　pipe arrangement　配管，配管図
　pipe bender　パイプベンダ【管曲げ工具】
　pipe coupling　管継手
　pipe cutter　パイプカッタ【管切断工具】
　pipe friction coefficient (factor)　管摩擦係数
　pipe joint　管継手
　pipe line　管路
　pipe orifice　管[内]オリフィス
　pipe tee　T字管
　pipe wrench　パイプレンチ【スパナの一種】
pipette　ピペット，移液管
pipework　配管
piping　配管　⑩ pipe arrangement
　piping diagram　配管図
　piping work　配管工事
piston　ピストン
　piston acceleration　ピストン加速度
　piston area　ピストン面積
　piston clearance　ピストンすきま
　piston cooling fresh water pump　ピストン冷却清水ポンプ
　piston crown　ピストンクラウン
　piston displacement　行程容積
　piston engine　容積形機関，ピストン機関
　piston groove　ピストンみぞ
　piston head　ピストンヘッド
　piston pin　ピストンピン
　piston pin bearing　ピストンピン軸受
　piston pressure　ピストン圧[力]
　piston pump　ピストンポンプ
　piston ring　ピストンリング
　piston ring gap　ピストンリングギャップ
　piston ring groove　ピストンリングみぞ
　piston ring land　ピストンランド
　piston rod　ピストン棒
　piston skirt　ピストンスカート　関 skirt：すその部分
　piston slap　ピストンスラップ【ピストンとライナのすきまが大きいときに生じるたたき音】
　piston speed　ピストン速度
　piston spring　ピストンばね
　piston stroke　ピストン行程
　piston top surface　ピストン頂面
　piston travel　ピストンの行程
　piston valve　ピストン弁
pit　名ピット，穴，くぼみ【溶接不良の一種】　動穴を開ける
pitch　ピッチ，程度，度合い，(音の)高低，縦揺れ，刻み，勾配
　pitch angle　ピッチ角
　pitch circle　ピッチ円
　pitch cone　ピッチ円すい
　pitch cone angle　ピッチ円すい角
　pitch diameter　ピッチ円直径
　pitch diameter of thread　ねじの有効径
　pitch gauge　ピッチゲージ【ねじのピッチを測定するゲージ】
　pitch line　ピッチ線
　pitch point　ピッチ点
　pitch ratio　ピッチ比

pitch surface　ピッチ面
pitching　ピッチング，縦揺れ
pitchometer　ピッチ計
Pitot state tube　ピトー静圧管
Pitot tube　ピトー管，静圧管
pitsaw file　半丸やすり
pitting　ピッチング，点食，孔食【腐食による点状のきず】
　pitting abrasion　まだら摩耗
　pitting corrosion　点食
　pitting of gear tooth　歯の点食
pivot　名ピボット，回転軸，中心点，枢軸，旋回軸【軸や板が旋回傾斜運動するとき支軸となる小さな突起部】動回転させる，軸を付ける
　pivot bearing　ピボット軸受【スラスト軸受の一種】
　pivot journal　ピボットジャーナル
　pivot suspension　ピボット支え
pixel　画素
place　名位置，場所
　動置く，配置する，設置する
　関 in place：適所に，定位置に／in place of ～：～の代わりに
plain　形明白な，平坦な
　副はっきりと，率直に，平らに
　plain bearing　平軸受，滑り軸受
　plain bracket　平ブラケット
　plain cylindrical furnace　平形炉筒
　plain washer　平座金
plan　名プラン，計画，予定，設計，図面，平面図 動計画する，考案する，設計する，予定する
　plan equation　馬力計算式，plan方程式
　plan view　平面図
planation　平面化，平坦化作用
Planck constant　プランク定数
plane　平面，平面図，かんな，面
　関 line：線／point：点
　plane bearing　平軸受
　plane cam　平面カム
　plane curve　平面曲線
　plane motion　平面運動
　plane of projection　投影面
　plane strain　平面ひずみ

　plane stress　平面応力
　plane view　平面図
planet　惑星，遊星
　planet gear　遊星歯車
planetary　形惑星の，軌道を回る，遊星形の 名遊星伝動(歯車)装置
　planetary drive　遊星歯車駆動装置
　planetary electron　軌道電子
　planetary gear　遊星歯車装置
　planetary motion　遊星運動
　planetary pinion gear　遊星小歯車
　planetary reduction gear　遊星減速装置，遊星減速歯車
planimeter　プラニメータ，面積計
planing　平削り
　planing machine　平削盤
planish　動平らにする
planning　計画
plant　プラント，設備，工場，装置
plasma　プラズマ
　plasma machining　プラズマ加工
plastic　名プラスチック，合成樹脂 形塑性の，可塑性の
　plastic deformation　塑性変形
　plastic fluid　塑性流体
　plastic forming　塑性加工
　plastic refractory　プラスチック耐火物
　plastic section modulus　塑性断面係数
　plastic strain　塑性ひずみ
　plastic working　塑性加工
plasticating capacity　可塑化能力
plasticity　塑性，可塑性
plasticization　可塑化
plasticize　動可塑化する
plasticizer　可塑剤
plasticizing efficiency　可塑化効率
plastics　プラスチック，可塑性物質【可塑性をもつ高分子物質】
plastification　可塑化
plastigauge　プラスチゲージ【軸受の間隙を測定するプラスチック製の細線】
plate　名プレート，板，平板，陽極 動めっきする，覆う

plate cam　板カム
plate clutch　板クラッチ
plate current　陽極電流
plate efficiency　段効率
plate fan　プレート送風機
plate-fin type heat exchanger　プレートフィン式熱交換器
plate gauge　板ゲージ
plate spring　板ばね
plate type heat exchanger　プレート式熱交換器
plate valve　板弁
plate voltage　陽極電圧
plated circuit　基板回路
platform scale　台ばかり
plating　めっき，被覆
platinum　プラチナ，白金　㊘Pt
platinum resistance thermometer　白金抵抗温度計
play　遊び，がた
play a part in　動〜に関係(与)する
plenty　形多くの，豊富な，十分な　関plenty of〜：十分な〜
pliability　柔軟性，たわみ性
pliers　プライヤ，やっとこ
plot　名プロット，作図　動作図する
plotter　プロッタ
plow　動掘り起こす
plow bolt　プラウボルト，回り止め付き頭ボルト
plowing　[摩擦面の]掘り起こし
plug　名プラグ，栓　動ふさぐ，埋める，詰める，栓をする
plug flow　プラグ流れ，栓流
plug gauge　穴ゲージ
plug socket　コンセント，差込栓
plug welding　プラグ溶接
plugger　充填器
plugging　プラギング，逆転制動
plugging relay　逆回転防止継電器
plumb　おもり
plumb line　鉛直線
plumbago　黒鉛，石墨
plumber block　中間軸受
plumber work　管工事
plummer　軸受台，軸受

plummer block　プランマブロック，中間軸受，軸受台
plunge　動突っ込む，押し込む
plunger　プランジャ【径に対して長い棒状のピストン】
plunger pump　プランジャポンプ
plus　形正の，陽の，プラスの　名プラス，正数，剰余　前〜を(に)加えて，プラスして，そしてまた
plus thread　プラスねじ
plutonium　プルトニウム　㊘Pu
PN junction　PN接合
pneumatic　形空気の，空気力の
pneumatic brake　空気ブレーキ
pneumatic tube　気送管
pneumatic controller　空気圧式制御装置
pneumatic cylinder　空気圧シリンダ
pneumatic drive　空気圧駆動
pneumatic impact wrench　インパクトレンチ
pneumatic motor　空気圧モータ
pneumatic pump　空気[圧]ポンプ
pneumatic spring　空気ばね
pneumatic starting valve　空気始動弁
pneumatic tube　気送管
PNP transistor　PNP型トランジスタ
PNPN transistor　PNPNスイッチ素子，サイリスタ
pocket　ポケット，受け口，くぼみ
point　名ポイント，箇所，点，先端，接点　動指さす，示す
point angle　先端角
point contact　点接触
point of contact　接点，接触点
point of inflection　反曲点
point out　動指示(摘)する
Point To Point control　PTP制御，位置決め制御
pointer　指針
pointer galvanometer　指針検流計
pointing tool　面取りバイト
poise　名ポアズ【流体の粘性率のCGS単位】，平衡，釣り合い　動平衡を保つ
poison　毒，有害物質

poison gas　毒ガス
poisonous　形有毒な，有害な
poisonous substances　毒物
Poisson's number　ポアソン数
Poisson's ratio　ポアソン比【軸方向に垂直荷重が加わったときの横ひずみと縦ひずみとの比】
polar　名極の，極性の
polar coordinates　極座標
polar modulus of section　極断面係数
polar moment of inertia　慣性極モーメント
polar moment of inertia of area　極断面二次モーメント
polar moment of resistance　抵抗極モーメント
polarity　極性
polarity of transformer　変圧器の極性
polarization　分極，偏光
polarization microscope　偏光顕微鏡
polarized light　偏光
polarizing microscope　偏光顕微鏡
pole　ポール，極，電極，磁極，柱
pole face　磁極面
pole piece　極片，磁極片
policy　政策
polish　名磨き　動磨く
polisher　純水装置
polishing　つや出し
pollutant　汚染物質
pollute　動汚す，汚染する
pollution　汚染，公害
　関 environmental pollution：環境汚染
pollution problem　公害問題
poly-　接頭ポリ「多い」「重い」の意
polybasic　形多塩基の
polycyclic　形多環式の
polycrystal　多結晶
polyester　ポリエステル
polyethylene　ポリエチレン
polygon　多角形
polygonal line　折れ線，屈折線
polyhedron　多面体
polymer　ポリマ，重合体，高分子
polymeric semiconductor　高分子半導体

polymerization　重合
　関 thermal polymerization：熱重合
polynomial　多項式
polyphase　形多相の
polyphase commutator machine　多相整流子電機
polyphase power　多層電力
Poly Tetra Fluoro Ethylene　PTFE，ポリテトラフルオロエチレン，テフロン
polytropic　形ポリトロープの
polytropic change　ポリトロープ変化
polytropic compression　ポリトロープ圧縮
polytropic efficiency　ポリトロープ効率
polytropic index　ポリトロープ指数
polytropic process　ポリトロープ過程
polyurethane　ポリウレタン
polyvinyl chloride　塩化ビニル樹脂
pool　［水］たまり
pool boiling　プール沸騰
poor　形不十分な，乏しい，劣った，わずかな
poor combustion　燃焼不良，不完全燃焼
poor compression　圧縮不良
poor conductor　不良導体
poor contact　接触不良
poor control　制御不良(不十分)
poor ground　アース不良
poor quality　低品質
poppet　ポペット
poppet valve　ポペット弁，きのこ弁【きのこ形の弁】
popping pressure of safety valve　安全弁吹出し圧力
popping stop pressure of safety valve　安全弁吹止まり圧力
popularity　需要
porcelain　磁器
pore　気孔，細い穴
porosity　気孔率，多孔性，空隙率
porous　形多孔性の

porous bearing 多孔質軸受
porous chrome ポーラスクロム
porous chrome plated cylinder liner ポーラスクロムメッキシリンダライナ
porous chromium plating ポーラスクロムめっき
porous membrane 多孔膜
porous structure 多孔質構造
port 名ポート, 口, 孔, 窓, 左舷, 港 形左舷の 副左舷に
port scavenging ポート掃気
portable 形携帯用の
portable fire extinguisher 持運び式消火器
portable fitting 取外し付属品
portable instrument 携帯測定器
portable type 移動型
porting diagram ポート線図
portion 一部, 部分
position 名ポジション, 位置, 姿勢 動～を置く, ～の位置を占める, ～の位置を定める
 関 in position：所定の場所に, 適所に
position angle 位置角
position head 位置水頭
position indicator 位置指示器
position sensor 位置センサ
position vector 位置ベクトル
positioner ポジショナ【位置を決める機器】
positioning 位置調整, 位置決め
positioning control 位置決め制御
positive 形正の, 正(陽)電気の, 陽性の, 確実な, 完全な, 明確な, 積極的な, 役に立つ
positive bias 正バイアス
positive blower 押込み送風機
positive catalyzer 正触媒
positive charge 陽電荷, 正電荷
positive displacement motor 容積形モータ
positive [electric] charge 陽電荷, 正電荷
positive electricity 陽電気

positive electrode 正極, 陽極
positive electron 陽電子
positive feed back 正帰還, 正フィードバック
positive hole 正孔
positive ion カチオン, 陽イオン
positive motion cam 確動カム
positive number 正数
positive plate 陽極板
positive pole 正極, 陽極
positive pressure 正圧, 陽圧
positive reactance 正相リアクタンス
positive sign 正符号
positive strain 正ひずみ
positively 副明確に, 確かに, まったく, 正(陽)電気を帯びて
positron 陽電子
possess 動所有する, 持つ
possibility 名可能性
possible 形可能な, 起こりうる
 関 make possible：可能にする
possibly 副あるいは, 多分
post 柱, くい
post- 頭「後の」「次の」の意
 反 pre-：「あらかじめ」「…以前の」「…前部にある」の意
post-brake ポストブレーキ
post-heating 後熱
post-ignition 遅延点火, 遅れ点火
Post Meridiem P.M., 午後
post-purge ポストパージ【ボイラ消火後の残留可燃ガスを排除すること】
postpone 動延期する
potassium カリウムの英語名 記 K
potassium hydroxide 水酸化カリウム
potassium nitrate 硝酸カリウム
potassium sulfate 硫酸カリウム
potential 名ポテンシャル, 電位 形位置の, 潜在する
potential circuit 電圧回路
potential device 変圧装置
potential difference 電位差
potential divider 分圧器

potential energy ポテンシャルエネルギ, 位置エネルギ
potential equivalent temperature 相当温度
potential flow ポテンシャル流れ
potential head 位置水頭, 位置ヘッド
potential heat 保有熱
potential transformer 計器用変圧器
potential wake 流線伴流
potentiometer ポテンショメータ, 電位差計, 分圧器
potentiometry 電位差測定法
pottery 陶磁器
pound ポンド【重さの単位：1lb＝約453.6g】
pour 動流れ出る, 注ぐ 名注入, 流出, (絶え間のない)流れ
pour point 流動点【油が流動しうる最低温度】
pour point depressant 流動点降下剤
pour test 流動点試験
pouring 鋳込み
pouring temperature 鋳込み温度
pouring time 鋳込み時間
powder 名粉, 粉末, 粉体, 火薬 動粉にする, ふりかける
powder coupling 粉体継手
powder emery 金剛砂
powder lubricant 粉末潤滑剤
power パワー, 動力, 出力, 電力, 仕事率, べき, 能 動動力を供給する, 動かす
power actuated control 他力制御 ⓢ power assisted control
power actuated governor 他力式調速機
power amplifier 電力増幅器
power board 配電盤室
power brake 動力ブレーキ
power cable 電力ケーブル
power circuit 電力回路, 動力回路
power cord 電源コード
power converter 電力変換装置
power cycle 出力サイクル
power cylinder パワーシリンダ
power delivered coupling 動力伝達継手
power diode パワーダイオード, 電力用ダイオード
power distribution 配電
power driven 形動力駆動の
power factor 力率, 出力係数
power factor meter 力率計
power failure 電源喪失
power gain 電力利得
power gas 動力ガス
power generation 発電
power house 動力室
power law 指数法則
power loss 電力損, 出力損
power method べき乗法
power operated control 他力制御
power outage 停電, 電源異常
power output 出力, 電力出力
power panel 電力盤
power plant 発電所, 動力装置
power rate 出力率
power ratio 出力比
power rectifier 順変換装置, 大電力整流器
power relay パワーリレー, 電力継電器
power requirement 所要動力, 電源条件
power shortage 電力不足
power source 電源
power station 発電所
power stroke 動力行程, 仕事行程
power supply 電源
power supply circuit 電源回路
power switch 電源スイッチ
power transformer 電源トランス
power transistor パワートランジスタ, 電力用トランジスタ
power transmission 送電, 動力伝達, 動力伝達装置, 伝動
power transmission shaft 伝動軸
power turbine 出力タービン
power unit パワーユニット, 動力装置
power winding 出力巻線

powered 形エンジン付の，〜を動力源とした　関 gasoline-powered engine：ガソリン内燃機関

ppb (Parts Per Billion)　十億分の一

ppm (Parts Per Million)　百万分の一

practicable　形実行できる，実用的な

practical　形実際の，事実上の，実用的な，現実的な

practically　副事実上，実際には

practice　名練習，習慣，実際，実地，実行　動実行する，練習する　関 in practice：実際には

Prandtl number　プラントル数【略語：Pr.】

pre-　接頭「プレ」「予め」「前の」「前部の」「…以前の」の意

precaution　警戒，用心，予防策

precede　動〜に先立つ，〜に優先する

precession　歳差運動，すりこぎ運動

precessional torque　歳差トルク

precious　形貴重な

precipitation　析出，沈殿
  precipitation hardening　析出硬化

precise　形正確な，精密な

precisely　副正確に

precision　精密さ，精度，正確
  precision gearing　精密かみあい
  precision machine　精密機械
  precision made　形精密仕上げの

preclude　動［予め］防ぐ，排除する

pre-coat　プレコート，下塗り

pre-coating agent　プレコート剤

pre-combustion chamber　予燃焼室

pre-cooler　予冷器

predetermine　動あらかじめ決める

predicate　動断定(言)する

predict　動予想(測)する

prediction　予測，予報

prefer　動好む，選ぶ

preferably　副もしできれば，むしろ

preference trip　優先遮断

preferred number　標準数

pre-harden steel　プレハーデン鋼【熱処理済み鋼】

pre-heat　動予熱する

pre-heater　予熱器

pre-heating　予熱

pre-ignition　過早着火，早期着火

premature　形早まった
  premature burst　過早破裂
  premature ignition　過早着火

pre-mixed combustion　予混合燃焼

preparation　準備，処理，前処理

prepare　動準備する

pre-purge　プレパージ【ボイラ点火前の可燃ガス排除】

prescribe　動指示する，規定する

presence　存在，出席　関 in the presence of 〜：〜に直面して

present　形現在の，緊急の，存在して　名現在，贈り物　動現れる，贈呈する，提出する

presentation　提示，発表，提案説明

preservation　保存

preservative　防腐剤

preserve　動保存する，保護する

preset　プリセット【前もって調整や設定をすること】

prespringing　弾性逆ひずみ

press　動押し付ける　名プレス，圧搾
  press working　プレス加工

pressed drop oiling　圧入滴下給油

pressing crack　圧縮割れ

pressure　圧力【圧力の単位：パスカル[Pa]】
  pressure angle　圧力角
  pressure atomizing burner　圧力噴霧バーナ
  pressure circuit　電圧回路
  pressure coefficient　圧力係数
  pressure coil　電圧コイル
  pressure compensated flow control valve　圧力補償流量制御弁
  pressure compound turbine　圧力複式タービン
  pressure connection terminal　圧着端子
  pressure container　圧力容器
  pressure control valve　圧力制御弁
  pressure converter　圧力変換器

pressure curve　圧力曲線
pressure difference　圧力差
pressure distillation　加圧蒸留
pressure distribution　圧力分布
pressure drag　圧力抵抗, 圧力抗力
pressure drop　圧力損失, 圧力降下
pressure energy　圧力エネルギ
pressure enthalpy diagram　P–H線図
pressure fan　押込み送風機
pressure feed lubrication　圧力注油, 圧力潤滑, 強制潤滑, 押込み注油
pressure fit　圧入
pressure gauge　圧力計
pressure gradient　圧力勾配
pressure head　圧力ヘッド, 圧力水頭
pressure inclination gauge　傾斜圧力計
pressure increase　圧力上昇
pressure indicator　圧力測定用インジケータ
pressure loss　圧力損失
pressure lubrication　強制潤滑
pressure measurement　圧力測定
pressure membrane　圧膜
pressure pump　圧力ポンプ
pressure ratio　圧力比
pressure recovery factor　圧力回復率
pressure reducing valve　減圧弁
pressure regulating valve　圧力調整弁
pressure regulator　圧力調整器
pressure relay　圧力継電器
pressure relief valve　圧力逃し弁
pressure resistance　圧力抵抗
pressure ring　圧力リング【ピストンリングの一種】
pressure rise　圧力上昇
pressure sensitive diode　感圧ダイオード
pressure sensor　圧力センサ
pressure stage　圧力段〈蒸気タービン〉
pressure switch　圧力スイッチ
pressure tank　圧力タンク
pressure terminal　圧力端子
pressure test　圧力試験
pressure thermometer　圧力式温度計
pressure transducer　圧力変換器
pressure transmission　圧力伝達
pressure turbine　圧力タービン
pressure type liquid level gauge　圧力液面計
pressure velocity compound turbine　圧力速度複式タービン
pressure vessel　圧力容器
Pressure Volume diagram　P–V線図, 圧力-容積線図
pressure wave　圧力波
pressure welding　圧接
pressureless sintering　常圧焼結
pressurize　動圧力をかける
Pressurized Water Reactor　PWR, 加圧水型原子炉
prestrain　逆ひずみ, 予備ひずみ
prestress forming　クリープテンパ【低温焼戻し処理の一種】
pretend　動〜のふりをする
prevent　動防ぐ, 阻止する
　prevent A from 〜 ing　動 A が〜することを防ぐ
preventative maintenance　予防保全
prevention　予防, 防止
preventive　形予防の
previous　形先の, 前の
　previous to 〜　前〜の前に
previously　副あらかじめ
prewhirl vane　予旋回羽根
primarily　副第一に, 主として
primary　形最初の, 一次の, 第一の, 主要な, 根本の, 一段の
　primary accelerator　主促進剤
　primary air　一次空気
　primary axis　一次軸
　primary battery (cell)　一次電池
　primary circuit　一次回路
　primary coil　一次コイル
　primary creep　第一期クリープ
　primary crystal　初晶
　primary current　一次電流
　primary energy　一次エネルギ
　primary flow　主流
　primary mol　基本モル

primary pinion　一段小歯車
primary quenching　一次焼入れ
primary shield　一次遮蔽
primary standard　一次標準, 原器
primary stress　一次応力
primary superheater　一次過熱器
primary voltage　一次電圧
primary wheel　一段大歯車
primary winding　一次巻線

prime　形主な, 最初の, 素数の
動[ポンプに]呼び水をする
prime fuel　始動燃料
prime mover　原動機
prime number　素数

primer　プライマ【塗料の一種】

priming　プライミング, 水けだち, 沸水, 呼び水
priming cock　呼び水コック
priming pump　始動ポンプ
priming valve　始動弁

primitive function　原子関数

principal　形主な, 主要な
principal axis　主軸
principal axis of stress　主応力軸
principal component　主成分
principal coordinate　主座標
principal dimension　主要寸法
principal particulars　主要目
principal plane of stress　主応力面
principal shearing strain　主せん断ひずみ
principal shearing stress　主せん断応力
principal strain　主ひずみ
principal stress　主応力

principally　副主に, 主として, 第一に

principle　原理, 原則, 法則
関 in principle：原理的には
principle of Archimedes　アルキメデスの原理
principle of causality　因果律
principle of conservation of energy　エネルギ保存の法則
同 law of conservation of energy
principle of continuity　連続の法則
同 law of succession
principle of electromagnetic induction　電磁誘導の原理
principle of superposition　重ね合わせの原理

print　名プリント, 印刷物
動押す, 印刷する

printed circuit　プリント回路

printed plate board　プリント基板

printer　プリンタ

printing　印字

prior　形先の, 先立つ
prior to ～　前～より前に

priority　優先順位
priority load　優先負荷【遮断されにくい負荷】

prism　プリズム, 角柱, 柱体, 柱

private　形個人の

pro oxidant　酸化促進剤

probability　確率
probability distribution　確率分布

probable　形起こりそうな
probable cause　推定原因
probable error　確率誤差

probably　副多分, おそらく

probe　プローブ, 探触子, 精査

problem　問題, 故障

procedure　手順, 行動, 手続き, 手法

proceed　動進行する, 前進する, 続行する, 実施する・される, 処置する, 行われる

process　動処理する, 加工する
名プロセス, 過程, 処理, 工程, 手順, 方法
関 absorption process：吸収法
process annealing　中間焼なまし
process chart (drawing)　工程図
process control　プロセス制御【制御量をプログラムにしたがって変化させる制御】
process design　プロセス設計, 工程設計
process instrumentation　プロセス計装
process monitoring　プロセス監視

processability　加工性

processing 加工，処理
　processing method　加工方法
processor プロセッサ，処理装置
prodigious 形膨大な，強大な
produce 動造る，製造する，生ずる，生産する，もたらす
product 生成物，製品，積
　関 logic product：論理積
　product of inertia　慣性相乗モーメント
　products of inertia of the area　断面相乗モーメント
　products of combustion　燃焼生成物
production 生産，製造，発生
　production control　生産管理
　production drawing　製作図
productivity 生産性
profile 名形状，輪郭，側面図，外形　動外形を描く
　profile control　ならい制御
　profile drag　形状抗力，翼形抗力
　profile friction loss　翼面摩擦
　profile gauge　輪郭ゲージ
　profile loss　翼断面損失
　profile shift　転位
　profile shifted gear　転位歯車
profilometer 粗さ計，粗面計
program 名プログラム，計画　動プログラムを作る，計画する　同 programme《英》
　program analysis　プログラム解析
　program control　プログラム制御
programming プログラミング
　programming language　プログラム言語
progress 名進行，前進，発展　動前進する，発達する
　progress of work　製造工程
progressive 形進行性の，進歩的な，革新的な，前進的な
progressively 副次第に，漸次
prohibit 動禁止する，妨げる
project 名計画，研究課題　動計画する，投影する，放出する，突出する
projected area 投影面積

　projected area ratio　投影面積比
projected blade area 羽根投影面積
projected plan 投影図
projecting 形突き出た
　projecting type　突出型
projection 投影[図]，突起，射出，計画，工夫
　projection drawing　投影図
　projection line[s]　寸法補助線
projector プロジェクタ，投光器，投影器
prolong 動長くする・なる
promote 動促進する，助長する
promoter 促進剤
prompt 形素早い，即座の，機敏な
promptly 副迅速に，早期に，ぴったり，ちょうど
prone to 動する傾向がある
Prony dynamometer プロニー動力計【摩擦動力計の一種】
proof 名プルーフ，試験，耐力，証明　形耐えられる，試験(検査)済みの　動検査(試験)する，防水加工する，耐えるようにする
　proof load　保証荷重
　proof mark　検印
　proof resilience　最大弾性エネルギ
　proof stress　耐力
　proof test　保証試験，耐力試験
propagate 動伝わる，伝播する
propagation 伝搬，伝播
　関 flame propagation：火炎伝播
propane プロパン
propel 動推進する，前進させる
propellant 推進剤，推進燃料
propeller プロペラ，推進器
　propeller aperture　プロペラアパチャ，プロペラすきま
　propeller blade　プロペラ羽根
　propeller boss　プロペラボス
　propeller diameter　プロペラ直径
　propeller disc area　プロペラ全円面積
　propeller efficiency　プロペラ効率
　propeller efficiency in open water　単独プロペラ効率

propeller expanded area プロペラ展開面積
propeller experiment プロペラ試験
Propeller HorsePower PHP, プロペラ馬力
propeller hub プロペラハブ
propeller leading edge プロペラ前縁
propeller performance プロペラ性能
propeller pitch プロペラピッチ
propeller pressure side プロペラ前進面
propeller propulsion プロペラ推進
propeller pull-up length プロペラ押込み量
propeller pump プロペラポンプ, 軸流ポンプ
propeller push up-distance プロペラ押込み量
propeller runner プロペラ羽根車
propeller shaft プロペラ軸
propeller shaft bearing プロペラ軸軸受
propeller singing プロペラ鳴音
propeller slip プロペラスリップ
propeller suction side プロペラ後進面
propeller tailing edge プロペラ後縁
propeller thrust プロペラスラスト
proper 形適当な, 正しい, 特有の, 固有の
proper motion 固有運動
properly 副適当に, 適切に, 正確に, きちんと, 厳密に
property 性質, 特性, 性状, 状態量, 物性値, 固有性
prop-jet プロップジェット【ジェットエンジンの一種】
proportion 比例, 割合
関 considerable proportion of：大部分の／in proportion to：〜に比例して
proportional 形比例の
関 be proportional to 〜：〜に比例する
proportional control 比例制御
proportional [control] action P動作, 比例動作
proportional element 比例要素
proportional gain 比例ゲイン
proportional limit 比例限度
Proportional plus Integral action PI動作, 比例積分動作
Proportional plus Integral plus Derivative action PID動作, 比例積分微分動作
proportional position action 比例動作
proportional shifting 比例推移
proportionate 形つり合った, 比例した 動つり合わせる, 比例させる
proportioning pump 定量ポンプ, 流量調節ポンプ
propose 動提案する, 申し込む, 企てる
propulsion 推進, 推進力
propulsion machinery 主機
同 main engine
propulsion turbine 推進タービン
propulsive 形推進力のある
propulsive coefficient 推進係数【主機の図示馬力と有効馬力との比】
propulsive efficiency 推進効率
prosecute 動遂行する, 行う
protect 動保護する
protecting tube 保護管
protection 保護, 防食
protection circuit 保護回路
protection current 防食電流
protective 形保護する, 保護用の
protective clothing 保護衣, 防護服
protective device 保安装置, 保護装置
protective gear 保護具
protective relay 保護継電器
protest 名海難報告書, 抗議, 異議, 反対
動主張する, 表明する, 抗議する
protocol プロトコル, 規約, 協定, 議定書
proton 陽子 関 neutron：中性子
prototype 原型, 試作機, 試作品, 原器, 見本
protractor 分度器

protrude 動はみ出る，突き出る
protrusion 突起部
prove 動証明する
provide 動備える，与える，供給する，準備する
provided [**that**] ～ 接～という条件で
provision 食料，供給，準備，条項
  熟 make provision for：～に備える
  provision store 食料庫
provisional design 仮設計
proximity switch 近接スイッチ
PS 馬力【1PS = 735.5W】
psi (**Pounds per Square Inch**) ポンド毎平方インチ
psychrometer 乾湿計
psychrometric chart 湿り空気線図，湿度線図
psychrometric ratio 乾湿比，湿り比
psychrometric saturated steam 湿り飽和蒸気
PTFE (**Poly Tetra Fluoro Ethylene**) テフロン
public nuisance 公害
pull 動引く，引っ張る，引き抜く，除去する
  pull-in torque 引入れトルク
  pull-out 同期はずれ
  pull-out test 引抜き試験
  pull-out torque 脱出トルク
pulley プーリ，滑車
  pulley motor プーリモータ
pulling speed 引張り速度
pulp パルプ
pulsate 動脈打つ，脈動する
pulsating 脈動
  pulsating combustion 振動燃焼，息つき燃焼，脈動燃焼
  pulsating current 脈動電流
  pulsating flow 脈流
  pulsating load 片振り荷重
  pulsating stress 片振り応力
pulsation 脈動
  pulsation effect 脈動効果
pulse パルス【信号の一種】
  pulse amplitude modulation パルス振幅変調
  pulse encoder パルスエンコーダ
  pulse generator パルス発生器
  pulse interval パルス間隔
  pulse motor パルスモータ
  pulse oscillator パルス発振器
  pulse operation 動圧過給，パルス動作，パルス運動
  pulse shape パルス波形
  pulse transfer function パルス伝達関数
  pulse wide method パルス幅変調方式
  Pulse Width Modulation control PWM制御，パルス幅電圧制御
pump 名ポンプ
  動送り出す，ポンプで注入する，ポンプでくみ上げる
  pump barrel ポンプ胴
  pump casing ポンプケーシング
  pump efficiency ポンプ効率
  pump head 揚程
  Pump HorsePower PHP，ポンプ馬力
  pump into 動注入する
  pump out 動汲み出す
  pump plunger ポンププランジャ
pumping ポンピング，排気
  pumping loss ポンプ損失
  pumping work ポンプ仕事
punch 名ポンチ 動打ち抜く
punching 打抜き
puncture test 破壊試験
pure 形純粋の
  pure iron 純鉄
  pure metal 純金属
  pure substance 純粋物質
  pure water 純水
purge 動～を除去する，～を追い出す 名パージ
purification 浄化，精製
purifier ピューリファイヤ，清浄機
purify 動清浄にする
purity 清浄，純正，純度
purpose 目的
push 動押す，突く
  名プッシュ，押し，突き

- push button 押しボタン
- push-down close 正栓
- push-down open 逆栓
- push fit 押込みばめ
- push out 動押し出す,排除する
- push rod プッシュロッド,押し棒,突き棒
- push up 動押し付ける
- push valve 押込み弁

put 動置く
- put away 動片づける
- put in 動取り付ける
- put on 動動かす,付ける,接続する
- put out 動消す,外す
- put together 動組み立てる,合計する

putty パテ【接合剤の一種】

- puzzle 動困らせる,当惑する

P-V (Pressure-Volume) diagram P-V線図,圧力-容積線図

PWM (Pulse Width Modulation) control PWM制御,パルス幅電圧制御

pycnometer ピクノメータ,比重びん【比重計の一種】

pyramid ピラミッド,角すい

pyro bearing 高温用軸受

pyroelectric material 焦電材料

pyrogenic reaction 高温反応

pyrolysis 熱分解

pyrometer 高温計

pyrophoric alloy 発火合金

pythagorean theorem ピタゴラスの定理

Q-q

QA (Quality Assurance) 品質保証
QC (Quality Control) 品質管理
quadrangle 四角形, 四辺形
quadrant コドラント, 象限, 四分円, 四分儀
quadrate 形正方形の, 四辺形の 名正方形
quadratic 形二次の, 正方形の 名二次式
 quadratic curve 二次曲線
 quadratic equation 二次方程式
 quadratic function 二次関数
 quadratic polynomial 二次多項式
 quadratic residue 平方剰余
 quadratic surface 二次曲面
quadrature 直角位相, 求積法
 quadrature component 直角成分
 quadrature formula 求積公式
 quadrature modulation 直角変調
 quadrature phase 直角位相
quadric 名二次曲面, 二次関数 形二次の, 二次式の
 quadric crank mechanism 四節回転機構
quadrilateral 名四辺形 形四辺形の
quadrillion 10の15乗
quadruple 形4倍の, 4重の
 quadruple expansion engine 四段膨張機関
 quadruple screw vessel 4軸船
quadrupole 4極, 四端子, 四極子
quake 動震える, 揺れる, 振動する 名震え, 揺れ, 地震
quakeproof 形耐震の, 耐震性のある
qualification 資格, 技能, 適性, 制限, 条件
 qualification test 認定試験
qualified 形正規の
qualify 動資格を与える, 制限する, 修飾する
qualitative 形性質上の, 質的な, 定性的な
 qualitative analysis 定性分析
quality クオリティ, 品質, 特性, 高級, 良質, 性質, 良否, 長所, 乾き度
 quality analysis 品質分析
 quality assessment 品質アセスメント, 品質評価
 Quality Assurance QA, 品質保証
 quality confirmation 品質確認
 Quality Control QC, 品質管理
 quality evidence 品質保証
 quality level 品質レベル, 品質基準
 quality specification 品質規格
 quality standard 品質基準
 quality verification 品質検証
quantification 定量化, 数量化
 quantification method 定量化法
 quantification theory 数量化理論
quantitative 形量の, 量的な
 quantitative analysis 定量分析
quantity 量, 数量, 分量
 関 a small quantity of ～ : 少量の～
 quantity control governing 流量加減調整
 quantity governing 分量調整
 quantity meter 定量計【流量計の一種】
 quantity of electric charge 電荷量
 quantity of electricity 電気量
 quantity of evaporation 蒸発量
 quantity of flow 流量
 quantity of fuel 燃料量
 quantity of heat 熱量
 quantity of light 光量
 quantity of magnetism 磁気量
 quantity of motion 運動量
 quantity of oil 油量
 quantity of state 状態量
 quantity of thermal flow 熱流量
quantization 数量化, 定量化, 量子化
quantize 動量子化する
quantum 量, 量子, 定量, 数量, 分量
 quantum electrodynamics 量子電

磁力学
　quantum mechanics　量子力学
　quantum number　量子数
　quantum of action　プランク定数
quarantine　検疫, 隔離
quark　クォーク【素粒子の構成粒子】
quarter　图4分の1, 居住区域
　動4等分する
　quarter grain　板目
　quarter master　操舵手
quartz　クォーツ, 石英, 水晶
　quartz clock　水晶時計
　　同 quartz chronometer
　quartz glass　石英ガラス
　quartz oscillator　水晶発振器
　quartz plate　水晶板
　quartz sand　珪砂
quasi-　「類似」「半～」「準～」の意を表す連結形
　quasi-equilibrium　準平衡
　quasi-peak value　準ピーク値
　quasi-static process　準静的過程
　quasi-stationary state　準定常状態
quaternary　形4要素から成る, 4の, 四元数の　图4要素, 四元数
quay　埠頭, 波止場, 岸壁
quench　图急冷, 焼入れ, 消火
　動急冷する, 焼入れする, 消す
　quench aging　焼入れ時効
　quench and temper　焼入れ・焼戻し
　quench annealing　急冷焼なまし
　quench crack　焼割れ【焼入れしたときに生じる割れ】
　quench distortion　焼入れひずみ, 焼入れ変形
　quench hardening　急冷硬化, 焼入れ硬化
quenching　クエンチング, 焼入れ, 急冷, 消光, 消火, 消滅
　quenching and tempering　焼入れ・焼戻し
　quenching area　冷却域
　quenching liquid　焼入れ液, クーラント
　quenching medium　焼入れ剤
　quenching oil　焼入れ油
　quenching rate　急冷速度
　quenching temperature　焼入れ温度, 急冷温度
　quenching treatment　焼入れ処理
question　图質問, 問題, 疑義
　動質問する
questionnaire　質問事項, アンケート
queue　キュー, 待ち行列
　queue control　待ち行列制御
queuing algorithm　待ち行列アルゴリズム
queuing theory　待ち行列理論
quick　形急速な, 短時間の
　quick acting relay　速動継電器
　quick charge　急速充電
　quick chilling　急冷
　quick disconnect coupling　急速継手
　quick freezing　急速冷凍, 急速凍結
　quick lime　生石灰
　quick response excitation　速応励起
　quick runner　高速度羽根車
　quick traverse　早送り
quicker　「quick」の比較級
quickly　副早く, 急に
quiescent　形静止した, 穏やかな
　quiescent point　静止点
　quiescent time　静止時間, 休止時間
quiet　形静かな, 穏やかな　副静かに, 穏やかに　图静けさ, 静寂　動静かにさせる
quieter　防音装置
quieting curve　緩衝曲線
quietly　副静かに, 平穏に, そっと
quill　クイル, 管, 中空の軸
　quill gear　クイル歯車
　quill shaft　クイルシャフト, たわみ軸, 中空シャフト
quilled　形管状の
quintet　5人組, 5個組, 5重奏
quintic　形五次の　图五次方程式
quintillion　10の18乗
quit　動やめる, 放棄する
quite　副全く, 完全に, 事実上, すっかり, かなり
quiz　图試験, クイズ
　動試験を行う, 詳しく質問する

quota 分け前, 割り当て
quotation 引用, 引用句, 引用文
  quotation mark　引用符【記号：
   " "】

quote 動引用する, 引用符で囲む
quotient 商, 指数, 比率

R-r

rabble 攪拌(かくはん)棒
race 名レース【軸受においてボールやころを挟む輪】動空転する
raceway 軌道, 水路, 導水路
  raceway diameter 軌道径
  raceway groove 軌道みぞ
  raceway surface 軌道面
  raceway track 軌道
racing レーシング, 空転, 急転, 乱調
rack ラック, 歯ざお, 歯板
  rack and pinion adjustment ラック＆ピニオン装置, 粗動装置
racking ユニオン式ラック
  racking stress 圧潰力
radar レーダ, 電波探知機
radial 形ラジアル, 放射状の, 半径の, 星形の, ふく射の
  radial acceleration 半径方向加速度
  radial ball bearing ラジアル玉軸受
  radial bearing ラジアル軸受
  radial blade clearance 翼の半径方向すきま
  radial clearance 半径方向すきま
  radial drilling machine ラジアルボール盤
  radial engine 星形機関
  radial equilibrium 半径方向つり合い
  radial factor ラジアル係数
  radial fan ラジアルファン, プレートファン, 径向き羽根
  radial feed 半径送り
  radial flow turbine 半径流タービン
  radial force ラジアル力, 半径方向力
  radial impeller 遠心羽根車, 半径方向羽根車
  radial internal clearance ラジアルすきま
  radial inward flow turbine 内向き半径流タービン
  radial load ラジアル荷重, 径方向荷重
  radial outward flow turbine 外向き半径流タービン
  radial piston pump ラジアルピストンポンプ, 星形ピストンポンプ
  radial play 径隙, 半径方向の遊び
  radial plunger pump ラジアルプランジャポンプ
  radial rotary engine 星形回転機関
  radial stress 半径方向応力
  radial thrust 半径方向スラスト
  radial valve gear ラジアル動弁装置
  radial vane 半径方向羽根, 放射羽根
  radial velocity 半径方向速度
radian ラジアン【角度の単位：rad.】
radiance 発光, 放射輝度
radiant 形放射の, 放射状の
  radiant boiler 放射ボイラ
  radiant efficiency 放射効率
  radiant energy 放射エネルギ, ふく射エネルギ
  radiant exitance 放射発散度
  radiant heat 放射熱, ふく射熱
  radiant heat transfer 放射伝熱
  radiant heating surface 放射伝熱面
  radiant intensity 放射強度, ふく射強度
  radiant pyrometer ふく射高温計
  radiant quantity 放射量
  radiant ray 放射線, ふく射線
  radiant superheater 放射過熱器
  radiant transmittance 放射透過率
radiate 動放射する, 発射する, 発散する
radiating pipe 放熱管
radiating surface 放射表面, 放熱面
radiation 放射, ふく射, 放射線
  radiation boiler 放射ボイラ【放射伝熱面のみのボイラ】
  radiation catalysis 放射[線]触媒作用
  radiation contact superheater 放射接触過熱器

radiation cooling　放射冷却
radiation damping　放射減衰
radiation decrement　放射減衰率
radiation efficiency　放射効率
radiation heat loss　放熱損失
radiation heating surface　放射伝熱面
radiation impedance　放射インピーダンス
radiation inspection　放射線探傷法【放射線を用いた材料内部欠陥の検知法】
radiation loss　放射損失
radiation pyrometer　放射高温計【放射エネルギによる温度上昇によって生じる熱起電力を測定し温度を計測する】
radiation resistance　放射抵抗
radiation superheater　放射過熱器
radiation thermometer　放射高温計, 放射温度計
radiation thickness gauge　放射線厚み計
radiative heat transfer　放射伝熱
radiator　ラジエータ, 放熱器, 冷却器, 放射器, 放射体
radical　ラジカル, 基, 遊離基, 根号
　radical polymerization　ラジカル重合
　radical reaction　ラジカル反応, 遊離基反応
radii　「radius」の複数形
radio　ラジオ, トランシーバ, 無線
　radio broadcast[ing]　ラジオ放送, 無線放送
　radio buoy　ラジオブイ
　radio chemical analysis　放射化学分析
　radio communication　無線通信
　radio control　ラジコン, 無線制御, 無線操縦
　RAdio Detecting And Ranging　レーダ, 電波探知機
　radio frequency　無線周波数, 高周波
　radio heating　高調波加熱
　radio isotope　放射性同位元素
　radio navigation　無線航法
　radio pliers　ラジオペンチ

radio relay system　無線中継方式
radio station　無線局
radio system　無線方式
radio telegram　無線電報
　㊤ radiogram
radio telegraphy　無線電信
radio telephone　無線電話
radio transmitter　無線送信機
radio wave　電波, ラジオ波
radio wave propagation　電波伝搬
radioactivation　放射化
radioactive　形 放射性の, 放射能のある
　radioactive atom　放射性原子
　radioactive constant　放射性崩壊定数
　radioactive contamination　放射能(性)汚染
　radioactive element　放射性元素
　radioactive equilibrium　放射平衡
　radioactive isotope　放射性同位元素
　radioactive ray　放射線
　radioactive substance　放射性物質
　　㊤ radioactive material
　radioactive thickness gauge　放射線厚み計
　radioactive tracer　放射性トレーサ
radiograph test　放射線透過検査
radiographic　形 X線撮影の, X線写真の
　radiographic test　放射線透過試験
radioisotope　放射性同位元素
radiological　形 放射性物質の, 放射線医学の
radiometer　ラジオメータ, 放射計
radiometry　放射測定
radium　ラジウム　㊤ Ra
radius　半径　㊀ diameter：直径
　radius gauge　ラジアスゲージ, 丸み測定用ゲージ
　radius of action　航続距離
　radius of convergence　収束半径
　radius of curvature　曲率半径
　radius of gyration　回転半径
　radius of gyration of area　断面二次半径
　radius vector　動径

raft いかだ
rag ウエス，ぼろきれ
  rag bolt　鬼ボルト
  rag wheel　鎖歯車，スプロケット
Rahmen ラーメン【骨組みの一種】
  rahmen structure　ラーメン構造【柱と梁で建物を支える構造】
rail 手すり，レール
  関 common rail：コモンレール【燃料を高圧で各インジェクタに供給する共通配管】
  rail clamp　レールクランプ【クレーンの安全装置】
raise 動上げる，揚げる，上昇させる　名増加
rake レーキ，傾斜
  rake angle　レーキ角，すくい角
  rake face　すくい面
  rake ratio　傾斜比
ram 名ラム【シリンダ内を往復する円柱】動衝突する，突き当てる
  ram balancer　ラムバランサ【ラム保持器】
  ram jet　ラムジェット【ジェットエンジンの一種】
ramification 分岐，支流，支脈
rammer 突き棒
ramp 斜道
  ramp function　ランプ関数
  ramp input　ランプ入力
  ramp response　ランプ応答【入力が一定速度で変化し続ける応答】
random 形でたらめの，無原則な，不規則な
  random error　偶然誤差
  random failure　偶発故障
  random number　乱数
  random signal　不規則信号
  random variable　確率変数
  random vibration　不規則振動
range レンジ，範囲，領域，限界
  range finder　測距機，距離計
  range index　距離指標
  range of combustion　燃焼範囲
  range of explosion　爆発範囲
  range resolution　距離分解能
  range scale　距離範囲
  range spring　レンジばね
rank 階数，ランク，階級
Rankine cycle ランキンサイクル【熱力学的サイクルの一種】
rap 名コツンとたたくこと　動コツンとたたく
rapid 形急速な，すばやい，瞬間の
  rapid access　高速アクセス
  rapid charge　急速充電
  rapid combustion　急激燃焼
  rapid cooling　急冷
  rapid cure adhesive　速乾接着剤
  rapid fall　急低下
  rapid flow　射流
  rapid solidification　急速凝固
  rapid start　急速起動
  rapid traverse　早送り
rapidity 急速，速度，高速性
rapidly 副すみやかに，迅速に
rapping bar 型抜き棒
rare 形まれな，珍しい，希薄な
  rare earth element　希土類元素
  rare gas　希ガス【アルゴン，ヘリウムなど】
  rare metal　レアメタル，希少金属
rasp 名荒やすり　動やすりをかける
  rasp cut file　鬼目やすり，石目やすり，わさび目やすり
ratch つめ，歯止め，つめ車
ratchet ラチェット，つめ歯，つめ歯車，つめ車
  ratchet and pawl　つめ歯と爪
  ratchet gearing　つめ歯装置
  ratchet handle　ラチェットハンドル
  ratchet jack　ラチェットジャッキ
  ratchet wheel　つめ車，一つ目車
rate 名割合，[比]率，速度，量，級　動見積もる，評価する，定格する
  関 at the rate of ～：～の割合で
  rate action　割合動作，レート動作
  rate control　比率制御
  rate conversion　速度変換
  rate of absorption　吸収率
  rate of burning　燃焼速度

rate of change　変化率
rate of charge　充電率
rate of climb　上昇率
rate of collection　回収率
rate of combustion　燃焼率, 燃焼速度
rate of conduction of heat　熱伝導率
rate of contact　接触比, かみ合い率
rate of creep　クリープ速度
rate of deposition　溶着率
rate of diffusion　拡散速度
rate of dilution　希釈度
rate of discharge　噴射率, 流量
rate of evaporation　蒸発率
rate of excess air　過剰空気率
rate of expansion　膨張率, 膨張比
rate of explosion　爆発率, 爆発比, 最高圧力比
rate of filtration　濾(ろ)過速度
rate of flow　流速, 流量
rate of fuel consumption　燃料消費率
rate of fuel injection　燃料噴射率, 噴射率
rate of gasification　ガス化率
rate of heat absorption　熱負荷
rate of heat release　熱発生率
rate of heat release in combustion chamber　燃焼室熱発生率
rate of heat transfer　伝熱速度
rate of heating　加熱速度
rate of operation　稼働率, 操業率
rate of outflow　流出流量
rate of pressure rise　圧力上昇率
rate of radiant heat transfer　放射伝熱速度
rate of reaction　反応速度, 反応率
rate of recovery　回収率
rate of sedimentation　沈殿速度
rate of settling　沈降速度
rate of shrinkage　収縮率
rate of spring　ばね定数
rate of temperature rise　温度上昇速度
rate of transformer　変圧比
rate of turnover　回転率
rate time　微分時間

Rateau turbine　ラトータービン【衝動タービンの一種】
rated　形 定格の
　rated capacitance　公称静電容量
　rated capacity　定格容量, 定格能力, 定格出力, 定格荷重
　rated continuous　定格連続
　rated current　定格電流
　rated dissipation　定格電力
　rated frequency　定格周波数
　rated horsepower　定格馬力, 定格出力
　rated impedance　定格インピーダンス
　rated insulation voltage　定格絶縁電圧
　rated load　定格負荷, 定格荷重
　rated load operation　定格負荷運転
　rated output　定格出力
　rated power　定格電力, 定格出力
　rated power consumption　定格消費電力
　rated power factor　定格力率
　rated pressure　定格圧力
　rated resistance　定格抵抗値, 公称抵抗値
　rated revolution　定格回転速度
　rated speed　定格速度, 定格回転数
　rated supply voltage　定格電源電圧
　rated temperature　定格温度
　rated torque　定格トルク
　rated value　定格値
　rated voltage　定格電圧
rather　副 かなり, むしろ, いくぶん
　rather A than B (A rather than B)　BよりむしろA
　rather than　〜よりむしろ, かえって
rating　レーティング, 定格, 定格出力, 評価, 率, 格付け, 等級
　rating life　定格寿命
　rating plate　銘板
ratio　比, 割合
　ratio control　比率制御
　ratio gear　比歯車
　ratio meter　比率計
　ratio of contact　接触率

ratio of current transformation　変流比
ratio of expansion　膨張比
ratio of specific heat　比熱比
ratio of the circumference of a circle to its diameter　円周率
ratio of transformation　変成比, 変圧比
rational　形有理の, 理性のある, 合理的な
rational function　有理関数
rational number　有理数
　㋝ irrational number：無理数
rationalization　合理化
rationalize　動合理化する, 有理化する, 正当化する
raw　形生(なま)の, 未加工の, 原材料の
raw data　未処理のデータ
raw material　原料, 素材
raw water　原水, 生水
ray　光線, 放射線, [円の]半径
ray velocity　電波速度, 光線速度
Rayleigh number　レイリー数【流体中における伝熱に関する無次元数】
rayon　レーヨン
reach　動到達する, 届く, 達する
react　動反応する
reactance　リアクタンス, 誘導抵抗【交流の抵抗】
reactance coupling　リアクタンス結合
reactance drop　リアクタンス降下
reactance modulator　リアクタンス変調器
reactance relay　リアクタンス継電器
reactance transistor　リアクタンストランジスタ
reactance voltage　リアクタンス電圧
reaction　反応, 反作用, 反力, 反動
　㋝ degree of reaction：反動度
reaction blade　反動羽根
reaction coil　再生コイル
reaction condenser　再生コンデンサ
reaction force　反力
reaction motor　反動電動機, 反作用電動機
reaction rate　反応速度
reaction stage　反動段
reaction torque　反動トルク
reaction turbine　反動タービン
　㋝ impulse turbine：衝動タービン
reaction velocity　反応速度
reaction zone　反応帯
reactive　形反動的な, 無効の
reactive component　無効成分
reactive cross current　無効横流
reactive current　無効電流
　㋙ active current：有効電流
reactive metal　活性金属
reactive power　無効電力
　㋙ active power：有効電力
reactive power compensation　無効電力補償
reactive power relay　無効電力継電器
reactivity　反応度, 反応性
reactor　リアクトル, 反応器, 原子炉
read　動読む, 示す, 校正する
readily　副たやすく, 直ちに
readiness　用意, 支度
reading　読み, 示度, 度数, 記録
readout　表示器, 表示装置, 計器, 読み出し
ready　形用意(準備)のできた, 迅速な
　㋝ be ready for：いつでも～する状態になっている
ready mixed paint　調合ペイント
reagent　試薬
real　形真の, 実際の, 実在する
real component　実数成分
real function　実関数
real gas　実在ガス
real image　実像
real number　実数
　㋝ imaginary number：虚数
real part　実部
　㋝ imaginary part：虚数部
real root　実根
real slip　真の失脚(スリップ)
real slip ratio　真のスリップ比
real time　実時間, 即時間

real time operation 実時間動作
real value expression 瞬時値表示
real value function 実変数関数
reality 現実, 実在, 真実
㈱ in reality:実は, 本当に
realization 実現, 現実化
realize 動十分に理解する, 実現する
really 副実際は, 本当に, とても
reamer リーマ【穴の内面を仕上げる切削工具】
reamer bolt リーマボルト【リーマ孔に通すボルト】
reaming リーマ仕上げ, リーマ通し【リーマを用いた仕上げ加工のこと】
reaming machine リーマ盤
rear 名後部, 後方, 背面, 背後
形後ろの
rear axle 後車軸
rear engine リアエンジン
rear hub 後ハブ
rear rim 後リム
rear spring 後ばね
rear view 背面図
rear wheel 後[車]輪
rearrangement 転移, 再配置
reason 理由
reasonable 形合理的な, ほどよい
reasonably 副合理的に, 無理なく, 適度に
reassemble 動再び組み立てる, 再び集める
reassembly 再組立
reattachment 再付着
Reaumur 形レ氏(列氏)目盛の
Reaumur thermometer 列氏温度計【水の沸騰点を80℃, 氷点を0℃とした温度計】
rebound 動はね返る 名反発, 回復
rebound leaf 押えばね板
rebound test 反発弾性試験
rebuild 動組み立てる, 再建する
recede 動後退する, 減退する
receive 動受ける, 受け取る, こうむる, 受信する
received power 受信電力

receiver レシーバ, 受信器, 受液器, 容器, 受器, 受け, 〜溜め, 受話器
receiver relay 受信継電器
recent 形新しい, 近頃の, 最近の
recently 副近頃, 最近
receptacle ソケット, コンセント
㊀ socket
reception 受信
receptor レセプタ, 受容体
recess 名くぼみ, みぞ
動くぼんだところに置く, くぼんだところを作る
recharge 動充放電する
rechargeble battery 充電式電池
reciprocal 形逆数の, 相互の, 相反する, 逆方向の
reciprocal vector 逆ベクトル
reciprocate 動交換する, 往復運動する ㈱ rotate:回転する
reciprocating 名往復動
形往復運動をする
reciprocating air compressor 往復空気圧縮機
reciprocating compressor 往復圧縮機
reciprocating engine レシプロエンジン, 往復機関
reciprocating motion 往復運動
㈱ rotary motion:回転運動
reciprocating pump 往復ポンプ
reciprocator 往復機関
reciprocity 相互作用
recirculate 動再循環する・させる
recirculation 再循環
recirculation flow control 再循環流量制御
recirculation pump 再循環ポンプ
recirculator 再循環器
reckoning 船位(の算出), 勘定, 見込み
reclaim 動取り戻す, 再生する, 改良する, 再利用する
reclaimed oil 再生油
reclaiming 再生
recline 動もたれかかる
reclosing relay 再閉路継電器

recognize 動識別する，認める
recoil 動はね返る，反動する 名はね返り，反動
recombination 再結合，組換え
　recombination coefficient　再結合係数
　recombination repair　再結合修復，組換え修理
　recombination velocity　再結合速度
recombiner 再結合器
recommend 動推奨する
recommendation 推奨，勧告
recompress 動再圧縮する
recondition 動修理する，修復する，回復させる
reconnect 動再接続する
reconstruct 動再建する，再構成する
reconstruction 再生，再構成
record 動記録する，表示する 名記録，証拠
recorder レコーダ，記録員，記録計，受信器，録音（録画）機
recording 名記録，録音，録画 形記録する，録音する，録画する
　recording instrument　記録計器　＝ recording meter
　recording media　記録媒体
　recording secretary　記録係　＝ recordist
　recording spot　画素
　recording thermometer　記録温度計
recover 動取り戻す，回復する
recovery 回収，回復，復旧，再生
　recovery factor　回復率
　recovery time　回復時間
recrystallization 再結晶
　recrystallization temperature　再結晶温度
recrystallize 動再結晶する・させる
recrystallized structure 再結晶組織
rectangle 長方形，矩形
rectangular 形長方形の，矩形の，直角の，直交する
　rectangular coordinates　直交座標
　rectangular duct　矩形管
　rectangular fin　矩形フィン
　rectangular hyperbola　直角双曲線
　rectangular section　長方形断面
　rectangular solid　直方体
　rectangular triangle　直角三角形
　rectangular wave　矩形波
rectification 整流，精留
rectifier 整流器，順変換装置
　rectifier action　整流作用
　rectifier circuit　整流回路
　rectifier diode　整流ダイオード
rectify 動整流する，修正する
rectifying action 整流作用
rectifying column 精留塔
rectilinear 形直線の，直線で囲まれた，直進する
　rectilinear guide　直動案内
　rectilinear motion　直線運動
recuperative heat exchanger 伝熱式熱交換器
recuperator 回収熱交換器，復熱装置
recurrence 再発，再起，再現，回帰
　recurrence relation　漸化式
recurrent 形再発する，回帰的な，周期的に起こる
　recurrent orbit　回帰軌道
recurring 形循環する，繰り返し発生する
　recurring decimal　循環小数
recursion 回帰，帰納
recursive 形帰納的な，再帰的な
　recursive function　帰納的関数
recut 動再切断する
recycle 名再循環，リサイクル 動再循環させる，再生利用する
recycling 再生利用
　recycling technology　リサイクル技術
red 形赤い，赤色の
　red globe lamp　紅球灯
　red heat　赤熱
　red hot shortness　赤熱高温脆（もろ）さ
　red lead　鉛丹，光明丹，赤鉛
　red lead paint　鉛丹ペイント
　red light　紅灯
　red oxide　べんがら【さび止め塗料】

red shortness　赤熱脆(もろ)さ
red side light　紅色げん灯
red stain　赤さび
redox　レドックス，酸化還元
redrawing　再絞り
redress　動仕上げを見直す
reduce　動減らす，軽減する，下げる，縮小する，還元する
reduced　形減じた，縮小した，不完全な，還元した，誘導の
　reduced capacity　性能低下
　reduced frequency　換算振動数
　reduced insulation　低減絶縁
　reduced iron　還元鉄
　reduced mass　換算質量
　reduced modulus　相当弾性係数
　reduced power　出力低下
　reduced pressure　減圧
　reduced pressure distillation　減圧蒸留
　reduced speed　減速
　reduced temperature　換算温度
reducer　径違い継手
reducibility　希釈性
reducing　還元，減ずること，縮小すること
　reducing agent　還元剤
　reducing coupling　径違い継手
　reducing cross　径違い十字
　reducing elbow　径違いエルボ
　reducing flame　還元炎
　reducing gauge　減圧計
　reducing harmonic current　高周波低減
　reducing joint　径違い継手
　reducing socket　径違いソケット
　reducing speed　減速
　reducing valve　減圧弁
reduction　減少，還元，縮小，絞り，減速　関 oxidation：酸化
　reduction flame　還元炎
　reduction gear　減速歯車，減速装置
　reduction gearing　減速機構
　reduction of area　絞り
　reduction ratio　減速比，縮小率
redundancy　余分，過剰，冗長性

　redundancy check　冗長検査【連続した情報の誤りを検出する検査のこと】
redundant　形余分な，過剰な，冗長な
　redundant system　冗長系
Redwood　レッドウッド【油の動粘度の単位】
　Redwood visco[si]meter　レッドウッド粘度計
　Redwood viscosity　レッドウッド粘度
reed relay　リードリレー
reed switch　リードスイッチ【近接スイッチの一種】
reed valve　リード弁【吸込み弁の一種】
reeding　細いたて溝
reel　リール，糸車，[機械の]回転部
reemphasize　動再強調する
reentrant　形内曲した，凹角の　名凹部
　reentrant angle　凹角
re-establish　動復旧(元)する
re-evaporation　再蒸発
re-expansion　再膨張　同 reinflation
refer　動参照する，呼ぶ，適用する，関係する，言及する，調べる，言う，引用する
　refer to ～ as　動 ～を…と呼ぶ
reference　参考，参照，基準，照合，関連，指示，参考文献
　reference condition　標準状態
　reference frame　座標系
　reference frequency　基準周波数
　reference fuel　標準燃料
　reference gauge　検定ゲージ
　reference input　基準値，基準入力
　reference junction　基準接点【熱電対の2接点のうち，基準温度に保たれている接点】
　reference level　基準レベル
　reference line　基準線
　reference number　部品番号，照合番号
　reference plane　基準面

reference point　基準点
reference signal　基準信号, 参考信号
reference standard　標準ゲージ, 基準ゲージ
reference surface　基準面
reference temperature　基準温度
reference vector　基準ベクトル
reference voltage　基準電圧
referential input　基準入力
refill　動再び満たす, 補充する
refine　動精製する, 精錬する
refined　形精練された, 微細な
refined oil　精製油
refinery　製油所
refit　動取り付け直す, 修理する
reflect　動反射する, 反響する
reflectance [ratio]　反射率
reflected ray　反射光線
reflected wave　反射波
reflecting microscope　反射顕微鏡
reflecting plate　反射板
reflection　反射
　関 incidence：入射
　reflection angle　反射角
　reflection coefficient　反射係数
　reflection factor　反射率
reflective　形反射する, 反省する
reflectivity　反射率, 反射力, 反射性
reflectometer　反射率計
reflector　反射鏡, 反射器, 反射材, 反射体
reflex　名反射　形反射的な　動反転させる, そらせる
reflux　還流, 逆流
　reflux condenser　還流凝縮器
　reflux ratio　還流比
reform　動改良する, 改善する
reformate　改質油
reformation　再構成, 再編成
reformed gas　改質ガス
refract　動屈折する・させる
refracted ray　屈折光線
refracted wave　屈折波
refracting telescope　屈折望遠鏡
refraction　屈折, 透過

refraction error　屈折誤差
refractive angle　屈折角
refractive index　屈折率
refractometer　屈折計
refractometry　屈折判定法
refractor　屈折レンズ
refractoriness　耐火性, 耐火度
refractory　形耐火性の, 耐熱性の　名耐火物
　refractory alloy　耐火合金
　refractory body　耐火物
　refractory brick　耐火れんが
　refractory material　耐火材
　refractory mortar　耐火モルタル
refrangibility　屈折性
refresh　動回復する, 新たにする, 補給する　名リフレッシュ, 再生
　refresh velocity　回復速度, 再生速度
refreshable　形再生可能な
refrigerant　名冷媒, 寒剤, 冷却剤　形冷却の
　refrigerant compressor　冷媒圧縮機
　refrigerant gas　冷媒ガス
　refrigerant liquid　液冷媒
　refrigerant mixture　混合冷媒
refrigerate　動冷却する, 冷蔵(冷凍)する
refrigerated carrier　冷凍船
refrigerating　冷蔵
　refrigerating agent　冷媒, 冷凍剤
　refrigerating capacity　冷凍能力
　refrigerating chamber　冷蔵室, 冷凍室
　refrigerating cycle　冷凍サイクル
　refrigerating effect　冷凍効果
　refrigerating machine　冷凍機
　refrigerating machine oil　冷凍機油
　refrigerating plant　冷凍装置, 冷凍・冷蔵装置
　refrigerating ton　冷凍トン
refrigeration　冷凍
　refrigeration cycle　冷凍サイクル
　refrigeration cycle of two stage compression　二段圧縮冷凍サイクル
　refrigeration system　冷却(凍)装置

refrigeration ton 冷凍トン
refrigerator 冷凍機, 冷蔵庫
refuse 名くず(屑), ごみ, 廃棄物 動断る, 拒絶する
regain 動回復する, 取り戻す
regard 動～と考える, ～とみなす, ～に注意を払う 名注意, 関心, 尊敬 関 as regards：～に関しては／without regard for (to)：～に関係なく, ～を無視して
regarding ～ 前～に関しては, ～の点では
regardless of ～ 前～にも関わらず, ～に関係なく
regenerate 動更生する, 再生する
regeneration 再生, 回収, 復興
regenerative 形再生させる, 蓄熱式の, 改造する
  regenerative air heater 再生式空気加熱器
  regenerative air preheater 再生式空気予熱器
  regenerative and reheat cycle 再生再熱サイクル
  regenerative apparatus 再生装置
  regenerative brake 回生ブレーキ
  regenerative brake equipment 回生制動装置
  regenerative braking 回生制動
  regenerative condenser 再生コンデンサ
  regenerative control 回生制御
  regenerative cycle 再生サイクル, 回生サイクル
  regenerative furnace 蓄熱炉
  regenerative heat exchanger 蓄熱式熱交換器
  regenerative pump 再生ポンプ
  regenerative reheating cycle 再生再熱サイクル
  regenerative signal 再生信号
  regenerative turbine 再生タービン
 regenerator 再熱器, 再生器, 熱交換器, 蓄熱器
region 領域, 区間, 区画, 範囲, 分野
register 名レジスタ, 記録, 記録器, 記憶器 動示す, 表示する, 記録する
  register control 蓄積制御方式, 間接制御方式
registered horsepower 登録馬力, 公称馬力
registration 記載, 登録, [画像の]重ね合わせ
regress 名後退, 復帰, 退行 動後戻りする, 後退する
regression 後戻り, 回帰
  regression analysis 回帰分析
  regression curve 回帰曲線
regular 形通常の, 標準的な, 規則正しい, 定期的な, 正式の, 一定の, 等辺等角の
  regular body 正面体
  regular deviation 定常偏差
  regular error 規則的誤差
  regular inspection 定期検査
  regular interval 一定間隔
  regular polygon 正多角形
  regular polyhedron 正多面体
  regular reflectance 正反射率
  regular transmission 正透過
  regular transmittance 正透過率
  regular valve 正弁
regularly 副規則正しく
regulate 動調整する, 規定する
regulating 調整
  regulating apparatus 調整器
  regulating capacity 調整容量
  regulating ring 調整リング, 調整輪
  regulating rod 調節棒
  regulating switch 調整開閉器
  regulating transformer 負荷時電圧調整器
  regulating valve 調整弁, 調節弁
regulation 規則, 規制, 調整
  regulation pressure 定格圧力
regulator レギュレータ, 調整(節)器, 減圧弁
reheat 動再加熱する 名再熱, 再燃焼
  reheat chamber 再熱器
  reheat cracking 再熱割れ

reheat cycle 再熱サイクル
reheat factor 再熱係数
reheat turbine 再熱タービン
reheater 再熱器, 再燃焼装置
reheating cycle 再熱サイクル
reheating regenerative cycle 再熱再生サイクル
reheating turbine 再熱タービン
reinforce 〔動〕補強する
reinforced 〔形〕強化した, 補強した
reinforced concrete 鉄筋コンクリート
reinforced plastics 強化プラスチック
reinforcement 強化, 補強, 補強材
reinsert 〔動〕再び差し込む
reinstate 〔動〕元に戻す, 復帰する
reject 〔動〕捨てる, 拒絶する, 退ける
rejected material 不合格材料
rejection 排除, 拒否, 除去, 不合格
relate 〔動〕関係する, 関連する
〔関〕relating to：～に関して
relation 関係, 関連, 比較式
〔関〕in relation to：～に関しては, ～については
relational 〔形〕関係の
relationship 関係, 関連
relative 〔形〕関連した, 相対的な
〔関〕relative to ～：～に比べて
relative coordinate 相対座標
relative cylinder charge 充填比
relative density 比重, 相対密度
relative displacement 相対移動, 相対変位
relative entropy 相対エントロピ
relative error 相対誤差, 誤差率
relative exit angle 相対流出角
Relative Humidity RH, 相対湿度
relative measurement 比較測定
Relative Metabolic Rate RMR, エネルギ代謝率
relative motion 相対運動
relative permeability 比透磁率【真空の透磁率に対するある物質の透磁率の比】
relative permittivity 比誘電率【真空の誘電率に対するある物質の誘電率の比】
relative position 相対位置
relative pressure 相対圧力
relative rotative efficiency プロペラ効率比
relative speed (velocity) 相対速度
relative temperature 相対温度
relative value 相対値
relatively 〔副〕相対的に, 比較的
relativity theory 相対性理論
relax 〔動〕緩める, 解除する
relaxation 減衰, 緩和, 弛緩
relaxation oscillation 緩和振動
relaxation time 緩和時間
relay 〔名〕リレー, 継電器 〔動〕中継する
relay contact 継電器接点
relay governor 間接調速機
relay operated control 他力制御
relay station 中継局
relay valve リレー弁
release 〔動〕解放する, 解除する, 放出する 〔名〕解放, 解除, 放出
release of drawing 出図
relevant 〔形〕関連がある, 適切な
reliability 信頼性, 信頼度
reliability decline 信頼性低下
reliability engineering 信頼性工学
reliability provision 信頼度規定
reliability standard 信頼度標準
reliable 〔形〕信頼できる, 頼りになる
relief 除去, 軽減, 逃げ, 救援
relief angle 逃げ角
relief of residual stress 残留応力除去
relief ring frame 調圧環
relief valve リリーフ弁【主として圧力を一定に保つために作動】, 逃し弁, 安全弁
relieve 〔動〕救助する, 軽減する, 逃がす, 解職する
relieving レリービング, リリーフ加工, 二番取り
relocatable 〔形〕再配置可能な
relocate 〔動〕再配置する, 配置し直す
relocation 再配置

reluctance リラクタンス，磁気抵抗
 reluctance motor　リラクタンスモータ，反作用電動機
 reluctance torque　リラクタンストルク
reluctivity　磁気抵抗率
rely　[動]頼る，当てにする
 rely on (upon) ~　[動]~を信頼する
remain　[動]残る，~のままでいる，~し続ける　[名]残り，残りもの
remained strain　残留ひずみ
remainder　残り，余り
 remainder theorem　剰余定理
remaining　[形]残っている，残りの
 remaining air　残留空気
 remaining fuel　残油
 remaining heat　余熱
 remaining portion　未使用部分
 remaining stress　残留応力
remanence　残留磁気
remanent　[形]残された，残留する
 remanent magnetic flux　残留磁束
 remanent magnetization　残留磁気，残留磁化
remark　[動]述べる，感想を言う
remarkable　[形]著しい，目立った
remedy　[形]改善，矯正　[動]直す，元に戻す
remember　[動]思い出す，覚えている，記憶する・させる
remind　[動]思い出させる，注意する
remote　[形]遠く離れた，遠方の
 remote batch processing　遠隔一括処理
 remote control　リモコン，遠隔制御，遠隔操作
 remote control circuit　遠隔制御回路
 remote control device　遠隔制御装置
 remote indication　遠隔指示
 remote measuring　遠隔測定
 remote sensing　遠隔探査，間接測定，リモートセンシング
 remote supervisory equipment　遠隔表示装置
 remote switch　リモートスイッチ，遠隔スイッチ
 remote terminal　リモートターミナル【端末装置】
 remote water level gauge　遠隔水面計
remotely　[副]遠隔で，遠く離れて
 remotely operated valve　遠隔操作弁
removal　除去，移動，移転
remove　[動]除去する，取り外す，取り除く，移動する，分解する
removing device　取外し用具
rend　[動]裂く，分裂させる
render　[動]状態にする，提供する，表現する
renew　[動]新替えする，一新する
renewable　[形]再生可能な，更新できる，回復できる
 renewable energy　再生可能エネルギ
renewal　新替え，再生，更新
renovate　[動]刷新する，修復する，活気づける
rent　[名]裂け目　[動]「rend」の過去・過去分詞
repair　[名]修理，修繕　[動]修理する
 repair expense　修繕費
 repair work　修繕工事，改修工事，補修工事
repay　[動]払い戻す，返金する
repeat　[動]反復する，繰り返す　[名]リピート，繰返し
 repeat count　繰返し回数
repeatability　反復性，繰返し性，繰返し精度
repeated　[形]繰り返された，たびたびの
 repeated load　繰返し荷重，片振り荷重
 repeated stress　繰返し応力
repeater　中継器，リピータ
repeating coil　中継コイル
repeating decimal　循環小数
repeating installation　中継装置
repetition　繰返し，反復，再現
repetitive　[形]繰り返される，反復性の
 repetitive control　繰返し制御
replace　[動]交換する，取り替える，代わる

replaceable

replace ～ with… 動～を…に置き換える
replaceable 形交換可能な
replacement 交換, 置換, 更新, 取替
replacer リプレッサ【脱着用具のこと】
replenish 動補充する, 補給する, 満たす
replenishment oil plan 補油計画
reply 動応答する, 答える 名応答, 回答, 返事
report 名リポート, 報告, 記録, 報告書 動報告する, 伝える, 記録する, 出頭する
represent 動意味する, 相当する, 描く, 表現する, 表す, 代表する
representation 表示, 表現, 代表
representative 名代表者 形代表的な, 典型的な
repress 動抑制する, 抑圧する, 再圧縮する
reprocessing 再処理
reproduce 動再生させる, 再現する, 複写する
reproducibility 再現性, 再現精度
reproduction 再生, 再現, 複写
repulsion 反発, 斥力, 反発性, 反発力 反 attraction:引力, 吸引力
　repulsion motor 反発電動機
repulsive 形反発する
　repulsive force 斥力
request 動頼む, 依頼する
require 動要求する, 必要とする
required 形必要な, 不可欠な, 必須の
　required horsepower 所要馬力
　required NPSH 必要有効吸込ヘッド
requirement 要求, 必要, 必要条件, 資格, 要件
rerun 動再実行する 名再実行
　rerun time 再実行時間
rescue 動救助する 名救助, 救出
　rescue boat 救命艇
research 名調査, 研究 動調査する, 研究する
　Research and Development R&D, 研究開発
　research boat (ship) 調査船, 海洋観測船
　research engine 試験研究用エンジン
　research octane number リサーチ法オクタン価【ガソリン機関のアンチノック性の指標】
reseat 動再び座らせる, 座部を取り替える
reseating pressure リシート圧力, 吹止まり圧力
resell 動再び売る, 転売する
reserve 動保存する, 保有する, 予約する 名保存, 予備, 保留
　reserve capacity 予備容量
　reserve feed water tank 予備給水タンク
　reserve oil bunker 予備燃料油タンク
　reserve [oil] tank 予備油槽, 貯蔵タンク
reserved word 予約語
reservoir レザーバ, (液体を入れる)容器, 油タンク, 貯蔵所, 貯蔵器, 貯水池
　reservoir tank リザーバタンク, 備蓄タンク
reset 動[計器などを]初期状態に戻す, リセットする, 復元する 名リセット
　reset action リセット作用, リセット動作
　reset time 復帰時間, リセット時間, 積分時間
residual 形残留の, 残りの 名残余, 残留, 残留物, 残渣
　residual capacity 残留容量
　residual carbon 残留炭素
　residual current 残留電流
　residual error 残差【一群の測定値とその平均値との差】
　residual flux 残留磁束
　residual fuel 残さ燃料, 重油
　residual gas 残留ガス

residual heat 余熱，残留熱
residual magnetism 残留磁気
residual oil 残油
residual pressure 残圧
residual resistance 残留抵抗
residual standard deviation 誤差標準偏差
residual strain 残留ひずみ
residual strain by machining 加工ひずみ
residual strength 残存強度
residual stress 残留応力
residual vibration 残留振動
residual voltage 残留電圧
residue 残さ，残留物，燃えがら，留数
residuum 残油，残留物
resilience 弾力性，弾性エネルギ
resilient 形弾力のある，回復力のある
resin レジン，樹脂
resinous 形樹脂の，樹脂製の
resintering 再焼結
resist 動抵抗する，耐える 名防食剤，絶縁塗料
resistance 抵抗，抵抗力，耐性
resistance bulb thermomcter 測温抵抗温度計
resistance curve 抵抗曲線
resistance drop 抵抗降下
resistance force 抵抗力
resistance heating 抵抗加熱
resistance meter 抵抗計
resistance pyrometer 抵抗高温計
resistance test 抵抗試験
resistance thermometer 抵抗温度計
resistance thermometer bulb 測温抵抗体
resistance value 抵抗値
resistance winding 抵抗巻線
resistant 形抵抗する，抵抗力のある，耐性のある，〜に耐える，耐〜の
関 abrasion resistant rubber：耐摩耗性ゴム／oil resistant rubber：耐油性ゴム

resisting moment 抵抗モーメント
resistivity 抵抗率，固有抵抗
resistor レジスタ，抵抗，抵抗器
resistor bulb 抵抗測温体
resizable 形サイズ変更可能な
resolution 分解，分解能，解像度，融解
resolution threshold 感度限界
resolve 動解決する，決定する，分解する 名決心，決意
resolver レゾルバ【角度検出用モータの一種】
resolving 分解
resolving power 分解能，解像力
resolving time 分解時間
resonance 共振，共鳴
resonance absorption 共鳴吸収
resonance amplitude 共振振幅
resonance characteristic 共振特性
resonance circuit 共振回路
resonance current 共振電流
resonance energy 共鳴エネルギ
resonance frequency 共振周波数，共振振動数
resonance phenomena 共振現象
resonance point 共振点
resonance tachometer 同調回転計
resonance voltage 共振電圧
resonant 形反響する，共鳴を起こす
resonant cavity 共振空洞
resonant coupling 共鳴結合
resonator 共振器，共鳴器
resource 資源，資産，財産
respect 動尊敬する，尊重する 名敬意，尊敬
関 in respect of 〜：〜に関しては／with respect to：〜に関して
respectable 形尊敬できる，りっぱな
respective 形それぞれの，各自の
respectively 副それぞれ，各自に，おのおの，別々に
respiration 呼吸作用
respirator レスピレータ，呼吸用マスク，呼吸用保護具，呼吸装置，防毒マスク
respire 動吸収する

respond 動応答する
response 反応，応答
　関 in response to：〜に反応（対応）して
　　response characteristic　応答特性
　　response curve　応答曲線
　　response time　応答時間
responsibility 責任，応答性
responsible 形責任のある
responsivity 応答度
rest 名休止，休息，静止，残り　動休止する，静止する，〜に基づく，〜に置く　関 at rest：静止して
　　rest mass　静止質量
　　rest on 〜　動〜に基づく，〜に頼る
restart 動再出発させる　名再出発，再始動
　　restart condition　再始動条件
　　restart point　再始動点
restitution 反発，復元
restoration 復旧，回復
　　restoration operation　復旧操作
　　restoration state　復旧状態
restorative 形復旧する，回復させる
restore 動もとに戻す，復旧させる
restoring force 復原力，復元力
restoring moment 復原モーメント
restrain 動抑制する，制限する
restrained packing ring 制限リング
restrict 動限定する，制限する
restricted 形制限した
restriction 制限，限定，絞り
　　restriction flowmeter　絞り流量計
　　restriction of mapping　制限写像
restrictor リストリクタ，絞り，絞り弁【流れを制限する装置】
result 名結果，成績　動結果として生じる，起因する，という結果になる，帰着する
　関 as a result：結果として，結局
　　result from 〜　動〜に原因する，〜に起因する，〜の結果として生ずる
　　result in 〜　動〜に終わる，〜に帰着する，〜という結果になる
resultant 形合成した，結果として生じる　名合力
　　resultant error　合成誤差
　　resultant force　合成力，合力
　　resultant stress　合応力
　　resultant wave　合成波形
resume 動再開する，戻る，取り戻す，要約する　名レジュメ，要約，概要，履歴書
　　resume function　レジューム機能
retail 名小売り　形小売りの　動小売りする
retain 動〜を保つ，保持する
retainer 保持器，リテーナ
　　retainer lock　割環
retaining circuit 保持回路
retaining nut 止めナット
retaining ring 保持リング，止め輪
retard 動遅らせる，妨害する
retardation 減速度，遅延，阻止，妨害
　　retardation coil　塞流コイル
　　retardation method　減速法
　　retardation time　遅延時間
retarded ignition 遅れ点火
retarded motion 減速運動
retarding force 減速力
retention 貯留，保留
retentivity 保持力，残磁性
retire 動退く，引退する，引き下がる
retort レトルト【蒸留や乾留に用いる実験器具／食品を加圧・加熱・殺菌する装置】
　　retort carbon　レトルトカーボン【電池，電極の原料】
　　retort pouch food　レトルト食品
retract 動引っ込む，収縮させる，引っ込ませる
retreat 名返却，退却，後退　動返却する，後退する
retreater リトリータ【温度計の一種】
retrieval 回復
retrieve 動取り返す，回収する，回復する，救出する，償う，訂正する
retrieving head リトリービングヘッド【連結装置】

retroaction 逆動
retrofit 图改造, 改装
  働改造する, 改装する
retry 働再び試みる, 再試験する
return 働戻る, 返却する
  图リターン, 戻り, 帰還, 復帰
  形帰りの, 元に戻る, 戻りの
  return bend　返しベンド, リターンベンド
  return flow burner　リターンフローバーナ
  return gear　追従装置
  return line　戻り管路
  return mechanism　復帰装置
  return oil pipe　戻り油管
  return oil pressure　戻り油圧力
  return stroke　戻り行程
  return type burner　還流式バーナ
  return value　戻り値【関数を評価したときに返される値のこと】
reunion 再結合
reusability 再利用性
rev 回転(動)回転速度を上げる
rev/min rpm, 毎分回転数
reveal 働現れる, 見せる, 暴露する
reverberator 反射器, 反射鏡, 反射灯, 反射炉
reverberatory 形反射する, 反響する
  reverberatory furnace　反射炉
reversal 反転, 逆転
  reversal scavenge　反転掃気
reverse 图逆転, 反対　形逆の
  働入れ替える, 逆にする, 反対にする, 逆転させる, 転換する
  reverse and reduction gear　逆転減速装置
  reverse current　逆電流, 逆流
  reverse current circuit breaker　逆流遮断器
  reverse current protection　逆電流保護(装置)
  reverse current relay　逆流継電器
  reverse drawing　逆絞り
  reverse engineering　分解工学
  reverse feedback　負帰還
  reverse flow　逆流
  reverse gear　逆転歯車, 後退歯車
  reverse osmotic membrane　逆浸透膜
  reverse power　逆電力
  reverse power protection　逆電力保護
  Reverse Power Relay　RPR, 逆電力継電器
  reverse response　逆応答
  reverse voltage　逆電圧
reversed 形逆にした, 逆転した, 反対の, 裏返しの, 左巻きの
  reversed phase　逆相
  reversed stress　両振り応力
reversibility 可逆性, 取り消し可能性
reversible 形可逆の, 逆(反対)にできる, もとの状態に戻れる
  reversible adiabatic change　可逆断熱変化
  reversible booster　可逆ブースタ
  reversible cell　可逆電池
  reversible change　可逆変化
  reversible converter　可逆変換装置
  reversible cycle　可逆サイクル
  reversible engine　可逆機関
  reversible motor　可逆電動機
  reversible pitch propeller　可逆ピッチプロペラ
  reversible process　可逆プロセス, 可逆過程
  reversible reaction　可逆反応
reversing 逆転
  reversing clutch　反転クラッチ
  reversing cycle engine　可逆サイクルエンジン
  reversing device　逆転装置
  reversing engine　逆転機関
  reversing gear　逆転装置, 逆転機
  reversing handle　逆転ハンドル
  reversing lever　逆転レバー, 逆転てこ
  reversing relay　リバーシングリレー
  reversing shaft　逆転軸
  reversing starter　可逆始動器
  reversing test　逆転試験
  reversing turbine　後進タービン
  reversing valve　逆転弁
review 图再調査, 再検討, 批判,

査読 動再調査する，批評する，査読する，復習する
revolute 形外巻きの，弓なりになった
　revolute joint　回転関節
　revolute pair　回転対偶
　revolute robot　多関節ロボット
revolution 回転，旋回
　revolution counter　積算回転計
　revolution indicator　回転計，回転表示器
　Revolution Per Minute　RPM, rpm, 毎分回転数
revolve 動回転する，循環する
revolved sectional view 回転断面図
revolver 回転装置
revolving 形回転する，旋回する，回転装置の 名回転
　revolving armature type　回転電機子形
　revolving field　回転界磁
　revolving field type motor　回転界磁形モータ
　revolving magnetic field　回転磁界
　revolving rotor　回転ロータ
　revolving speed　回転速度
revving 回転，「riv」の現在分詞 関 sound of revving engines：エンジンの回転音
rewire 配線し直す
rewrite 書き直す
Reynolds' law of similarity レイノルズの相似則
Reynolds number レイノルズ数
Reynolds stress レイノルズ応力
rhenium レニウム 記 Re
rheology 流動学，レオロジ【物質の変形および流動一般に関する学問分野のこと】
rheometer レオメータ，電流計，検流計，流量計
rheostat 加減（可変）抵抗器，摺動抵抗器
rheostatic braking 発電ブレーキ，抵抗制動
rheostatic control 抵抗制御

rhodium ロジウム 記 Rh
rhombic 形ひし形の，斜方形の
　rhombic distortion　ひし形ひずみ
rhythm リズム，調子，周期，調律
rib 名リブ，肋骨，肋材，力骨，つば 動〜に肋骨（肋材）をつける
ribbed tube リブド管【つば形または管の内面にらせん溝のある管】
ribbon microphone リボンマイクロフォン【マイクロフォンの一種】
rich 形金持ちの，豊富な，濃厚な
　rich mixture　濃厚ガス
rickety 形ぐらついた，倒れそうな
rid 動取り除く，自由にする
　関 get rid of 〜：〜を取り除く
riddle 動〜にふるいをかける，しみ込む
　riddle condition　浸水状態
rider ライダ【調整用の分銅】
ridge うね，筋，隆起部，尾根
　ridge on the piston ring　ピストンリングのカラー
ridging リッジング【歯車表面に生じる損傷の一種】
riffler 波形やすり
rig 名三脚，索具，装備，用具，リグ 動〜に[索具を]装備する，設置する
right 名右，正当 形右側の，正しい，正当な 副正しく，完全に，まったく，すぐに 動正常にする，まっすぐになる，矯正する
　right angle　直角
　right [angled] triangle　直角三角形
　right circular cone　直円すい
　right circular cylinder　直円柱
　right couple　復原偶力
　right hand rule　右手の法則
　right handed propeller　右回りプロペラ
　right handed screw rule　右ねじの法則
　right now　まさに今
　right projection　直角投影
righting arm (lever)　復原てこ
righting couple　復原偶力

righting moment 復原力, 復原モーメント
rigid 形硬直した, 堅固な, 固定式の
　rigid bearing　固定軸受
　rigid body　剛体
　rigid coupling　固定継手
　rigid flanged coupling　固定フランジ継手
　rigid frame　ラーメン【鉄骨構造の一種】
　rigid inclusion　剛体介在物
　rigid joint　剛節
　rigid motion　剛体運動
　rigid plastic　硬質プラスチック
　rigid rotor　剛性ロータ
　rigid shaft　剛性軸
　rigid shaft coupling　固定軸継手
　rigid type　剛型
rigidity　剛さ, 剛性
rigorous　形厳密な, 精密な, 正確な
rim　名リム, 縁
　動～にへり(縁)をつける
rimmed steel　リムド鋼【Mnで脱酸した鋼】
ring　名リング, 輪, 環　動鳴る・らす, 鳴り響く, ～らしく聞こえる
　ring analysis　環分析
　ring back tone　呼出し音
　ring bolt　リングボルト, 環付きボルト, つりボルト
　ring bulkhead　環状隔壁
　ring bus　環状母線
　ring connection　環状結線, 環状接続
　ring current　環状電流
　ring end gap　リング合口すきま
　ring expander　リングエクスパンダ【ピストンリング挿入用具】
　ring flutter　リングフラッタ
　ring gear　リングギア, 輪歯車【円盤状歯車】
　ring groove　リングみぞ
　ring life buoy　救命ブイ, 浮輪
　ring lubricator　リング注油器
　ring modulator　リング変調器【整流器形振幅変調器の一種】
　ring nut　リングナット【円形状のナット】
　ring oiled bearing　リング注油軸受
　ring spring　輪ばね
　ring tube resister　環状抵抗器
　ring valve　リング弁, 蛇の目弁, 輪形弁
　ring winding　環状巻
Ringelmann chart　リンゲルマン濃度表【煤煙の濃度表】
ringing　リンギング, 呼出し信号
　ringing circuit　リンギング回路, 信号回路
　ringing signal　鳴動信号
　ringing tone　呼出し音
ripping　リッピング【歯面に生じる損傷の一種】
ripple　名さざ波, 波紋, 脈動
　動さざ波を立てる, 波紋を起こす
　ripple current　脈動流, リップル電流
　ripple factor　脈動率
　ripple frequency　リップル周波数
　ripple voltage　リップル電圧
rise　動上がる, 上昇する, 現れる
　名上昇, 増加, 出現　熟 give rise to ～ : を起こす, を生ずる
　rise above　動超える
　rise time　立ち上がり時間
rising　形上昇する, 増加(大)する
　rising process　上昇過程
risk　名リスク, 危険, 危険度, 賭け
　動危険にさらす, 危険を冒す, 賭ける
　risk analysis　リスク分析
　risk assessment　リスクアセスメント, 危険性事前評価
　risk management　リスクマネジメント, リスク管理
river pattern　リバーパターン【破面の模様の一種】
rivet　名リベット　動リベットで留める, くぎづけにする
　rivet holder　当て盤
　rivet joint　リベット継手
robot　ロボット
robust　形がっちりした, 丈夫な

robust control　ロバスト制御【制御したい対象の特性が変化しても，安定に動作する事を保証する制御のこと】

rock　图岩，岩石
　動ゆする，振動させる
　rock wool heat insulating material　ロックウール保温材

rocker　ロッカー，揺れ腕
　rocker arm　揺れ腕，ロッカアーム
　rocker fulcrum shaft　支点軸

rocking　ロッキング，揺れ，揺動
　rocking arm　揺れ腕，ロッキングアーム
　rocking attitude control　揺動姿勢制御
　rocking bar　揺れ棒
　rocking lever　揺りてこ
　rocking shaft　揺れ軸
　rocking valve　ロッキング弁，揺れ弁

Rockwell hardness　ロックウェル硬さ

rod　ロッド，棒

roentgen　レントゲン【X線・γ線照射量の単位】
　roentgen ray　X線

roil　動［液体など］をかき回す，かき濁す

role　役割，役目

roll　動転がる，回転する，圧延する　图ころがり，横揺れ
　roll control　回転制御
　roll grinder　ロール研削盤
　roll turning lathe　ロール旋盤

rolled　形圧延された
　rolled core type　ロールコア型【変圧器などの鉄心構造】
　rolled material　圧延材
　rolled steel　圧延鋼
　rolled steel for general structure　一般構造用圧延鋼材
　rolled thread　転造ねじ
　rolled tube　圧延管

roller　ローラ，ころ
　roller arm　ローラ腕
　roller bearing　ころ軸受
　roller chain　ローラチェーン
　roller clearance　ローラすきま
　roller electrode　ローラ電極
　roller follower　ころ従動節
　roller hardening　ローラ加工
　roller leveler　ひずみ取り機
　roller thrust bearing　スラストころ軸受
　roller wedge piece　ローラくさび片

rolling　回転，圧延，転造，（船の）横揺れ
　rolling bearing　ころがり軸受，ころ軸受
　rolling circle　ころがり円，母円
　rolling contact　ころがり接触
　rolling experiment　横揺れ試験
　rolling friction　ころがり摩擦
　rolling indicator　横揺れ指示器
　rolling ladder　移動はしご
　rolling mill　圧延機，圧延工場
　rolling moment　横揺れモーメント
　rolling reduction　圧下量
　rolling resistance　ころがり抵抗
　rolling roll　圧延ロール
　rolling surface　ころがり面

rollover　ころがり，転倒

room　室，部屋
　room temperature　室温

root　图根，平方根，歯車の歯元，根元，底　動根付く，定着する
　root angle　歯底円すい角
　root bend　裏曲げ
　root circle　歯元円，歯底円
　root clearance　谷底すきま
　root diameter　谷径，歯元円直径
　root face　ルート面
　root length　歯元の長さ
　root mean square　二乗平均
　root mean square value　実効値，二乗平均値
　root of thread　ねじの谷底
　root of tooth　歯元
　root opening　ルート間隔
　root radius　ルート半径
　root running　裏溶接
　root sign　根号

root valve　元弁
Roots blower　ルーツ送風機
Roots pump　ルーツポンプ
rope　名なわ，綱，ロープ　動縄で縛る
　rope drive　ロープ駆動
　rope gearing　ロープ伝動装置
　rope off ～　動～を仕切る，囲む
　rope pulley　ロープ車
　rope way　ロープウェー，索道
rosette　ローゼット【運動部分を制御するカム】
rot　動腐る，朽ちる
rotameter　ロータメータ，浮き子式流量計【流量計の一種】
rotary　形回転する，回転式の
　rotary actuator　回転式アクチュエータ【流体エネルギを利用した回転装置】
　rotary blower　回転送風機
　rotary burner　ロータリバーナ，回転式バーナ
　rotary compressor　回転圧縮機
　rotary converter　回転変流器
　rotary encoder　ロータリエンコーダ【回転角度検出器】
　rotary engine　ロータリエンジン，回転式発動機，回転機関
　rotary indicator　回転表示器
　rotary joint　回転結合，回転継手
　rotary machine　回転機械
　rotary motion　回転運動
　　関 reciprocating motion：往復運動
　rotary oil burner　回転式[油]バーナ
　rotary piston engine　回転ピストン機関
　rotary press　輪転機，輪転印刷機
　rotary pump　ロータリポンプ，回転ポンプ
　rotary scavenging valve　回転掃気弁
　rotary stabilizer　回転揺れ止め
　rotary switch　ロータリスイッチ，回転スイッチ
　rotary valve　回転弁【方向制御弁の一種】
　rotary vane　回転羽根

　rotary vane meter　回転羽根式流量計
rotate　動回転する・させる，回る，循環する・させる
rotating　名回転　形回転運動をする
　rotating amplifier　増幅発電機
　rotating blade　回転羽根
　rotating body　回転体
　rotating disc　回転円板
　rotating field　回転磁界
　rotating field type　回転磁界形
　rotating gear　回転歯車
　rotating magnetic field　回転磁界
　rotating mass　回転質量
　rotating meter　回転計
　rotating part　回転部分
　rotating shaft　回転軸
　rotating stem valve　回転弁
rotation　回転，循環，[天体の]自転
　rotation angle　回転角
　rotation loss　回転損失
　rotation moment　回転モーメント
　rotation spectrum　回転スペクトル
　rotation speed　回転速度
　rotation test　回転試験
rotational　形回転する，回転の
　rotational energy　回転エネルギ
　rotational flow　渦流れ
　rotational motion　回転運動
　rotational shaft center　回転軸中心
　rotational speed　回転速度
　rotational viscometer　回転粘度計
rotator　回転子
rotodynamic pump　ターボポンプ
rotor　ロータ，回転子
　関 stator：固定子
　rotor blade　回転翼羽根，動翼，回転翼
　rotor core　回転子鉄心
　rotor drum　ロータ胴
　rotor shaft　ロータ，回転軸
　rotor winding　ロータ巻線，回転子巻線
　rotor yoke　回転子継鉄
rouge　ルージュ，ベンガラ【艶だし剤】
rough　形粗い，加工をしない，大ざっぱな

rough arrangement 概略配置図
rough cutting 荒削り
rough finishing 荒仕上げ
rough forging 荒地
rough machining 粗加工
rough map (drawing) 略図
rough running 不安定な動作
rough sea 荒天
rough surface 粗面
roughen 動粗くする, ざらざらにする
roughened surface 粗面
roughing 荒加工, 粗削り
roughing cut 荒削り
roughing tool 荒削りバイト
roughly 副おおよそ, 大まかに
roughness 粗さ, 粗度
roughness curve 粗さ曲線
roughness of thread ねじ表面粗さ
round 形丸い, 円形の 名円, 回転 動丸くする, 丸みをつける
round angle 周角, 円周角, 360°
round bar 丸棒
round bracket 丸かっこ
round down to ～ 動～を切り捨てる
round end 回転端
round key 丸キー
round-off error 丸め誤差
round off to ～ 動～を四捨五入する
round [screw] thread 丸ねじ
round steel 丸鋼
round up to ～ 動～を切り上げる
rounded end 回転端
rounding 丸み, アール
rounding error 丸め誤差
roundness 真円度
route 名経路 動経路を定める
routine 名手順, 日常業務, 日課 形日常の, 決まりきった
routine adjustment 定期調整
routine inspection 定期点検
routine maintenance 定期整備作業
routine procedure 決まりきった手順
routine test 定期試験
row 列, 並び, 行
row of rivet リベット列
RPM (Revolution Per Minute) rpm, 毎分回転数
RPR (Reverse Power Relay) 逆電力継電器
rub 動こする, 摩擦する, 塗りつける
rubber ラバー, ゴム
rubber belt ゴムベルト
rubber buffer ゴム緩衝器
rubber bush ラバーブッシュ
rubber insulated gloves 電気用ゴム手袋
rubber isolator 防振ゴム
rubber mat ゴムマット
rubber packing ゴムパッキン
rubber tape ゴムテープ
rubber tube ゴム管
rubbing こすること, 摩擦
rubbing surface こすれ面, 摩擦面
rubbing velocity こすり速度
rubbish くず, がらくた
rubidium ルビジウム 記Rb
ruck がらくた, くず, しわ
rudder ラダー, 舵, 方向舵
rudder angle indicator 舵角指示器
rugged 形頑丈な, 起伏の多い
rule 名ルール, 規則, 規定, 定規, 公式 動支配する, 治める, 定規で線を引く
ruler ルーラ, 定規, 平定規, 物差し
Rules of the Roads 海上衝突予防法
run 動運転する, 流す, 流動する, 走る, 延びる, 続く 名走行, 行程, 運転
run back 動戻る
run on 動続ける, 動かす, 進行する
run out 名振れ, 心振れ 動絶やす, 不足する, 尽きる
run out of ～ 動～がなくなる
run up 動急上昇する
runaway ship 便宜置籍船
runaway speed 無拘束速度

runner 運転者，羽根車，湯道
running 名運転，稼働，流れ
　形動いている，運転中の，流動する，連続する
　running average　移動平均
　running block　動滑車
　running cost（charge）　運転費，ランニングコスト
　running expense　運転費，経常費
　running fit　動きばめ
　running gear　駆動装置
　running hours　運転時間
　running-in　すり合わせ運転，ならし運転，なじみ運転
　running lamp　航海灯
　running-out　使い果たすこと
　running repairs　簡単な[応急]修理
　running resistance　走行抵抗
　running rigging　動索
　running test　走行試験，運転検査
　running time　運転時間
rupture　名ラプチャ，破壊，破裂，破断　動破裂する
　rupture elongation　破断伸び
　rupture plate　破裂板
　rupture strength　ラプチャ強さ
　rupture test　破壊試験
rupturing capacity　遮断容量
rush　動勢いよく流れる，突進する，急いでする
rushing charge　急速充電
rust　名さび　動さびる，腐食する
　rust blister　さびぶくれ
　rust inhibitor　防錆剤，さび止め
　rust preventing grease　さび止めグリース
　rust prevention　さび止め
　rust-proof　形さび止めの，防食の
　rust resisting paint　さび止めペイント
　rust scale　さびスケール
rustless steel　ステンレス鋼，不しゅう鋼
rusty　形さびた，さび付いた

S-s

Sabathe cycle サバテサイクル，複合サイクル
sacrifice 名犠牲 動犠牲にする
saddle 鞍(くら)，サドル【支持台の一種】
  saddle key　サドルキー，鞍キー
safe 形安全な
  safe limit　安全限度
  safe load　安全荷重
  safe working load　安全使用荷重
safeguard 安全装置，予防策
safely 副安全に
safety 安全，安全性，無事
  safety analysis　安全解析
  safety assessment　安全性評価
  safety awareness　安全意識
  safety belt　救命帯，安全ベルト
  safety cap　安全キャップ
  safety control　安全管理
  safety device　安全装置，保安装置
  safety engineering　安全工学
  safety evaluation　安全評価
  safety factor　安全率，安全係数
  safety fuse　安全ヒューズ(可溶片)
  safety glass　安全ガラス
  safety glasses　保護メガネ
  safety lamp　安全灯
  safety line　安全索
  safety lock　安全錠，安全装置
  safety management　安全管理
  safety measure　安全対策
  safety planning　安全計画
  safety precaution　安全対策
  safety relief valve　リリーフバルブ，安全逃がし弁
  safety review　安全審査
  safety shoes　安全靴
  safety signal　安全信号
  safety signplate　安全標識
  safety valve　安全弁【圧力過昇の緊急時に作動】
  safety zone　安全地帯
sag 動たるむ，たわむ 名たわみ【竜骨中央のたわみ】
sagging サギング【船体の曲がりの一種】
sail 名帆 動帆走する，航海する
sailor 船員，水兵
salient 形目立った，突出した，突角の 名凸部
  salient angle　凸角
  salient pole machine　突極電機
  salient pole synchronous induction motor　巻線型起動同期電動機
salinity 塩分
salinometer 塩分計，検塩計
salt 名塩，食塩 形塩気のある
  salt water　塩水，海水
saltation 躍動，激変【砂泥粒子が水や空気の流れによって跳ねながら運ばれる現象のこと】
salvage 名海難救助 動海難から救助する，(船舶)を引き揚げる
same 形同様な，同じ 関about the same as：と同じである／in the same way (manner)：同様に，同じ方法で
sample サンプル，見本，標本，試料
  sample function　標本関数
  sample number　サンプル数
sampled-data control サンプル値制御
sampling サンプリング，試料採取，抜取り，調査
  sampling action　間欠動作
  sampling distribution　標本分布
  sampling function　サンプリング周波数，標本化周波数
  sampling inspection　抜取り検査
  sampling point　試料採取場所
  sampling rate　サンプリング速度
  sampling test　抜取り試験
  sampling time　サンプリング時間
sand 砂
  sand blast　砂吹き【金属などの表面を磨いたりすること】

sand box 砂箱, 砂場
sand casting 砂型鋳造［物］
sand cloth 研磨布, 布やすり
sand mark 砂傷
sand paper 研磨紙, 紙やすり
sandblasting サンドブラスト【表面処理の一種】
sander サンダ【研磨剤で物体の表面を磨く装置】
sanding machine サンダ
sandwich construction サンドイッチ構造
sanitary 形衛生上の, 清潔な
sanitary fixture 衛生器具
sanitary pump サニタリポンプ, 衛生ポンプ
saponification 加水分解, 鹸(けん)化
Sargent cycle サージェントサイクル【熱力学的サイクルの一種】
satellite 名衛星, 人工衛星 形衛星のような
satellite broadcasting 衛星放送
satellite communication 衛星通信
satellite navigation 衛星航法
satisfactory 形満足な, 適切な
satisfy 動満足する, 満たす
saturable reactor 可飽和リアクタ
saturate 動浸す, 飽和する, 飽和状態にする
saturated 形飽和した, ぬれた, 浸透した
saturated air 飽和空気
saturated boiling 飽和沸騰
saturated compound 飽和化合物
saturated humidity 飽和湿度
saturated liquid 飽和液
saturated liquid line 飽和液線
saturated magnetic flux 飽和磁束
saturated solution 飽和溶液
saturated steam 飽和蒸気
saturated steam pressure 飽和蒸気圧力
saturated temperature 飽和温度
saturated vapor 飽和蒸気
saturated vapor line 飽和蒸気線
saturated vapor pressure 飽和蒸気圧力
saturated water 飽和水
saturated water line 飽和水線
saturation 飽和, 飽和度, 飽和状態
saturation adiabat 飽和断熱線
saturation characteristic 飽和特性
saturation current 飽和電流
saturation curve 飽和曲線
saturation degree 飽和度
saturation factor 飽和率
saturation liquid 飽和液
saturation liquid line 飽和液線
saturation magnetization 飽和磁化
saturation point 飽和点
saturation power 飽和出力
saturation pressure 飽和圧力
saturation region 飽和領域
saturation state 飽和状態
saturation steam 飽和蒸気
saturation temperature 飽和温度
saturation vapor line 飽和蒸気線
saucer 受け皿
save 動救う, 節約する, 保存する
saw 名のこぎり 動のこで引く, のこぎりを使う, 「see」の過去形
saw dust おがくず, のこくず
sawing 切断
sawing machine のこ盤
sawtooth 名のこ歯 形ギザギザの
sawtooth current のこぎり歯状電流
sawtooth oscillator のこぎり波発振器
sawtooth voltage のこぎり歯状電圧
sawtooth waveform のこぎり波形
say 動言う, 述べる, 表現する, 〜と書いてある
Saybolt visco[si]meter セイボルト粘度計
scalar スカラ【方向を持たない量のこと】 関vector：ベクトル
scalar control スカラ制御
scalar matrix スカラ行列
scalar product スカラ積
scalar quantity スカラ量
scale 名スケール, 目盛り, 階級,

尺度, はかり, 堆積物, 湯あか, さび 動天秤(はかり)で計る, 縮尺で〜を書く
  scale coefficient 汚れ係数
  scale deposit スケールデポジット, スケール付着(堆積)
  scale mark 目盛線
  scale of reliability 信頼性の尺度
  scale plate 目盛板
scalene 名不等辺三角形 形不等辺の, 斜軸の
scaling スケーリング, スケール生成, さび皮落とし, 基準化, 拡大縮小, 位取り, 焼損, 剥れ
  scaling circuit 計数回路
  scaling hammer チッピングハンマ, さび打ちハンマ
scan 動細かく調べる, 走査する 名スキャン, 視界, 走査
scanner スキャナ, 走査器
scanning スキャニング, 走査
  scanning direction 走査方向
  scanning line 走査線
  scanning pattern 走査パターン
  scanning point (spot) 走査点
  scanning speed 走査速度
  scanning system 走査系
scantling 材料寸法, 角材
scar 名傷, 傷あと 動傷になる, 傷あとを残す
scarf 動接合する, そぎ継ぎにする
  scarf joint そぎ継ぎ, すべり刃継ぎ
scarfing スカーフィング, 相欠き
scatter 動ばらまく, 散乱させる 名まき散らすこと, 散乱
  scatter factor 散乱係数
scattered 形散乱した, 分散した
  scattered light 散乱光
  scattered wave 散乱波
scattering 名分散, 散乱 形分散した, ばらばらの
  scattering angle 散乱角
  scattering cross section 散乱断面積
  scattering layer 散乱層
scavenge 動掃気する, 取り除く, 清掃する 名掃気

  scavenge air passage 掃気通路
  scavenge blower 掃気送風機
  scavenge factor 掃気係数
  scavenge port 掃気ポート, 掃気孔
  scavenge pump 排油ポンプ
scavenging スカベンジング, 掃気, 掃気作用
  scavenging action 掃気作用
  scavenging air 掃気
  scavenging air pressure 給気圧
  scavenging air receiver 掃気空気だめ
  scavenging blower 掃気送風機
  scavenging efficiency 掃気効率
  scavenging manifold 掃気マニホールド
  scavenging port 掃気ポート
  scavenging pressure 掃気圧
  scavenging process 掃気過程
  scavenging pump スカベンジングポンプ, 掃気ポンプ, 掃気送風機
  scavenging ratio 給気比
  scavenging stroke 掃気行程
  scavenging valve 掃気弁
  scavenging with perfect mixing 完全混合掃気
SCC (Stress Corrosion Cracking) 応力腐食割れ
scene 現場, 場面
schedule 名予定[表], 工程 動予定する, 表を作る
  schedule drawing 工程図
scheduled maintenance 定期保守
scheduled outage 作業停止
schema スキーマ, 概要, 図表
schematic 形概要の, 図解の, 模型の 名図表, 回路図, 図式, 概要, 図解
  schematic diagram 概要図, 図表, 略図 ◎ schematic drawing
scheme 名計画, 体制, 要項 動計画を立てる, たくらむ
  scheme drawing 計画図
science 科学, 自然科学
scientific 形科学的な
scintillate 動閃光を発する
scintillation シンチレーション, 閃

光，火花
scissor 動はさみで切る
scissors はさみ
scleroscope 反発硬度計
scoop ひしゃく，スコップ
scope スコープ【テレスコープ（望遠鏡），マイクロスコープ（顕微鏡）】
score 名引っかき傷，割れ目 動〜に刻み目をつける，〜に印をつける，傷をつける
scoring スコーリング，かじり【傷の一種】
Scotch boiler スコッチボイラ【船用ボイラの一種】
SCR (Silicon Controlled Rectifier) シリコン制御整流器
scramble スクランブル【電波の暗号化】
scram-jet スクラムジェット【超音波の気流を利用して燃料を燃焼させて推力を得るラムジェット】
scrap スクラップ，くず，廃物
　scrap iron　くず鉄
scrape 動削る，こすり落とす
scraper きさげ，スクレーパ
　scraper ring　油かきリング，オイルリング
scraping きさげ仕上げ
scratch 動引っかく，こする 名スクラッチ，かき傷
　scratch hardness　引っかき硬さ
　scratch test　引っかき硬さ試験
scratching スクラッチング，引っかき傷
screen 名スクリーン，仕切り，ついたて，画面，遮蔽，ふるい，金網 動仕切りをする，遮へいする，選別する，審査する
　screen tube　スクリーン管，仕切管
screened cable シールドケーブル
screening effect 遮へい効果
screw 名スクリュ，ねじ，プロペラ 動ねじ込む，ねじで調整する
　screw aperture　プロペラ孔
　screw bolt　ねじボルト
　screw brake　ねじブレーキ
　screw cap　ねじ込み口金
　screw compressor　スクリュ圧縮機，ねじ圧縮機
　screw coupling　ねじ継手，ねじ連結器
　screw cutting　ねじ切り
　screw ditch　らせん溝
　screw driver　ドライバ，ねじ回し
　screw gear　ねじ歯車
　screw hole　ねじ穴
　screw motion　ねじ運動
　screw pair　ねじ対偶
　screw propeller　スクリュプロペラ，らせん推進器
　screw pump　ねじポンプ
　screw shaft　プロペラ軸
　screw socket　ねじ込みソケット
　screw spanner　自在スパナ
　screw stay　ねじ控え
　screw stud　植込みボルト
　screw tap　ねじタップ
　screw thread　ねじ，ねじ山
　screw thread gauge　ねじゲージ
　screw type joint　ねじ形継手
　screw wrench　自在スパナ
screwed type pipe fitting ねじ込み形管継手
screwed valve ねじ込み弁
scribble 動走り書きする，落書きする
scribe 動けがく，画線器で線を引く
scriber けがき針
scribing block トースカン【けがき工具の一種】
scroll 名渦巻，渦形 動巻く，スクロール（画面移動）する
　scroll compressor　スクロール圧縮機
scrub 動こする，ごしごし洗う
scrub[bing] brush 洗濯ブラシ，たわし
scrubber スクラバ，洗浄器，清浄器
　scrubber dedustor　スクラバ脱塵装置
scuffing スカフィング【細いかき傷】

scum 浮きあか，あく，皮膜
  scum valve 水面吹出し弁
scupper 排水口，スカッパ
  scupper pipe 排水管
scuttle 小窓
sea 海，海洋
  sea chest シーチェスト，海水箱【海水取入口】
  sea clutter 海面反射
  sea cock 船底弁
  sea condition 海面状態
  sea gauge 喫水，気圧測深器
  sea margin シーマージン
  sea mile 海里【1海里 = 1852m】
  sea speed 航海速力，対水速力
  sea trial 海上試運転
  sea water 海水
  sea water cooling 海水冷却
  sea water lubricating stern tube 海水潤滑船尾管
  sea water protection law 海洋汚染防止法
  sea water service pump 海水サービスポンプ
  sea water temperature 海水温度
  sea worthiness 堪航性
seal 名シール，密封，密閉，印，封印 動シールする，密閉する
  seal air シールエア【炉内ガスの漏れを防止するために炉壁貫通部に導入される加圧空気】
  seal groove シールみぞ
  seal ring シールリング
  seal tape シールテープ
  seal weld 漏れ止め溶接
sealed 形密封した
  sealed bearing シール軸受
  sealed container 密閉容器
sealing シーリング，密封，密閉
  sealing action 密閉作用
  sealing device 密閉装置
  sealing surface シール面
  sealing water 封水
seam 名縫い目，縫合 動縫い合わせる
  seam set シームセット【[金属板，革細工など]継ぎ目をならす道具】
  seam welding シーム溶接，縫合わせ溶接【重ね抵抗溶接の一種】
seamen's competency certificate 海技免状
seamless 形継目のない
  seamless pipe (tube) 引抜管，継目無し管
search 動探す，調べる 名捜索，調査
  search for ～ 動～を探す，求める
  search into ～ 動～を調査する
searcher すきまゲージ
searchlight 探照灯，サーチライト
season 名季節，時季 動乾燥させる，味付けする
  season crack 置割れ，時期割れ
seasonal 形季節の，季節的な
seasoning シーズニング，枯らし，乾燥
seat 名シート，座部，台，台座 動据えつける，座る，着席させる
  関 valve seat：弁座
seating face 座面
seating ring 座環
seaway 海路，船脚，荒海
second 形第二の 副二番目に 名秒
  second hand tap 二番タップ，中タップ
  second law of motion 運動の第2法則
  second law of thermodynamics 熱力学の第2法則
  second moment of area 断面二次モーメント
  second order lag element 二次遅れ要素
  second stage 第2段
secondary 形第2の，二次の，次の，従属的な
  secondary air 二次空気
  secondary battery (cell) 二次電池，蓄電池，充電式電池
  secondary burning 二次燃焼
  secondary circuit 二次回路
  secondary coil 二次コイル

secondary combustion 二次燃焼
secondary combustion chamber 二次燃焼室, 補助燃焼室
secondary current 二次電流
secondary flexure 二次たわみ
secondary flow 二次流れ
secondary loss 二次損失
secondary pinion 二段小歯車
secondary stress 二次応力
secondary voltage 二次電圧
secondary winding 二次巻線

secondly 副第二に, 次に

section セクション, 断面, 切片, 薄片, 断面図, 部分, 区域, 分割
熟 in section：断面で
section area　断面積
section line　切断線
section modulus　断面係数
section paper　方眼紙
section steel　形鋼

sectional 形区間の, 組合わせ式の, 断面(図)の
sectional area　断面積
sectional boiler　組合わせボイラ
sectional drawing　断面図
sectional form　断面形状
sectional header　組合わせ管寄せ
sectional size　断面寸法
sectional view　断面図

sector セクタ, 部門, 区域, 扇形
sector display　部分表示
sector gear　セクタ歯車, 扇形歯車

secular 形長年の, 永続の
secular change　経年変化, 経年変形

secure 形安全な, 確かな　動しっかり締める, 確保する, 安全にする

securely 副しっかりと

security 安全確保, 保護, 保証
熟 in security：安全に, 無事に

sediment 沈殿物, 沈降物

sedimentation 沈殿, 沈降作用
sedimentation equilibrium　沈降平衡
sedimentation separation　沈降分離
sedimentation tank　沈殿タンク
sedimentation velocity　沈降速度

see 動見る, 見える, 調べる

seek 動(探し)求める, (達成しようと)努める

Seebeck effect ゼーベック効果

seem 動〜のように見える
熟 It seems that 〜：〜のように思われる, 〜のようだ

segment 名セグメント, 整流子片, 部分, 区分, 切片, 弧, 弓形, (機械などの)扇形部分
動分ける, 分割する
segment gear　扇形歯車
segment rack　扇形ラック
segment wheel　扇形歯車

segmental 形部分の・からなる, 区分の・からなる, 弓形の

segregation 分離, 偏析【融金属が凝固する際に, 成分が一部にかたよる現象】

seize 動焼き付く, 動かなくする, 膠着する

seizing シージング【結索法の一種】

seizure 焼付き

select 動選ぶ, 選択する
形選んだ, 抜粋した

selection 選択, 選別
selection signal　選択信号

selective 形選択的な, 選択の
selective absorption　選択吸収
Selective Catalytic Reduction system　SCR, 選択接触還元システム
selective diffusion　選択拡散
selective emission　選択放出
selective gear　選択かみ合い継手
selective hardening　局部硬化
selective protection　選択(遮断)保護
selective quenching　局部焼入れ
selective radiation　選択放射
selective resonance　選択共振

selectivity 選択度, 選択性

selector 選択装置, セレクタ
selector channel　選択チャンネル
selector panel　転換盤
selector valve　切替弁

Selenium rectifier セレン整流器

self- 「自己」「自分で」「自動的な」の意を表す連結形

- self-absorption 自己吸収
- self-actuated control 自力制御
- self-aligning bearing 自動調心軸受, 心合わせ軸受
- self-annealing 自己焼なまし
- self-balancing instrument 自己平衡計器
- self-capacity 自己容量
- self-contained 形内蔵型の, 全て揃った
- self-cooled 形自冷式の
- self-demagnetization 自己減磁
- self-demagnetizing force 自己減磁力
- self-discharge 自己放電
- self-evaporation 自己蒸発
- self-excitation 自励, 自己励磁
- self-excited 形自励[式]の
- self-excited AC generator 自励交流発電機
- self-excited generator 自励発電機
- self-excited motor 自励電動機
- self-excited oscillation (vibration) 自励振動
- self-focusing 自己集束
- self-hardening 自硬性, 自硬化
- self-hardening steel 自硬鋼
- self-healing 自己回復
- self-heating 自己加熱
- self-holding 自己保持
- self-ignition 自己着火, 自己点火
- self-impedance 自己インピーダンス
- self-inductance 自己インダクタンス
- self-induction 自己誘導
- self-limitation 自己制限, 自主規制
- self-loading 定格荷重
- self-locking nut 戻り止めナット
- self-lubricating 形自動注油[式]の, 自己潤滑の
- self-lubricating bearing 自己注油軸受
- self-luminous 形自己発光[性]の
- self-operated control 自力制御
- self-operating 形自動[式]の, 自動制御の
- self-priming pump 自吸ポンプ
- self-propulsion 自力推進
- self-protection 自己防衛, 自衛
- self-purification 自然浄化, 自浄
- self-reactance 自己リアクタンス
- self-regulation 自己規制, 自動制御
- self-replication 自己再生
- self-reset 自己復帰
- self-reversing 自己逆転
- self-scattering 自己散乱
- self-sealing 形自己密封式の
- self-shielding 自己遮へい
- self-similarity 自己相似性
- self-starter 自動始動機, 自動スタータ
- self-steering 形自動操舵の
- self-sufficient robot 自立ロボット
- self-supporting 形自己支持の
- self-supporting cable 自己支持形ケーブル
- self-timer 自動タイマ, セルフタイマ
- self-ventilated 形自己通風の
- self-winding 自動巻きの
- **selsyn** (**self-synchronizing**) セルシン【同期式変位伝送装置】, シンクロ
- selsyn motor セルシン電動機
- **semi-** 接頭「半分の」,「部分的な」,「準〜」,「不完全な」の意
- semi-automatic 形半自動式の, 半自動化された
- semi-automatic ship 半自動化船
- semi-axial pump 半軸流ポンプ
- semi-axis [双曲線などの]半軸
- semi-built[-up type] 半組立式
- semi-built up crank shaft 半組立型クランク軸
- semi-circular 半円
- semi-closed cycle 半密閉サイクル
- semi-closed cycle gas turbine 半密閉サイクルガスタービン
- semi-conductor 半導体
- semi-conductor chip 半導体素子
- semi-conductor diode 半導体ダイオード

semi-conductor integrated circuit　半導体集積回路
semi-conductor laser　半導体レーザ
semi-conductor processing equipment　半導体製造装置
semi-conductor rectifier　半導体整流装置
semi-cylinder　半円筒
semi-diameter　半径
semi-elliptic spring　弓形ばね
semi-finishing　中仕上
semi-floating axle　半浮動軸
semi-fluid　[形]半液体の
semi-flush type　半埋込形
semi-infinite body　半無限体
semi-infinite solid　半無限固体
semi-insulated substrate　半絶縁性基盤
semi-killed steel　セミキルド鋼
semi-logarithm　片対数
semi-major axis　長半径
semi-metal　半金属
semi-minor axis　短半径
semi-open system　半開放式
semi-permanent　[形]半永久的な
semi-permeable membrane　半透膜
semi-portable fire extinguisher　半移動式消火器
semi-radiation boiler　半放射ボイラ
semi-rigid type　半剛型
semi-solid　半固体
semi-steel　鋼性鋳鉄, セミスチール
semi-transparent　[形]半透明の
send　[動]送る, 届ける
sending current　送信電流
sending end output　送信端出力
senior　[形]上級の
sennit　センニット, 組みひも
sensation　知覚
sense　[名]センス, 感覚, 意識, 意義　[動]検知する, 探知する, 感知する　[関] in a sense：ある意味では, ある点では
sensibility　感度, 感覚, 感受性
sensible　[形]分別のある, かなりの, 感じられる

sensible heat　顕熱
　[反] latent heat：潜熱
sensible heat factor　顕熱比
sensing　センシング【センサを用いて圧力や温度などを計測・判別すること】
sensing device (instrument)　検出装置
sensing element　検出素子, 検出部
sensitive　[形]敏感な, 感度の高い, 感覚による
sensitive load　高感度負荷
sensitive paper　感光紙, 印画紙
sensitivity　感度, 敏感度, 感受性
sensor　センサ, 感知装置, 検出器
separate　[動]分離する, 隔てる, 引き離す　[形]離れた, 別の, 独立した, 個々の
separate excitation　他励
separated　[形]分離された
separated printing　別刷り, 抜き刷り
separately　[副]単独に, 別に, 別々に, 離れて
separately excited generator　他励発電機
separately excited motor　他励電動機
separating　分離, 析出
separation　分離, はく離, 独立
separation efficiency　分離効率
separation rupture　分離破壊
separator　セパレータ, 分離器
sequence　[名]シーケンス, 連続, 順序, 連鎖
　[動]順番に並べる, 整理する
sequence check　順番検査
sequence circuit　シーケンス回路
sequence control　シーケンス制御
sequence diagram　シーケンス図
sequence of numbers　数列
sequence valve　シーケンス弁
sequential　[形]連続する, その結果として生ずる
sequential access　順次アクセス
sequential circuit　シーケンス回路
sequential control　シーケンス制御, 逐次制御

sequential decoding 逐次復号
sequential scanning 順次走査
sequential trace type scheme いもづる方式
serial 形連続的な, 直列の
serial access storage 順次アクセス記憶装置
serial addition 直列加算
serial number シリアルナンバ, 通し番号, 製造番号, 認識番号
serial operation 直列運転
serial printer シリアルプリンタ【1字ごと印字するプリンタ】
serial tap 増径タップ
serial to parallel conversion 直列・並列変換
serial transmission 直列伝送
series 連続, 系列, 直列, 級数
関 a series of：一連の／in series：直列に(で)
series branch 直列分岐
series capacitor 直列コンデンサ
series circuit 直列回路
series coil 直列コイル, 直巻きコイル
series compensation 直列補償
series connection 直列接続
series development 級数展開
series dynamo (generator) 直巻発電機
series motor 直巻電動機
series resistance 直列抵抗
series resonance 直列共振
series spot welding 直列点溶接
series transformer 直列変圧器
series winding 直巻, 直列巻[線]
series wound 形直巻きの
series wound generator 直巻発電機
series wound motor 直巻電動機
serious 形重大な, 深刻な, まじめな
seriously 副まじめに, 深刻に
serrated shaft セレーション軸
serration セレーション, のこ歯切欠き
serve 動奉仕する, 仕える, 果たす, 供給する, 役立つ

service 名サービス, 点検, 操作, 修理, 業務, 奉仕, 供給, 貢献
関 in service：運行されていて, 使われていて 動点検・修理をする, 提供する, 整備する
service area 有効範囲, 供給区域
service bolt 仮止めボルト
service cable 引込みケーブル
service condition 航海状態, 使用状態
service entrance 引込み口
service hours 運転時間
service interruption 不通, 停電
service life 使用期間, 耐用年数
service pipe 引込み管
service pump サービスポンプ, 雑用ポンプ
service speed 航海速力
service tank サービスタンク, 常用タンク, 供給タンク
service telephone 専用電話
service voltage 供給電圧
servicing サービス, 修理
servo 形サーボ機構の, サーボ制御の
servo-actuated control 他力制御
servo-actuator サーボアクチュエータ
servo-amplifier サーボ増幅器
servo-control サーボ制御
servo-cylinder サーボシリンダ
servo-mechanism サーボ機構【物体の位置・速度・姿勢などを制御量とし, 電気や流体を用いて任意の変化に追従させる自動制御機構のこと】
servo-motor サーボモータ
servo starting motor 始動機
servo-system サーボ系, サーボシステム, サーボ機構
servo-valve サーボ弁
set 動取り付ける, 設置する, 定め, 設定する, 合わせる, 固まる 形固定した, 指定の
名セット, 設置, 残留ひずみ, 一組, 一式, 設定, 硬化, 凝結, 凝固

shaft

関 permanent set：永久ひずみ
set-back　そり
set-bolt　止めボルト
set-function　集合関数
set-point　設定値，設定点，目標値
set-point control　定値制御
set-pressure of safety valve　安全弁設定値
set-screw　止めねじ
set-square　三角定規
set-tap　組みタップ
set-time　整定時間
set up　動設定する，引き起こす，配置する，準備する，組み立てる
set-up　セットアップ，準備，設定，段取り，配置，組織，機構，装置　形組み上がった
set-up stress　組立応力
set-value　設定値
setting　セッティング，据付け，設定，調整，凝固，凝結，硬化
setting angle　取付角
setting drawing　据付図
setting point　凝固点
setting range　制定範囲
setting temperature　硬化温度，ゲル化温度，制定温度
setting time　整定時間
settle　動据える，置く，安定させる，沈殿させる，澄む・ませる，解決する
settling　静置，据付け，沈降[分離]，沈殿[物]
settling tank　セットリングタンク，澄ましタンク，沈殿タンク
settling time　整定時間
settling velocity　沈降速度
several　形
several thousand volts　数千ボルト
severally　副別々に，個別に
severe　形きびしい，激しい，厳格な，ひどい
severe condition　過酷な条件
severe wear　重摩耗
severely　副激しく，ひどく
sew　動縫う，縫い合わせる

sewage　下水，汚水，汚物
sewage disposal　下水処理，汚水処理
SFC（Specific Fuel Consumption）燃料消費率
shackle　シャックル，[南京錠の]掛け金
shade　名かさ，日陰
動隠す，日よけする
shading coil　くま取りコイル
shadow　名影，投影，陰影　動陰にする，暗くする，光をさえぎる
shaft　シャフト，軸，車軸
shaft alignment　軸系アライメント，軸心調整
shaft alley　軸路
shaft angle　軸角
shaft basis system of fits　軸基準はめあい
shaft bearing　軸受
shaft coupling　軸継手
shaft current　軸電流
shaft earthing device　軸系接地装置
shaft diameter　軸径
shaft end output　軸端出力
shaft gear　軸歯車
shaft generator　軸発電機
shaft governor　軸調速機
shaft grounding device　軸系接地装置
Shaft HorsePower　SHP，軸馬力，軸出力
shaft insulation　軸絶縁
shaft line　軸線
shaft liner　軸スリーブ，軸ライナ
shaft locking device　軸遊転防止装置
shaft output　軸出力
shaft power　軸動力
shaft rake　軸傾斜
shaft seal　シャフトシール，軸封
shaft sleeve　軸スリーブ，軸ライナ
shaft stool　軸受台
shaft tube　軸管
shaft tunnel　シャフトトンネル，軸路【プロペラシャフトの通路】
shaft washer　内輪

shafting

shaft work 軸仕事

shafting 軸系, 軸材
- shafting alignment 軸系アライメント, 軸心調整
- shafting arrangement 軸系装置

shake 動振り動かす, 揺れる 名振動, 動揺

shakedown シェークダウン, 機械の慣らし運転

shall 助～すべきである, 当然～だろう

shallow 形浅い

shank 柄[幹], 軸, 錨幹

shape 名形, 形状, 外形, 輪郭, 型, 姿 動形づくる, 具体化する
- shape coefficient 形状係数
- shape A into B　AをBに成型する, AでBを作る
- shape factor 形状係数, 形状因子
- shape memory alloy 形状記憶合金
- shape of section 断面形状
- shape steel 形鋼

shaped 形成形加工した
- 関 I-shaped：I 型／bow-shaped：弓形の

shapeless 形無形の

shapely 副形の良い, 均整のとれた

shaping 整形, 成形, 造形, 型削り
- shaping circuit 整形回路
- shaping die 成形型

share 動共有する, 分配する, 分け合う

sharp 形鋭い, 鋭利な, 急な, 激しい
- sharp crested orifice 刃形オリフィス
- sharp edged orifice 薄刃オリフィス
- sharp freezer 急速凍結機

sharpen 動研ぐ, 鋭くする

sharply 副すばやく, 鋭く, 激しく

sharpness 鋭さ

shave 動そる, 削る

shaving [金属などの]削りくず

shear 名せん断 動～を刈る, 切る, せん断する, 切断する
- shear angle せん断角
- shear bolt せん断ボルト
- shear deformation せん断変形
- shear flow せん断流
- shear fracture せん断破壊
- shear layer せん断層
- shear modulus せん断弾性係数
- shear pin シャーピン, せん断ピン
- shear plane せん断面
- shear stress せん断応力

shearing せん断, せん断加工
- shear[ing] force せん断力
- shearing force diagram せん断力図
- shearing load せん断荷重
- shearing machine せん断機
- shearing modulus 横弾性係数
- shear[ing] strain せん断ひずみ
- shearing strain energy せん断ひずみエネルギ
- shear[ing] strength せん断強さ
- shear[ing] stress せん断応力

sheath シース, 外装, 鞘, 被覆, 外筒
- sheath current シース電流
- sheath thermocouple シース熱電対
- sheath voltage シース電圧

sheave シーブ, 滑車, 綱車

shed 動発する, 取り除く

sheer 動向きを変える 名舷弧
- sheer plan 側面線図【船体側面の形状や甲板・水線の位置などを示す線図】

sheet シート, 板, 薄板
- sheet brass 薄黄銅板
- sheet copper 薄銅板
- sheet glass [薄]板ガラス
- sheet iron 薄鋼板, 鉄板
- sheet lead 薄鉛板
- sheet metal 薄板金, 金属薄板
- sheet rolling mill 薄板圧延機
- sheet steel 薄鋼板
- sheet stock 薄板材

shell シェル, 胴, 殻, 貝殻, 外形, 外観, ドーム状に湾曲した薄い板の構造物
- 関 upper (lower) bearing shell：上部(下部)軸受メタル
- shell and tube heat exchanger 円筒多管式熱交換器
- shell mark 貝殻模様

shell metal　裏金
shell plate　胴板, 外板
shell-type transformer　シェル型変圧器【鉄心がコイルを囲む構造】
shield　名シールド, 遮へい[物], 保護物　動防護する, 保護する
shield bearing　防じん軸受
shield driving method　シールド工法
shielding　遮へい
shielding angle　遮へい角, 遮光角
shielding efficiency　遮へい効果
shielding ring　シールドリング
shielding wire　シールド線, 遮へい線
shift　動移す, 移動させる, 動かす, 変える, 切り替える, 転位する　名シフト, ずれ, 変位, 変化, 転位, 移送, 交代, 変更
shifted gear　転位歯車
shifting load　移動荷重
shifting magnetic field　移動磁界
shifting pump　移送ポンプ
shim　シム, はさみ金, ワッシャ
shimmy　異常振動
shine　動輝く, 光を放つ　名光, 光沢, つや, 日光
ship　名船　動出荷する, 船に積む, 船に送る
ship inspection certificate　船舶検査証書
ship owner　船主
ship resistance　船体抵抗
ship service air　雑用空気
ship side valve　船体付き弁
ships particular　船体要目
shipyard　造船所
shock　名ショック, 衝撃　動衝撃を与える
shock absorber　ショックアブソーバ, 衝撃吸収装置, 緩衝器(装置)
shock excited oscillator　ショック発振器
shock load[ing]　衝撃荷重
shock pulse　衝撃パルス
shock resistance　耐衝撃性
shock test　衝撃試験

shock testing machine　衝撃試験機
shock valve　ショックバルブ, 防衝弁
shock wave　衝撃波, 爆風
shoot　動撃つ, 放つ　名発射
shop　工場, 作業場, 部門, 職場
shop gauge　工作ゲージ
shop test　工場試験
shore　名岸, 海岸, 支柱, 突っ張り　動上陸させる, 支柱で支える
Shore hardness　ショア硬さ
Shore hardness test　ショア硬さ試験
shore power　陸電
Shore scleroscope hardness test　ショア硬さ試験
short　名ショート, 短絡, 不足　形短い, 不十分な　副短く, 簡単に, 不足して　動ショートする・させる　熟 in short：要するに, 一言で言えば
short break　瞬断
short bunker　ショートバンカ【燃料不足】
short circuit　ショート, 短絡, 漏電
short circuit capacity　短絡容量
short circuit current　短絡電流
short circuit ratio　短絡比
short circuit ring　短絡環
short circuit test　短絡試験
short circuit winding　短絡巻線
short current relay　短絡継電器
short of　形〜不足して, 〜除いて
short of shipment　搭載不足
short pitch winding　短節巻き
short shunt　内巻き
short term demand forecast　短期需要予測
short term load forecasting　短期電力負荷予測
short time current　短時間電流
short time rating　短時間定格
short water　水面過降
short wave　短波
shortage　不足, 欠点, 欠陥
shortcoming　欠点, 欠陥
shorten　動縮める, 短くする, 減じる

shortly 副まもなく，手短かに
shot ショット，発射，発砲，弾丸
　shot blasting　ショットブラスティング【金属表面に鋼球を吹き付ける清浄法】
　shot effect　散弾効果
　shot peening　ショットピーニング【小球を当てる加工硬化法】
should 助～すべきである，～だろう，万一～ならば
shoulder 肩，[溶接の]ルート面
show 動見せる・える，示す，現れる，案内する　名ショー，展示会
shown 動「show」の過去分詞
SHP（**Shaft HorsePower**）軸馬力
shrink 動縮む，縮ませる，焼きばめする　名収縮
　shrink fit　焼きばめ【高温度で広がった所に部材を挿入し冷却して固定する】
shrinkage 縮小，縮み，収縮，減少
　shrinkage allowance　縮みしろ
　shrinkage crack　縮み割れ，収縮割れ
　shrinkage fit　焼きばめ
　shrinkage hole　収縮巣
　shrinkage rule　鋳物尺，伸び尺
　shrinkage stress　収縮応力，縮み応力
shroud ring シュラウドリング，囲い輪
shrouded blade シュラウド羽根
shrouding 囲い板
shrunk 動「shrink」の過去・過去分詞
　shrunk welded crank shaft　焼ばめクランク軸
shunt 動分路をつくる，切り替える　名シャント，分巻，短絡，分路，分流器
　shunt capacitor　分路コンデンサ
　shunt characteristic　分巻特性
　shunt circuit　分岐回路
　shunt coil　分巻コイル
　shunt effect　シャント効果，短絡効果
　shunt field current　分巻界磁電流
　shunt generator　分巻発電機
　shunt motor　分巻電動機
　shunt resistance　分路抵抗
　shunt winding　分巻巻線，分岐巻線
　shunt wound　形分巻の
　shunt wound generator　分巻発電機
shunting resistance 分巻抵抗
shut 動締める，閉鎖する　形閉じた　名閉鎖
　shut away　動隔離する
　shut down　動閉鎖する，停止する
　shut-down　運転停止，活動停止
　shut off　動止める，遮断する
　shut-off　締切，遮断
　shut-off cock　開閉コック
　shut-off head　締切揚程
　shut-off operation　締切運転
　shut-off valve　遮断弁
shuttle 名往復運転，シャトル　動往復する　形往復の
　shuttle service　往復(折り返し)運転
　shuttle valve　シャトル弁
SI（**Super Intendant**）船舶監督
SI unit 国際単位
　同 The International System of Units
side 名サイド，横，側，側面，面，片側，わき，辺　形横の，横方向の，側面[から]の，側面への
　関 one side：片方
　side anchor　振れ止め
　side bend　側曲げ
　side bunker　サイドバンカ(側燃料庫)
　side clearance　側すきま
　side effect　副作用
　side elevation　側面図
　side fillet weld　側面すみ肉溶接
　side force　横力
　side light　舷灯，側面灯
　side lobe　サイドローブ，副ローブ
　side pole　側柱
　side rake　横すくい角
　side rod　側棒
　side scatter　側方散乱
　side thrust　サイドスラスト，側方推力，側圧
　side thruster　サイドスラスタ【船体

sight 名視覚, 眺め, 見ること 動見つける, 認める, 観測する, 照準を合わせる
- sight glass サイトグラス, 点検窓, のぞき窓
- sight hole のぞき穴

sign 名サイン, しるし, 兆候, 標識, 看板, 記号, 信号 動署名する, 契約する
- sign board 掲示板, 看板
- sign character 符号文字
- sign digit 符号数字

signal 名印, 表示, 合図, 信号 動信号を送る, 合図する
- signal circuit 信号回路
- signal converter 信号変換器
- signal current 信号電流
- signal detection 信号検出
- signal gain 信号利得
- signal generator 信号発生器
- signal lamp 表示灯
- signal oscillator 信号発振器
- signal processing 信号処理
- signal speed generator 速度検出器
- signal transformation 信号変換
- signal transmitter 信号伝送器, 信号変換器
- Signal-to-Noise ratio SN比【信号電力と雑音電力との比】

signature 署名, サイン, 信号
significance 意義, 重要性
- significance level 危険率, 有意水準

significant 形重大な, 重要な, 意味のある, かなりの
- significant digit (figure) 有効数字

signify 動示す, 意味する
silence 名沈黙, 静けさ 動静かにさせる
silencer サイレンサ, 消音器
silent 形静かな, 沈黙した
- silent chain サイレント鎖【伝動用チェーンの一種】
- silent gear 低音歯車

silica シリカ, 二酸化けい素 (SiO_2)
- silica gel シリカゲル【主に乾燥剤として用いる】
- silica gel grease シリカゲルグリース
- silica glass 石英ガラス, シリカグラス
- silica sand けい砂

silicate けい酸塩
silicic acid けい酸
silicon シリコン, けい素 記Si
- silicon bronze けい素青銅
- silicon carbide 炭化けい素
- silicon chrome steel けい素クロム鋼
- Silicon Controlled Rectifier SCR, シリコン制御整流器
- silicon dioxide 二酸化けい素
- silicon fluoride フッ化けい素
- silicon rectifier シリコン整流器
- silicon resin シリコン樹脂
- silicon steel けい素鋼

silicone シリコン, けい素樹脂
- silicone grease シリコングリース
- silicone oil シリコンオイル
- silicone resin シリコン樹脂
- silicone rubber シリコンゴム

silk 絹, 絹糸
silver 名銀 記Ag 形銀の, 銀製の 動銀めっきを施す
- silver bromide 臭化銀
- silver chloride 塩化銀
- silver fluoride フッ化銀
- silver leaf 銀箔
- silver nitrate 硝酸銀
- silver plating 銀めっき
- silver solder 銀ろう【硬ろうの一種で硬くて耐熱性・伝導性にすぐれる】

similar 形類似の, 相似の
similarity 相似性, 類似性
- similarity law 相似則

similarly 副同様に
simple 形簡単な, 単純な, 単一の 名単体, 要素, 元素

simple average 単純平均
simple balance てんびん
simple beam 単純はり
simple bending 単純曲げ
simple calculation 単純計算
simple color 単色
simple equation 一次方程式
simple harmonic motion 単振動
simple pendulum 単振り子
simple shear stress 単純せん断応力
simple stress 単純応力
simple substance 単体
simple support 単純支点
simple support[ed] beam 単純支持はり
simple turbine 単式タービン
simple working stress 単純使用応力
simplex 形単純な，単一の，単信の 名単体
　simplex method 単体法，シンプレックス法
　simplex pump 単式ポンプ
simplicity 簡単，単純，平易
　simplicity radio 簡易無線
simplify 動簡単にする，平易にする，単純にする
simplistic 形余りにも単純(簡単，安易)な
simply 副簡単に，単純に，単に，平易に
Simpson's rule シンプソンの法則
simulate 動模倣する
simulation シミュレーション，模擬実験
simulator シミュレータ，模擬実験装置，模擬操縦装置
simultaneous 形同時に起こる，同時に存在する
　simultaneous calling 同時呼び出し
　simultaneous equation 連立方程式
　simultaneous linear equation 連立一次方程式
　simultaneous measurement 同時測定
　simultaneous operation 同時演算
　simultaneous start 一斉始動
　simultaneous transmission and reception 同時送受信
simultaneously 副同時に，一斉に
since ～ 接～して来ら，～だから 前～以来，～から 副それ以来
sine 正弦，サイン
　関 cosine：余弦，コサイン／tangent：正接，タンジェント
　sine curve 正弦曲線，サインカーブ
　sine theorem 正弦定理
　sine wave 正弦波
singing 鳴音
　singing propeller 鳴音プロペラ
single 形単一の，単独の，ただ1個の
　関 every single：一つ一つの
　single acting 形単動式の
　single acting engine 単動機関
　single acting pump 単動ポンプ
　single block 単滑車
　single bond 単結合
　single bus 単母線
　single cell 単一電池
　single coil 単コイル
　single contact 単接点
　single crystal 単結晶
　single current system 単流式
　single cylinder engine 単シリンダ機関
　single direction thrust bearing 単式スラスト軸受
　single element system 単要素式，一要素式
　single flow 片流れ
　single gauge 片口ゲージ
　single gear 一段歯車装置
　single geared drive 一段歯車駆動
　single helical gear 単はすば歯車
　single hole nozzle 単孔ノズル
　single layer winding 単層巻
　single operation 単独運転
　single pass cooler 単流冷却器
　single pass heat exchanger 単流熱交換器
　single phase 形単相の 名単相
　single phase AC 単相交流
　single phase active power 単相有効電力

single phase apparent power　単相皮相電力
single phase braking　単相制動
single phase circuit　単相回路
single phase flow　単相流
single phase generator　単相発電機
single phase induction motor　単相誘導電動機
single phase instantaneous current　単相瞬時電流
single phase instantaneous voltage　単相瞬時電圧
single phase motor　単相電動機
single phase power　単相電力
single phase reactive power　単相無効電力
single phase transformer　単相変圧器
single point tool　バイト
single pole　単極
single precision　単精度
single propeller　一軸プロペラ
single pulley drive　単ベルト駆動
single reduction gear　一段減速装置
single row　単列
single row angular contact radial ball bearing　単列アンギュラコンタクト型ラジアル玉軸受
single row ball bearing　単列玉軸受
single row roller bearing　単列ころ軸受
single screw vessel　一軸船
single shielded bearing　片シールド軸受
single sided impeller　片側吸込み羽根車
single stage　単段，一段
single stage air compressor　一段空気圧縮機
single stage pump　単段ポンプ
single suction pump　片吸込ポンプ
single threaded screw　一条ねじ
single throw switch　単投スイッチ
single unit　名単一ユニット　形一体型の
single whip　単滑車
singly　副一つ一つ，個々に，単独に

singular　形まれな，珍しい，特異な
　singular point　特異点
singularity　特異性
sink　動沈む，沈める，沈没する　名シンク，流し，吸込み
sinking pump　掘り下げポンプ
sinking speed　降下速度
sinter　名焼結物　動焼結する
　sinter compact　焼結体
sintered alloy　焼結合金
sintered magnet　焼結磁石
sintering　焼結
　sintering furnace　焼結炉
sinuous　形曲がりくねった，波形の
　sinuous header　波形管寄せ
sinusoid　正弦曲線
sinusoidal input current　正弦波入力電流
sinusoidal oscillator circuit　正弦波発振回路
sinusoidal wave　正弦波
siphon　サイフォン，吸上げ管
　siphon barometer　サイフォン式気圧計【U字形の管を持つ水銀気圧計のこと】
　siphon cup　注油サイフォン
　siphon gauge　曲管圧力計
　siphon lubricator　サイフォン注油器
siren　サイレン，号笛
sirocco fan　シロッコファン，多翼送風機
sit　動すわる，位置する
site　名敷地，場所　動設置する，置く
situate　動置く，位置させる，位置を定める
situation　状況，事態
six-four brass　六四黄銅
size　名サイズ，大きさ，寸法，かさ，型　動寸法で分類する
　size effect　寸法効果
skeleton　骨格，骨組
　skeleton construction　骨格(架構式)構造
　skeleton drawing　骨組図，構造線図
sketch　名下書き，略図　動略図を

作る，見取図を描く，写生する
sketch drawing　見取図
skew　形斜めの，非対称の
名ゆがみ，曲がり
skew back　スキューバック【プロペラ羽根の設計中心線と羽根先端とのずれの距離のこと】
skew bevel gear　はすばかさ歯車，食い違い軸歯車
skew curve　空間曲線，三次元曲線
skew distribution　非対称分布
skew gear　食い違い歯車
skew wheel　スキュー車【2軸が平行でもなく交わりもしていない回転伝達用の摩擦車のこと】
skewed slot　斜めスロット
skid　すべり，すべり止め，枕木，防舷(げん)材
skill　熟練，巧みさ，技能
skilled　形熟練した，熟練を要する
skilled crewmember　熟練船員
skin　皮膚，表皮
skin effect　表皮効果
skin friction　表面摩擦
skin horsepower　摩擦有効馬力
skin pass rolling　調質圧延
skirt　スカート，裾(すそ)
関 piston skirt：ピストンスカート
skylight　スカイライト，天窓
slab　スラブ，背板，平板，厚板，鋼片
slabbing cutter　平削りフライス
slack　形ゆるい　副ゆるく
名ゆるみ，たるみ
slacken　動（ねじなどが）緩む
slag　スラグ，鉱滓(こうさい)
slag furnace　スラグ炉【鉛鉱石を焙焼してスラグ化する炉】
slake　動［石灰を］消化する，消和する，［渇きを］満足させる
slaked lime　消石灰，水酸化カルシウム
slamming　スラミング，水面衝撃
slant　動斜めになる，傾斜する
名傾斜，斜面
slant distance　傾斜距離
slant face　傾斜面

slave　従属装置
slave cylinder　追従シリンダ
slave station　従局
sledge hammer　大ハンマ
sleeve　スリーブ，軸ざや
sleeve bearing　摺動ベアリング，スリーブ軸受
sleeve joint (coupling)　スリーブ継手
sleeve nut　締め寄せナット
sleeve valve　スリーブ弁，筒形弁
slenderness ratio　細長比
slice　名薄切り，スライス
動薄く切る，削り取る，切り分ける，分割する
slide　動すべる，滑走する，摺動する　名スライド，すべり，滑走
slide bar　すべり棒
slide bearing　すべり軸受
slide caliper　ノギス
slide fit　すべりばめ
slide guide　すべり座
slide head　スライドヘッド
slide rheostat　すべり抵抗器
slide rule　計算尺
slide spool valve　スライドスプール弁【方向制御弁の一種】
slide valve　すべり弁
slide way　すべり面
slider　スライダ，すべり子
slider crank mechanism　スライダクランク機構
sliding　形すべる，移動する
sliding bearing　すべり軸受
sliding cam shaft system　カム軸移動式
sliding contact　すべり接触
sliding crank mechanism　すべりクランク機構
sliding door　引き戸
sliding fit　すべりばめ
sliding friction　すべり摩擦
sliding jack　送りジャッキ
sliding joint　すべり継手
sliding key　すべりキー
sliding motion　すべり運動

sliding pair　すべり対偶
sliding pressure operation　変圧運転
sliding surface　すべり面
sliding vane compressor　ベーン圧縮機
sliding vane type rotary pump　すべり羽根式回転ポンプ
sliding velocity　すべり速度
slight　形わずかな(の)
slightly　副わずかに，少し
slime　ヘドロ，軟泥
sling　吊り索，吊り縄，スリング
　sling chain　掛け鎖
slinger　油切り
slip　動すべる，滑り落ちる　名スリップ，失脚，すべり，ずれ【slide は「長くすうっとすべる」，slip は「急につるりとすべる」】
　slip back　逆流
　slip casting　スリップ鋳造
　slip dynamic characteristic　すべり動特性
　slip factor　すべり率
　slip frequency　すべり周波数
　slip joint　すべり継手
　slip line　すべり線
　slip meter　すべり計
　slip-on type　差込フランジ
　slip plane　すべり面
　slip-proof　すべり止め
　slip ratio　スリップ比，スリップ率，すべり率，失脚比
　slip ring　スリップリング，集電環，滑動環
　slip stream　[プロペラの]後流
slippage　すべり[量]，[目標との]ずれ
slipper　すべり金
　slipper pad　すべり金受け
　slipper ring　すべり環
slippery　形すべりやすい
slit　動切り開く，細長く切る，切り目を入れる　名スリット，細長い切り口
slop　動[液体が]こぼれる，はねる
slope　名スロープ，傾斜，勾配，傾き，たわみ角　動傾斜する，傾く
sloshing　スロッシング，液面動揺
slot　名スロット，みぞ，みぞ穴，細長いみぞ　動～にみぞ[穴]をつける，はめ込む
　slot welding　スロット溶接，みぞ溶接
slotted　形細長い穴(みぞ)のついた
　slotted head set screw　すりわり付き止めねじ
　slotted link　みぞ付きリンク
　slotted nut　みぞ付きナット
slow　形遅い，低速の，緩慢な　副遅く，ゆっくり　動遅くする，遅らせる
　slow acting relay　緩動継電器
　slow cooling　徐冷
　slow down　動減速する
　slow filtration　緩速ろ過法
　slow runner　低速羽根車
　slow speed engine　低速機関
slowdown　減速
slower　「slow」の比較級
slowly　副おそく，ゆっくり
sludge　スラッジ，析出物，汚泥【排水処理タンク，ボイラの底などにたまる泥状の廃物のこと】
　sludge tank　スラッジタンク
slug　名スラグ，散弾　動弾を込める
　slug flow　スラグ流
sluice　水門　同 sluice gate
　sluice door　スルースドア，堰戸
　sluice valve　スルース弁，仕切弁
slurry　スラリ【泥，粘土，セメントなどの懸濁液のこと】
small　形小さい，少ない，小型の，わずかな，小規模の
　small diameter　小口径
　small end　[連接棒の]スモールエンド，小端　反 big end：大端
　small scale integrated circuit　小規模集積回路
smart　形冴えた，活発な　動うずく，ずきずき痛む　名痛み，苦痛
smith　金属細工人，鍛冶屋
　smith welding　鍛接

smith work　火造仕事
smithing　鍛冶，鍛造
smog　スモッグ【煙と霧が混合した状態】
smoke　名煙　動煙を出す，いぶる
　smoke and soot　ばい煙
　smoke box　煙室，煙道
　smoke chart　煙色図
　smoke consuming apparatus　無煙装置
　smoke detector　煙探知器
　smoke dust　煤塵（ばいじん）
　smoke exhaust　排煙
　smoke indicator　スモークインジケータ，ばい煙濃度計
　smoke meter　ばい煙濃度計
　smoke pipe　煙道管
　smoke tube　煙管
　smoke tube boiler　煙管ボイラ
smokeless　形煙を出さない，無煙の
smooth　形滑らかな，平らな　動滑らかになる，平滑化する
　smooth drum　平胴
　smooth nozzle　平ノズル
　smooth surface　平滑面
　smooth water area　平水区域
smoothing circuit　平滑回路
smoothing device　平滑装置
smoothing filter　平滑フィルタ
smoothing roll　くせ取りロール
smoothly　副なめらかに，円滑に
smother　動窒息させる，厚くおおう　名濃霧，いぶり灰
Sn base white metal　錫系ホワイトメタル
S-N curve　S-N 曲線
SN (Signal-to-Noise) ratio　SN 比，信号対雑音比
snagging　[鋳造品，鍛造品の]ばり取り
snap　動パチンと閉める，ポキンと折れる，機敏に動く　名締め金，留め金
　snap acting　スナップ作動
　snap gauge　はさみゲージ，軸用限界ゲージ
　snap ring　止め輪
　snap switch　ひねりスイッチ
snatch　動強奪する　名断片
　snatch block　開閉滑車，切欠き滑車
snifter (snifting) valve　[蒸気機関の]空気調節バルブ，漏らし弁
snub　動急に止める，抑える　形ずんぐりした　名冷遇，緩衝器
snubber circuit　スナバ回路【スイッチ回路において，スパイク電圧を防止する回路】
so　副そのように，そんなに，それほど，とても　接～するように，～するために，それで
　関 and so on：～など
　so as to ～　～するように，～するために
　so…as to ～　～するほど…
　so called　形いわゆる
　so far as ～　接～の限りでは
　so that ～　その結果～，だから～
　so ～ that…　非常に～なので…，…であるほど～
　so to speak (say)　いわば，まるで
soak　動浸す，吸収する　名浸透，しみ込み
　soak[ing] test　浸水試験
soaking pit　均熱炉【鋼塊を圧延するために均一な温度に加熱する炉のこと】
soap　石鹸，アルカリ金属塩
society　学会，協会，社会
　Society of Mechanical Engineers　機械学会
socket　ソケット，受け口
　socket and ball joint　玉継手
　socket and spigot joint　いんろう継手
　socket joint　ソケット継手
　socket outlet　コンセント
　socket pipe　ソケット管
　socket welding　差し込み溶接
　socket wrench　箱スパナ
soda　ソーダ，炭酸ソーダ，苛性ソーダ
　soda acid extinguisher　酸アルカリ消火器

soda ash　ソーダ灰
soda niter　ソーダ硝石
sodium　ナトリウム　㊋Na
　sodium bicarbonate　重炭酸ナトリウム，重そう
　sodium carbonate　炭酸ナトリウム，ソーダ灰，炭酸ソーダ
　sodium chloride　塩化ナトリウム，食塩
　sodium grease　ナトリウムグリース
　sodium hydroxide　水酸化ナトリウム，苛性ソーダ
　sodium ion　ナトリウムイオン
　sodium nitrate　硝酸ナトリウム
　sodium soap base grease　ソーダ石けん基グリース
　sodium sulfate　硫酸ナトリウム
soft　形柔軟な，やわらかい
　㊥hard：硬い
　soft coal　軟[質]炭，瀝(れき)青炭
　soft hammer　軟質ハンマ
　soft iron　軟鉄
　soft lead　軟鉛
　soft magnetic material　軟質磁性材料
　soft metal　ソフトメタル，軟金属，軟質金属
　soft packing　ソフトパッキン
　soft patch　ボルト締め当て金
　soft rubber　軟質ゴム
　soft solder　軟ろう，軟質はんだ【約370℃以下で溶融する鉛とスズの合金はんだ】
　soft steel　軟鋼
　soft water　軟水
　　㊥hard water：硬水
soften　動軟化する，軟水にする
softening　軟化，軟水法，軟水焼きなまし
　softening agent　軟化剤
　softening degree　軟化度
　softening point　軟化点
software　ソフトウェア【プログラムやその作成技術などの総称】
soil　名汚損，汚物
　動よごす，汚損する
　soil pipe　汚水管

soil pump　汚水ポンプ
sol　ゾル【流動性のコロイド溶液】
solar　形太陽の
　solar battery (cell)　太陽電池
　solar collector　太陽熱集熱器
　solar energy (power)　太陽エネルギ
　solar heat　太陽熱
　solar light　太陽光[線]
　solar panel　太陽電池パネル
SOLAS convention　SOLAS条約【海上における人命の安全のための条約】
solder　名はんだ
　動はんだ付けをする
　solder joint　はんだ継手
　solder plating　はんだめっき
soldering　はんだ付け
　soldering iron　はんだごて
　soldering lamp　トーチランプ，ブローランプ
　soldering paste　はんだペースト
solenoid　名ソレノイド，筒型コイル
　solenoid brake　電磁ブレーキ
　solenoid operation　電磁操作
　solenoid valve　ソレノイド弁，電磁弁
soleplate　台板，基礎板
solid　形固体の，濃い，堅い，立体の　名固体，固形物，立体
　solid angle　立体角
　solid bearing　一体軸受
　solid body (state)　固体
　solid cam　立体カム
　solid carburizing　固体浸炭法
　solid crank shaft　一体クランク軸
　solid drawn pipe　引抜管，継目無し管
　solid electrolyte　固体電解質
　solid film lubrication　固体膜潤滑
　solid forging　丸打ち，実体鍛錬
　solid friction　固体摩擦
　solid fuel　固体燃料
　solid injection engine　無気噴射機関
　solid insulator　固体絶縁物
　solid line　実線
　solid lubricant　固体潤滑剤

solid matter　固形体
solid of revolution　回転体
solid particle　固体粒子
solid phase　固相
　関 liquid phase：液相，gas phase：気相
solid piston　一体型ピストン
solid point　凝固点
solid propeller　一体型プロペラ
solid resistor　固体抵抗器
solid rotor　塊状回転子
solid shaft　中実軸
solid solution　固溶体
　関 α solid solution：α固溶体
solid solution hardening　固溶硬化
solid type　一体型
solidification　固形化処理，固化，固体化，凝固
solidify　動凝固する，凝固させる
solidifying　凝固
　solidifying point　凝固点
　同 solidification point, solid point, freezing point
solidity　固体性，弦節比，剛率
solidus line　固相線
solitary　形単独の，ただ一つの
　solitary wave　孤立波
solubility　溶解度，溶解性，可溶性
　solubility curve　溶解度曲線
　solubility limit　溶解限度
soluble　形溶ける，溶けやすい，溶解性の　名溶解分
　soluble anode　可溶陽極
　soluble oil　水溶性油
solute　溶質
solution　溶剤，溶解，溶液，解，解答
　solution gas　溶解ガス
　solution treatment　固溶化処理
solve　動解く，解決する，溶解する
solvent　名洗浄液，溶媒，溶剤　形溶かす，溶解力のある
　solvent extraction　溶媒抽出
　solvent insoluble matter　溶剤不溶分
some　形いくらかの，ある〜，何れかの〜，かなりの
　関 in some of：〜の一部において
　some of　〜のいくらか，一部の〜
someone else　ほかの誰か
something　重要なもの
sometimes　副ときどき，たまに
somewhat　副いくぶん，多少，やや
somewhere　副どこかに(へ)，いつか，およそ
sonic　形音速の
　sonic precipitator　音波脱塵装置
　sonic speed (velocity)　音速
soon　副間もなく，すぐに，早く
　関 no sooner 〜 than…：〜するや否や…
sooner or later　副遅かれ早かれ
soot　すす，ばい煙
　soot blower　スートブロア，すす吹き器
　soot blowing　スートブロー(すす吹き)
　soot fire　スートファイア
sophisticate　動精巧なものにする
sorbent　吸着剤，吸収剤
sorbite　ソルバイト【焼入れ組織の一つ】
sorption　収着，吸収，吸着
sorry　形気の毒で，すまないと思う
　関 I'm sorry that 〜：〜を残念に思う
sort　名種類
　動分類する，区分けする
sound　名音，響き　動響く，伝わる，水深を測る　形健全な，正直な
　sound absorbing material　吸音材
　sound bar　聴音棒
　sound effect[s]　音響効果
　sound insulation　遮音
　sound intensity　音の強さ
　sound level meter　騒音計，音量計
　sound pollution　騒音公害
　sound pressure　音圧
　sound proof chamber　防音室
　sound proof door　防音戸
　sound signal　音声信号
　sound speed (velocity)　音速
　sound volume　音量
　sound wave　音波
sounding　サウンディング，[タンク

の]測深
sounding pipe　測深管
sounding table　測深表
source　ソース，源，資源，動力源，光源，熱源，電源
source current　電源電流
source follower　ソースフォロワ【電界効果トランジスタ(FET)の電力増幅回路】
source of power　電源
source voltage　電源電圧
SOx　硫黄酸化物
関 NOx：窒素酸化物
space　名室，場所，空間，宇宙 動間隔をあける
space charge　空間電荷
space factor　占積率
space heater　スペースヒータ
space lattice　空間格子
space permeability　真空透磁率
space vector　空間ベクトル
spacer　スペーサ【間隔をあけるもの】
spacing　間隔，線間距離，字間
spacing signal　間隔符号
spall　動砕ける，剥離する
spalling　スポーリング，破砕，剥離
span　名スパン，期間，範囲，翼幅 動〜にかける，及ぶ，綱でしばる
spanner　スパナ
spar　円材
spar buoy　円柱浮標(ブイ)
spare　動思いやる，節約する 名予備品　形予備の，余分の
spare gear　予備品
spare machine　予備機器
spare parts　予備品，予備部品
spark　名スパーク，火花，閃光 動火花を出す，閃光を発する
spark arrester　火花防止装置
spark chamber　放電箱【荷電粒子の飛跡を観測する装置のこと】
spark[ing] coil　点火コイル
spark discharge　火花放電
spark gap　火花すきま
spark ignition engine　火花点火機関
spark[ing] plug　点火プラグ

spark sintering method　放電焼結法
spark test　火花試験
sparking　スパーキング，火花発生
sparking potential　破壊電位
sparking voltage　火花電圧
sparkle　名火の粉，閃光，きらめき 動火花を発する，輝く，きらめく
sparkless　形無火花の
sparse　形希薄な，まばらな
sparse matrix　疎行列
spatial　形空間の，空間的な
spatial angular frequency　空間角周波数
spatial coherence　空間コヒーレンス
spatter　スパッタ
speak　動話す，伝える，通信する
speaking circuit　通話回路
speaking tube　伝声管
special　形特別の，特殊な，専門の
special character　特殊文字
special function　特殊関数
special [heat] resistance steel　特殊耐熱鋼
special steel　特殊鋼
special survey　特別検査
special tool　専用工具
specialize　動専門にする，特化する
specialized form　特殊な形
specific　形特有の，固有の，特定の，比の
specific carrying capacity　比負荷容量
specific charge　比電荷
specific conductance　導電率
specific consumption　消費量
specific energy　比エネルギ
specific enthalpy　比エンタルピ
specific entropy　比エントロピ
Specific Fuel Consumption　SFC，燃料消費率
specific gravity　比重
specific gravity meter　比重計
specific heat　比熱
specific heat at constant pressure　定圧比熱
specific heat at constant volume

specifically

定容比熱，定積比熱
specific heat consumption 熱消費率
specific heat ratio 比熱比
specific humidity 比湿
specific humidity at saturation 飽和湿度
specific impulse 比推力
specific power 比出力
specific pressure 面圧
specific resistance 比抵抗，固有抵抗，抵抗率
specific response 比感度
specific sliding すべり率
specific speed 比速度，比較回転数
specific steam consumption 蒸気消費率
specific strength 比強度
specific surface area 比表面積
specific thermal conductivity 熱伝導度
specific thrust 比推力
specific torsion angle 比ねじれ角
specific velocity 見かけ速度
specific viscosity 比粘度
specific volume 比容積，比体積
specific weight 比重量

specifically 副特に
specification スペック，仕様，仕様書，諸元，設計書，性状表
specified 形規定の
specified temperature 規定温度
specified time relay 限時継電器
specified value 規格値
specify 動明記する，指定する，仕様書に記入する
specimen 試験片，見本，試料
spectral 形スペクトルの
spectral analysis スペクトル解析
spectral density スペクトル密度
spectral distribution スペクトル分布
spectral emissivity 分光放射率
spectral line スペクトル線
spectral radiant energy スペクトル放射エネルギ
spectral reflectance 分光反射率
spectral transmittance 分光透過率

spectrometer 分光計，分光器，分析計
spectrometry 分光測定法，分光分析
spectrophotometer 分光測定器，分光光度計
spectrophotometry 分光測光法，分光測色法
spectroradiometer 分光放射計
spectrum スペクトル，分光
spectrum amplitude スペクトル振幅
spectrum analysis スペクトル分析，分光分析
spectrum analyzer スペクトル分析器
spectrum locus スペクトル軌跡
specular 形鏡のような
specular reflection 鏡面反射
specular surface 鏡面
specular transmission 正透過
speculate 動思索する，推測する
speculum 金属鏡，反射鏡
speech ことば，話すこと，話し方
speech amplifier 音声増幅器
speech recognition 音声認識
speed 名スピード，速度，速力，回転数 動速度を上げる
speed [change] gear 変速装置，変速機
speed characteristic 速度特性
speed control 速度制御
speed counter 回転計数器
speed droop 速度低下
speed enhancement 増速機構
speed exchanger 変速器
speed factor 速度係数
speed governor 調速機
speed indicator 速力指示器
speed length ratio 速長比
speed limit 速度制限
speed limiting device 速度制限装置
speed meter (indicator) 速度計
speed of light 光速
speed of sound 音速
speed over the ground 対地速力 (O.G.)
speed range 速度範囲

speed ratio　変速比
speed reducer　減速機
speed reduction　減速，速度低下
speed regulation　速度調整，速度変動率
speed regulator　速度調整器
speed sensor　速度センサ
speed signal　速度信号
speed switch　速度スイッチ
speed test　速度試験，回転試験
speed through the water　対水速力 (STW)
speed torque characteristic curve　速度トルク特性曲線
speed trial　速力試験
speed variation　速度変動率
speed voltage　速度電圧
spell　動[話・字を]つづる，～を意味する　名魅力，期間，しばらくの間
spelter　スペルタ，亜鉛【不純物を数％含んだ亜鉛】
spelter solder　硬質はんだ
spend　動消費する，消耗する
spent　動「spend」の過去・過去分詞　形消費された，消耗した，疲れきった
sphere　球，球体，球面
spherical　形球形の，球の
spherical angle　球面角
spherical bearing　球面軸受
spherical block　球状ブロック
spherical body　球体
spherical cam　球面カム【立体カムの一種】
spherical coordinates　球[面]座標
spherical mechanism　球面運動機構
spherical mirror　球面鏡
spherical pair　球面対偶
spherical roller bearing　球面ころ軸受
spherical shell　球殻
spherical tank　球形タンク
spherical triangle　球面三角形
spherical valve　玉形弁
sphericity　球形度，真球度

spheroid　回転楕円体
spheroidal graphite cast iron　球状黒鉛鋳鉄，ノジュラ鋳鉄
spheroidizing treatment　球状化処理
spherometer　球面計
spider　三脚台，わく
spider of motor　電動機わく材
spider-web coil　くもの巣コイル
spigot　[樽などに差し込む]栓，差込み(ねじ込み)部
spigot joint　いんろう(はめ込み)継手
spill　動こぼれる，あふれる　名こぼれ，汚染
spill clearance　逃しすきま
spill loss　こぼれ損
spill port　逃し穴
spill valve　逃がし弁，加減弁
spillage　こぼすこと，流出
spillage oil　流出油
spin　動回す，紡ぐ，回転して作る　名スピン，回転
spin hardening　回転焼入れ
spindle　スピンドル，軸，心棒，主軸，回転軸
関 valve spindle　弁棒
spindle bush　主軸ブッシュ
spindle nose　主軸端
spindle oil　スピンドル油
spindle sleeve　主軸スリーブ
spindle speed　主軸速度
spinning　スピニング，回転
spiral　形螺旋(らせん)形の，旋回した　名スパイラル，らせん，渦　動らせん形にする，過巻状に進む
spiral angle　ねじれ角
spiral arrangement　らせん状配列
spiral bevel gear　はすばかさ歯車
spiral casing　渦形室
spiral coil　らせんコイル
spiral flute　ねじれ溝
spiral gasket　渦巻形ガスケット
spiral gear　ねじれ歯車，はすば歯車
spiral groove bearing　スパイラルグルーブ軸受
spiral pipe　スパイラル鋼管

spiral spring　渦巻ばね，ぜんまい
spiral stairway　らせん階段
spiral thermometer　らせん温度計
spiral vortex　渦巻流れ

spirit　形精神の，アルコールの　名工業用アルコール，蒸留酒，エッセンス，有効成分，魂
spirit thermometer　アルコール温度計
spirit varnish　揮発性ワニス
spirit[s] of salt　塩酸
spirit[s] of wine　エタノール
spirit[s] of wood　木精，メタノール

splash　動はね散らす，はねかける　名はねかけ
splash guard　はねよけ
splash lubrication　はねかけ注油，飛沫潤滑
splash proof　防滴

splice　動添え継ぎをする　名接続，添え継ぎ
splice plate　添え継ぎ台，継ぎ目板

splicer　継ぎ台，接続具
splicing machine　接合機
spline　名キー，キー溝，スプライン　動〜にキー[溝]をつける
spline curve　スプライン曲線
spline function　スプライン関数
spline shaft　スプライン軸
spline tooth　スプライン刃

split　動割れる，裂ける　名割れ目，ひび
split bearing　割り軸受
split cotter　割りコッタ【割りピンの一種】
split gear　割り歯車
split nut　半割りナット
split phase induction motor　分相誘導電動機
split phase protection　分相継電方式　同 split-phase relay scheme
split phase start　分相始動
split pin　割りピン
split pulley　割りベルト車

spoil　動害する，損なう
sponge　名スポンジ，吸収物，海綿　動吸い取る，スポンジで洗う

spontaneous　形自発的な，自然な
spontaneous combustion (ignition)　自然発火
spontaneous ignition temperature　自然発火温度

spool　スプール，糸巻き
spool valve　スプール弁

spot　名点，はん点，しみ　動〜をよごす
spot facing　座ぐり
spot test　滴下試験，スポット分析
spot welding　スポット溶接，点溶接

spotting　位置決定，すり合わせ，しみ，はん点

spout　名噴出，噴水　動〜を噴出する，吹き出す
spout velocity　噴出速度

spray　名スプレ，噴霧，散布　動噴霧する
spray angle　噴霧角
spray burner　スプレ式バーナ
spray combustion　噴霧燃焼
spray cooling　噴霧冷却
spray damping machine　霧吹き機
spray distribution　噴霧分布
spray dryer　噴霧乾燥器
spray gun　スプレガン
spray nozzle　噴霧ノズル
spray pattern　噴霧状態
spray penetration　貫通力

sprayer　霧吹き，噴霧器
spraying　噴霧，吹付け，溶射
spread　動広げる，広がる，伸ばす，散布する　名広がり，散布度
spreadability　分散性，広がること
spreading　延展，伸展，伝播，塗布，拡大
spreading coefficient　拡張係数
spreading method　展開法

spring　動跳ぶ，はねる　名スプリング，ばね，ぜんまい
spring balance[r]　ばねばかり
spring bearing (shoe)　ばね受け
spring bender　ばね曲げ
spring buffer　ばね緩衝器

spring capacity　ばね容量
spring constant　ばね定数
spring contact　ばね接点
spring control　ばね制御
spring cotter　割りピン
spring governor　ばね調速機
spring guide　ばね案内
spring index　ばね指数
spring key　ばねキー
spring line　斜め係留索
spring load　ばね荷重
spring-loaded　形ばね懸架式
spring-loaded centrifugal governor　ばね荷重式遠心調速機
spring motor　ばねモータ
spring pendulum　ばね振り子
spring pressure　ばね圧力
spring rigging　ばね装置
spring safety valve　ばね安全弁
spring shackle　ばね吊り
spring steel　ばね鋼
spring tension　ばね張力
spring washer　ばね座金
　同 spring lock washer
sprinkle　動まき散らす,振りかける
sprinkler　スプリンクラ
　sprinkler system　散水式,スプリンクラ装置
sprocket　スプロケット,鎖歯車の歯
　sprocket wheel　チェーン歯車,鎖歯車
sprue　スプル,[鋳造の]湯道
spur　名拍車　動〜に拍車をかける
　spur bevel gear　食違い歯車
　spur gear（wheel）　平歯車
sputtering　スパッタリング,溶射法
spy　名探偵,精査
　動見つけ出す,密かに調査する
　spy cock　小コック,調べコック
square　名正方形,二乗,直角定規　形正方形の,二乗の,四角な,直角の　動正方形にする,四角にする,二乗する
　square bar　角材
　square claw washer　つめ付き角座金
　square engine　方形機関

square key　角キー
square matrix　正方行列
square pole　四角柱
square root　平方根
square steel　角鋼
square thread　角ねじ
square wave　矩形波,方形波
squared paper　方眼紙
squaring shear　直刃せん断機
squeeze　動〜を圧搾する,〜を絞る,推し進める　名圧搾,はめ込み
　squeeze effect　スクイーズ効果
　squeeze molding machine　圧搾造型機
　squeeze packing　はめ込み型パッキン
squill vice　しゃこ万力
squirrel　名リスかご
　動蓄える,大切に保存する
　squirrel cage damper　かご形ダンパ
　squirrel cage induction motor　かご形誘導電動機
　squirrel cage motor　かご形電動機
　squirrel cage rotor　かご形回転子
　squirrel cage winding　かご形巻線
squish　スキッシュ
stability　安定性,復原力
　stability analysis　安定度解析
　stability criterion　安定判断
stabilization　安定,安定化
stabilize　動安定させる,固定させる
stabilizer　スタビライザ,安定化装置
stabilizing supply　安定化電源
stabilizing treatment　安定化処理【熱処理の一種】
stable　形安定な,安定した,静止した
　stable burning　安定燃焼
　stable equilibrium　安定つり合い
　stable state　安定状態
stack　名スタック,積み重ね,堆積,排気筒,山,煙突,立て筒　動積み重ねる
　stack gas　煙道ガス,煙突ガス
stacking fault　積層欠陥
staff　名スタッフ,職員

stage 名ステージ，段，段落，足場 動上場する，上演する，計画的に実施する
  stage diagram efficiency 段線図効率
  stage efficiency 段効率
  stage filter 階段ろ過器
  stage internal efficiency 段落内部効率
  stage pressure 段圧力
  stage pressure coefficient 段圧力係数
  stage temperature 段温度
  stage valve 段弁
stagger 動千鳥に配列する，食い違わせる，よろめく 名よろめき，千鳥足，波形配列 形ジグザグ配列の，千鳥形の
  stagger angle 食違い角
staggered arrangement 千鳥配列
staggered header 波形管寄せ
staggered tube layout 千鳥管配置
stagnation よどみ
  stagnation point よどみ点
  stagnation pressure 全圧力，よどみ点圧力
  stagnation temperature 総温度，よどみ点温度
stain 名しみ，さび 動さびる，汚れる，しみがつく
staining ステーニング【軸受表面に生じるしみの一種】
stainless 形さびない，汚れのない
  stainless steel ステンレス鋼
stake 名くい，支柱，賭け 動賭ける，くいを打って仕切る
stall 名失速，エンスト，停止 動失速する，止める
stalling 失速
  stalling angle 失速角
  stalling speed 失速速度
  stalling torque 停動トルク
stamp 動踏みつける，粉砕する 名打出し機，圧断機，突き棒
  stamp forging 型鍛造
stamping core plate 電機子鉄板
stanchion 名柱，支柱，スタンチョン 動～に支柱をつける
stand 動耐える，立つ，位置する 名スタンド，台
  stand alone 形独立した，それだけで動く
  stand alone power system 自立電源システム
  stand by 動待機する，用意する，スタンバイする
  stand-by 名スタンバイ，予備，待機状態 形控えの，待機の
  stand-by facility 予備設備
  stand-by machine 予備機器
  stand-by pump 待機ポンプ
  stand-by service 予備供給
  stand for 動～を表す，代表する
  stand idle 動使用されない状態である
standard 名標準，標準器，基準，規格 形標準の，規格にあった
  standard atmosphere 標準大気
  standard atmospheric pressure 標準大気圧
  standard capacitor 標準コンデンサ 同 standard condenser
  standard cell 標準電池
  standard condition 標準状態
  standard cycle 標準サイクル
  standard deviation 標準偏差
  standard electrode 標準電極
  standard error 標準誤差
  standard frequency 標準周波数
  standard gauge 標準ゲージ
  standard inductance 標準インダクタンス
  standard input 標準入力
  standard instrument 標準計器
  standard output 標準出力
  standard parts 標準部品
  standard pitch circle 標準ピッチ円
  standard resistance 標準抵抗
  standard signal generator 標準信号発生器
  standard solution 標準液
  standard spur wheel 標準平歯車
  standard temperature 標準温度

standard thread gauge 標準ねじゲージ
standard type 標準型
standard voltage 標準電圧
standardization 標準化, 規格化
standardize 動標準化する, 標準にあわせる, 規格化する
standardized normal distribution 標準化正規分布
standing 形直立の, 静止した, 持続的な
standing block 固定滑車
standing order 当直基準
standing rigging 静索
standing wave 定常波, 定在波
standstill 名停止, 静止, 足踏み 形据え置きの
star 名星, 星形 形星形の 動星印をつける
star connected device 星形接続機器
star connection 星形結線, Y結線
star-delta connection Y-Δ結線
star network 星状網
star-star connection Y-Y結線
star wheel 星形車
starboard 名右舷 形/副右舷の(に) 動[舵を]右に取る
stare 動じっと見つめる
start 動始める, 始動する 名スタート, 出発, 開始
start signal スタート信号
start-up 運転開始, 始動
startability 始動性
starter スタータ, 始動器
starter button 起動ボタン
starter circuit 起動回路
starter panel 起動パネル
starting 始動, 起動
starting air 始動空気
starting air bottle (receiver) 始動空気だめ
starting button 始動ボタン
starting cam 始動カム
starting characteristic 始動特性
starting circuit 始動回路
starting compensator 始動補償器
starting contactor 始動接触器
starting current 始動電流, 起動電流
starting device 始動装置
starting engine 始動機
starting friction 起動摩擦
starting handle 始動ハンドル
starting method 始動法, 起動方式
starting point 出発点, 起点, 原点
starting relay 始動継電器
starting resistance 始動抵抗
starting rheostat (resistor) 始動抵抗器
starting shaft 始動軸
starting speed 始動速度
starting system 始動装置
starting torque 起動トルク
starting transformer 始動変圧器
starting valve 始動弁
starting voltage 始動電圧
starve 動欠乏する
state 名状態, 事情 動述べる
state density 状態密度
state equation 状態方程式
state function 状態関数
state indicator 状態表示器
state variable 状態変数
state vector 状態ベクトル
static 形静的な, 静止した 名静電気
static balance 静的つり合い
static capacitor 静電コンデンサ, 電力用コンデンサ
  同 static condenser
static carrying capacity 静止負荷容量
static characteristic 静特性
  反 dynamic characteristic:動特性
static charge 静電荷
static control 定位制御
static coupling 静電結合
static discharge 静電放電
static electricity 静電気
static equilibrium 静つり合い
static fatigue 静疲労
static friction 静摩擦
static head 静[圧]水頭, 静落差

static induction　静電誘導
static induction transistor　静電誘導トランジスタ
static Leonard system　静止レオナード方式
static load　静[止]荷重
static machine　静電誘導起電機
static pressure　静圧
static stability　静的安定
static test　静的試験, 静止試験
static tube　静圧管
static unbalance　静的不つり合い
statical　形静的な, 静止した
statical balance　静的つり合い
statical deflection　静的たわみ
statical equilibrium　静つり合い
statical friction　静摩擦
statical head　位置水頭
statical moment of area　断面一次モーメント
statical stress　静応力
statically　副静的に
statically determinate beam　静定はり
statically indeterminate beam　不静定はり
statics　静力学
station　ステーション, 場所, 発電所, 本部, 部署, 位置
stationarity　定常性
stationary　形静止した, 固定の, 定常の, 不変の
stationary armature　固定電機子
stationary blade (vane)　固定羽根, 静翼　関 moving blade：動翼
stationary contact　固定接触子
stationary creep　定常クリープ
stationary current　定常電流
stationary fit　締まりばめ
stationary flow　定常流
stationary fluid　静止流体
stationary part　固定部(分)
stationary state　定常状態
stationary vector　固定ベクトル
stationary vibration　定常振動
stationary wave　定常波
statistical　形統計的な, 統計上の
statistical analysis　統計的分析
statistical error　統計誤差
statistical fluctuation　統計的変動, 統計的ゆらぎ
statistical mechanics　統計力学
statistical physics　統計物理学
statistics　統計学
stator　ステータ, 固定子
　関 rotor：ロータ, 回転子
stator armature　固定電機子
stator blade　静翼
stator coil　固定子コイル
stator core　固定子鉄心
stator current　固定子電流
stator frame　固定子枠
stator winding　固定子巻線
status　状態, 状況
statutory　形法定の, 法令の
stay　動留まる, 支持する　名ステー, 支え, 張り綱, 支索, 支線
stay bar　支え棒
stay bolt　控えボルト
STCW convention　STCW条約【船員の訓練, 資格証明及び当直に関する条約】
steadiness　安定
steady　形安定した, 定常の
steady analysis　定常解析
steady flow　定常流
steady load　不変負荷
steady motion　定常運動
steady pin　固定ピン
steady state　安定状態, 定常状態
steady state creep　定常クリープ
steady state creep rate　定常クリープ速度
steady state current　定常電流
steady state deviation　定常偏差
steady state error　定常偏差, 定常状態誤差
steadying resistance　安定抵抗
steam　名蒸気, 水蒸気
　動蒸発する　形蒸気の
steam accumulator　蒸気アキュムレータ
steam bleeding　抽気

steam boiler　蒸気ボイラ
steam brake　蒸気ブレーキ
steam calorimeter　蒸気熱量計
steam chart　蒸気線図
　⊜ steam diagram
steam chest　蒸気室　⊜ steam box
steam compression refrigerating machine　蒸気圧縮式冷凍機
steam consumption　蒸気消費量
steam cylinder　蒸気シリンダ
steam demulsibility number　蒸気抗乳化度
steam drum　蒸気ドラム
steam engine　蒸気機関
steam evaporator　蒸化器
steam explosion　水蒸気爆発
steam extraction (bleeding)　抽気
steam gauge　蒸気圧力計
steam generating tube　蒸発管
steam generator　蒸気発生器
steam gland　蒸気パッキン押え
steam hammer　蒸気ハンマ
steam heat　蒸気熱量
steam heater　蒸気暖房器
steam heating coil　蒸気加熱コイル
steam jet　蒸気噴射
steam jet [air] ejector　蒸気噴射式空気エゼクタ
steam jet oil burner　蒸気噴射式バーナ
steam line　蒸気線
steam pipe　蒸気管
steam pocket　蒸気だまり
steam point　水蒸気点
steam port　蒸気口
steam power　蒸気動力，汽力
steam power plant　蒸気動力プラント
steam pressure　蒸気圧力
steam pressure regulator　蒸気圧力制御器
steam radiator　蒸気放熱器
steam rate　蒸気消費率
　⊜ steam consumption ratio
steam reservoir　蒸気だめ
steam reheater　蒸気再熱器

steam seal　蒸気シール
steam separator　汽水分離器
steam ship　汽船
steam smothering system　蒸気消火装置
steam stop valve　蒸気塞止弁
steam strainer　蒸気こし
steam superheater　蒸気過熱器
steam table　蒸気表
steam temperature　蒸気温度
steam tight　気密
steam trap　蒸気トラップ【蒸気配管内から自動的に凝縮水を排出し除去する装置】
steam turbine　蒸気タービン
steam turbine generator　タービン発電機
steam valve　蒸気弁
steam whistle　汽笛
steam winch　蒸気ウインチ
steaming　蒸気洗い
steaming economizer　蒸気エコノマイザ
steaming test　汽醸試験
steel　鋼鉄，鋼（はがね）
　⑳ forged steel：鍛鋼
steel alloy　合金鋼
steel ball　鋼球
steel bar　鋼棒
steel brush　鋼製ブラシ
steel casting　鋳鋼，鋳鋼品
steel forging　鍛鋼
steel frame　鉄骨
steel ingot　鋼塊，インゴット
steel making　製鋼
steel pipe　鋼管
steel plate　鋼板，板金
steel scriber　けがきばり
steel tape　はがね巻尺
steel vessel　鋼船
steel wire　鋼線
steel wool　鋼綿，スチールウール
steep　動浸す，つける
　形険しい，急な
steeply　副急勾配に，急速に
steer　動舵をとる，操縦する

steering ステアリング，舵，舵取り装置
　steering angle　舵取り角
　steering engine　舵取り機
　steering gear (apparatus)　操舵装置
　steering motor　舵取り電動機
　steering system　操舵システム
　steering wheel　操舵輪

Stefan-Boltzmann's law　ステファン-ボルツマンの法則【電磁波のエネルギに関する物理法則】

stellite　ステライト【耐摩耗性・耐食性に富む合金】

stem　ステム，軸，心棒，船首，船首材　動由来する，止める
　stem bushing　弁棒ブッシュ

step　名ステップ，階段，足掛け　動踏む，踏み入れる
　step angle　ステップ角
　step bearing　ステップ軸受
　step bolt　段付ボルト
　step-down transformer　降圧変圧器
　step function　ステップ関数
　step input　ステップ入力
　step out　同期はずれ，脱調
　step response　ステップ応答
　step-up transformer　昇圧変圧器

stepped　形段の付いた，階段状の
　stepped drum　段付胴
　stepped gear　ずれば歯車
　stepped pulley　段車
　stepped screw　ステップねじ

step[ping] motor　ステッピングモータ，ステップ電動機，パルスモータ

stereograph　立体図

stereographic projection　立体射影，平射図法，ステレオ投影法

sterilization　滅菌，消毒

sterilizer　滅菌器

stern　船尾，とも
　stern bush　船尾管ブッシュ
　stern fast　船尾索
　stern tube　船尾管
　stern tube bearing　船尾管軸受
　stern tube gland　船尾管グランド押え
　stern tube sealing　船尾管シール装置
　stern tube shaft　船尾管軸

stick　動固着する，突き刺さる，食い込む　名スティック，棒
　stick circuit　保持回路

sticking　焼付き【摺動面が固着し動かなくなること】
　sticking friction　付着摩擦，固体摩擦

sticky　形粘着性の

stiff　形硬い

stiffen　動堅くなる，硬化させる

stiffener　補強材，硬化剤

stiffness　剛性，堅さ，剛さ，剛直性
　stiffness coefficient　剛性率，剛性係数

stifle　動窒息させる，[火などを]消す

still　副まだ，なお　名蒸留器　動蒸留する　形静止した
　still column　蒸留塔
　still kettle　蒸留がま
　still water　静水

stimulate　動刺激する，激励する

stir　動かき混ぜる，かき回す

Stirling engine　スターリングエンジン

stochastic　形確率論的な，推計学的な
　stochastic function　確率関数
　stochastic limits　推計限度
　stochastic method　確率的手法
　stochastic process　確率過程

stock　名貯蔵，蓄え　動仕入れる，蓄える
　stock fire　埋火

stoke　動[炉など]に火をたく，燃料をくべる

Stokes' law　ストークスの法則【流体中を運動する球と，それに働く抵抗に関する法則】

stone　名石，砥石　形石の，石製の　動砥石で磨く

stool　台，腰掛け，踏み台，軸受台

stop 動～を止める，～を妨げる 名停止
 stop ～ ing 動～することをやめる
 stop lamp（light） 停止灯
 stop ring pliers 止め輪用プライヤ
 stop signal 停止信号，ストップ信号
 stop to ～ 動立ち止まって～する
 stop valve 止め弁
stoppage 停止，障害，閉塞
stopped-up 形詰まった
stopper 停止装置，ストッパ，止め栓
storage 貯蔵，貯留，保管，記憶装置
 storage battery（cell） 蓄電池
 storage capacity 記憶容量
 storage device 記憶装置
 storage effect 蓄積効果
 storage element 記憶素子
 storage gain 積分利得
 storage heater 蓄熱ヒータ
 storage medium 記憶媒体
 storage principle 蓄積原理
 storage protection 記憶保護
 storage pump 揚水ポンプ
 storage tank ストレージタンク，貯蔵タンク
 storage time 蓄積時間
 storage tube 蓄積管
store 名店，貯蔵 動蓄える
 store room 貯蔵室，物置
stowage 積込み，積荷，格納
straddle 動またがる 名またがること
 straddle cutter またぎフライス
straight 形まっすぐな，一直線の，垂直な，水平の
 straight air brake 直通空気ブレーキ
 straight angle 平角，180度
 straight bed 通しベッド
 straight bevel gear すぐばかさ歯車
 straight blade 等厚翼
 straight circuit 直通回路
 straight compound 直型
 straight edge ストレートエッジ，直定規
 straight line 直線
 straight oil ストレートオイル

 straight polarity 正極性
 straight run 直留
 straight run distillation 直留蒸留
 straight run gasoline 直留ガソリン
 straight tail dog 回し金
 straight thread 平行ねじ
 straight type engine 直列型機関
 straight-way 形まっすぐに通す，直線的な
straighten 動まっすぐにする，整頓する，整理する
straightening ひずみ取り
 straightening roll くせ取りロール
 straightening vane 整流板，整流壁
straightness 真直度
strain 動変形させる，ひずませる，曲げる，引張る，濾（こ）す 名ひずみ，変形
 strain aging ひずみ時効
 strain circle ひずみ円
 strain energy ひずみエネルギ
 strain energy of dilation 膨張ひずみエネルギ
 strain energy of distorsion 形状ひずみエネルギ
 strain figure ひずみ模様
 strain gauge ひずみ計
 strain hardening ひずみ硬化【再結晶温度以下で塑性変形させることで，金属の硬さと強さが増大すること】
 strain meter ひずみ計
 strain sensor ひずみセンサ
strained glass 強化ガラス
strainer ストレーナ，こし器
straining pulley テンションプーリ，張り車
strand 名より糸，線 動［ロープの］よりを切る，座礁する
stranded wire より線
strange 形妙な，不思議な
strap 名帯金，目板 動革ひもで縛る，締め付ける
 strap brake 帯ブレーキ
strapped joint 当て継手
strategy 戦略，策略

stratified 形層化した
  stratified charge　層状給気
  stratified charge combustion　成層燃焼
  stratified flow　成層流れ
stratify 動層[状]にする，階層化する
stray 動さまよう，道に迷う
  形さまよっている
  stray current corrosion　電食，迷走電流腐食
  stray electron　迷走電子
  stray flux　漂遊磁束
  stray magnetic field　漂遊磁界
  stray voltage　浮遊電圧
streakline　流脈線
stream　名流れ，流線
  動流れる，流出する
  stream function　流れ関数
  stream line　流線
  stream line flow　流線流れ，層流
  stream line shape　流線形
  stream line wake　流線伴流
  stream surface　流面
  stream tube　流管
street　街路，通り
  street elbow　めすおすエルボ
strength　力，強さ，強度
  関 fatigue strength：疲れ強さ
  strength calculation　強度計算
  strength evaluation method　強度評価手法
  strength of materials　材料力学
  strength test　強度試験
  strength weld　耐力溶接
strengthen 動強くする，補強する
strengthened glass　強化ガラス
stress　名応力，圧迫，ストレス
  動圧力を加える
  stress concentration　応力集中
  stress concentration factor　応力集中係数
  stress corrosion　応力腐食
  Stress Corrosion Cracking　SCC，応力腐食割れ
  stress creep rate curve　応力-クリープ速度線図
  stress diagram　応力図
  stress fatigue　応力疲労
  stress function　応力関数
  stress intensity　応力度，応力強さ
  stress intensity factor　応力拡大係数
  stress relaxation　応力緩和
  stress relief　応力除去，応力緩和
  stress relief annealing　応力除去焼鈍し
  stress relief heat treatment　応力除去熱処理
  stress relieving　応力除去
  stress strain curve　応力-ひずみ曲線
  stress strain diagram　応力-ひずみ線図
stretch 動引き伸ばす，広げる 名ストレッチ，伸び，引張り，膨張
stretcher strain　ストレッチャストレイン【しま状のひずみ模様の一種】
stretching screw　調整ねじ
strict 形厳しい，厳格な，凡帳面な
strictly 副厳しく，厳密に
  strictly speaking　厳密に言えば
  同 speaking strictly
strike 動打つ，衝突する，攻撃する
striking energy　衝撃エネルギ
string　名ひも，糸，弦
  動糸に通す，弦を張る
  string electrometer　弦電位計
stringer　縦材，ストリンガ
strip 動はぐ，欠けさす，取り除く
  名細長い鉄片
  strip down　動分解する，解体する
  strip steel　帯鋼
stripe　ストライプ，縞(しま)
striped pattern　縞模様
stripped 形ねじ山(歯)をすり減らした
  stripped nut　ねじ山のつぶれたナット
stripper pump　残油ポンプ
stroboscope　ストロボスコープ【急速に運動する物を止まっているように観測(撮影)する装置】
stroke　ストローク，行程
  stroke bore ratio　行程内径比

stroke of piston　ピストン行程
stroke of slide　ストローク長さ
stroke volume　行程容積
stroll　動ぶらつく，さまよう
strong　形強い，強固な，かたい，丈夫な
strong acid　強酸[性]
　関 weak acid：弱酸
strong alkali　強アルカリ[性]
strong base　強塩基
strong base number　強塩基価
strong electrolyte　強電解質
strongly　副強く，強固に，かたく
strongly acidic resin　強酸性樹脂
strongly basic resin　強塩基性樹脂
strontium　ストロンチウム　記 Sr
strop　ストロップ，環索
structural　形構造の，構造用の
structural analysis　構造解析
structural drawing　構造図
structural formula　構造式
structural mechanics　構造力学
structural shape　構造用形鋼
structural steel　構造用鋼
structure　名構造，構造物，組織
　動構築する，組織化する
　関 ship's structure：船体
strut　支柱，圧縮材，突張り
stub　短い突出物
stub thread　低山ねじ
stub tooth　低歯
stuck　動「stick」の過去・過去分詞
stud　スタッド，植込みボルト
stud bolt　植込みボルト
study　名研究，勉強，検討
　動研究する，調査する，勉強する
stuff　名材料，要素
　動詰め込む，詰め物をする
stuffing　スタッフィング，詰め物
stuffing box　スタッフィングボックス，パッキン箱
stuffing gland　パッキン押え
stuffing nut　締付けナット
stuffing ring　スタッフィングリング
stumble　動つまずく，よろめく
sturdy　形頑丈な，しっかりした

style　名様式，型，技法，方式
　動称える，設計する
style of drawing　製図の表現法
stylus　触針
sub-　接頭「下」「半分」「亜」「副」「やや」の意
subassembly　サブアセンブリ，部分組立[物]
subassembly drawing　部分組立図
subcooled boiling　サブクール沸騰
subcooling　サブクーリング，亜冷却
　関 degree of subcooling：サブクール度
subcritical　形臨界未満の
subcritical pressure boiler　亜臨界圧ボイラ
subdivide　動細分する，解体する
subject　名主題，題目，主観，科目
　形受ける，受けやすい
　動さらす，～に当てる，支配する，受けさせる
　関 be subjected to：～を受ける
subject to ～　前～を条件として，～を仮定して
subjective　形主観の，主観的な
　反 objective：客観的な
sublayer　底層，副層，下層
sublimation　昇華
submarine　名潜水艦，海底動[植]物
　形海底の，海中で使う
submarine cable　海中電線，海底ケーブル
submerge　動水中に入れる
submerged arc welding　サブマージドアーク溶接，覆光溶接
submersible pump　水中ポンプ
submicron　サブミクロン
submicron particle　極微粒子
submit　動服従させる，屈服させる，提出する
subscribe　動署名する，承諾する，応募する
subscript　形下付きの
　名下付き文字　関 superscript：上付きの，上付き文字
subsequent　形次の

subsequently 副その後, 続いて
subsidence 陥没, 沈降
subsidiary 形補助の, 従属的な
　subsidiary equation　補助方程式
subsonic 形音速以下の, 亜音速の
　subsonic flow　亜音速流れ
subspace 部分空間
substance 物質, 物, 無機物, 内容
　関 flammable substance：助燃剤
substantial 形実在した, 現実の, 実質的な, しっかりした, 丈夫な
substitute 動代入する, 置換する　形代理の, 代わりの
　substitute character　置換文字
substitution 置換, 代入, 代用
　substitution method　置換法
subsystem 部分システム, 部分系
subtract 動減ずる, 引く, 引き算をする
subtraction 引き算, 減法, 引くこと
subtractive polarity 減極性
subtractor 減算器
subzero treatment サブゼロ処理【熱処理の一種】
succeed 動成功する, 継承する, 続く
success 成功
successful 形成功した, 好結果の
successfully 副うまく, 首尾よく
succession 連続, 継続
　関 law of succession：連続の法則
successive 形連続する, 継続的な
such 形そのような　副そんなに～な　代そのような人(物, 事)
　関 in such a case：そんな場合に(は)／in such a way as to：～するように, ～するような方法で
　such ～ as…　…のような～
　such ～ that…　非常に～なので…, …ほど～
　such as ～　～のような
suck 動吸う, 吸い込む
suction サクション, 吸入, 吸引, 吸込作用
　suction air　吸気
　suction blower　吸込送風機
　suction cam　吸気カム
　suction draft　吸込通風
　suction head　吸込水頭
　suction hose　吸入ホース
　suction manifold　吸気マニホールド
　suction pipe　吸入管, 吸気管
　suction port　吸気口
　suction pressure　吸入圧力
　suction pump　吸上げポンプ
　suction stroke　吸入行程
　suction temperature　吸入温度
　suction tube　吸込み管
　suction valve　サクションバルブ, 吸入弁
sudden 形突然の, 急な
suddenly 副突然に, 不意に
suffer (from) 動被る, 悩まされる
suffice 動十分である, 満足する
sufficient 形十分な
　sufficient accuracy　高精度
　sufficient condition　十分条件
sufficiently 副十分に, かなり
suggest 動提案する, 示唆する, 示す
suit 動適する, 好都合である
suitability 適合(性)
suitable 形適当な, 適切な, 適した, 似合う
suited 形適した, ふさわしい
sulfate 動硫酸で処理する, 硫酸化する　名硫酸塩　同 sulphate
　sulfate of alumina　硫酸アルミニウム
sulfide 硫化物
sulfur 名硫黄, 硫黄分　記 S　形硫黄の　動硫化する
　sulfur content　硫黄分, 硫黄含有量
　sulfur crack　サルファクラック, 硫黄割れ
　sulfur dioxide　SO_2, 亜硫酸ガス, 二酸化硫黄
　sulfur oxides　SOx, 硫黄酸化物
sulfureted hydrogen　硫化水素
sulfuric 形硫黄の
　sulfuric acid　硫酸
　sulfuric [acid] anhydride　三酸化硫黄, 無水硫酸
　sulfuric acid corrosion　硫酸腐食

sulfuric acid vapor　硫酸蒸気
sulfuric anhydride　SO₃, 無水硫酸
sulfurizing　浸硫処理
sulphide　⦿ sulfide
sulphur　⦿ sulfur
Sulzer boiler　ズルザーボイラ
Sulzer engine　ズルザー機関
sum　図合計, 総額, 和
　動合計する, 要約する
summary　図一覧, 概要
　形要約した, 手短な
summation　合計, 総和, 求和
summit　山頂, 頂上, [ねじの]山の頂
sump　油だめ, 排水だめ
　sump pump　排水ポンプ
　sump tank　サンプタンク, 集油タンク
sun and planet gear　遊星歯車装置
sun gear　太陽歯車
sunk　動「sink」の過去分詞
　形沈没した, 埋没した
　sunk head rivet　沈みリベット
　sunk key　沈みキー
super computer　スーパコンピュータ, 超高速計算機
super engineering plastic　超機能樹脂
superalloy　超合金
supercharge　動過給する, 与圧する　図過給
supercharged engine　過給機関
supercharger　過給機
supercharging　過給
　supercharging blower　過給送風機
　supercharging pump　過給ポンプ
　supercharging with constant pressure　静圧過給方式
superconduct　動超電(伝)導する, 超電(伝)導を起こす
superconducting　形超電導の
　superconducting coil　超電導コイル
　superconducting generator　超電導発電機
　superconducting magnet　超電導磁石
　superconducting stability　超電導安定性
　superconducting state　超電導状態

superconduction　超電導
superconductivity　超電導性
supercooling　過冷却　⦿ degree of supercooling : 過冷却度
supercritical　形臨界超過の
　supercritical fluid　超臨界流体
　supercritical pressure boiler　超臨界圧ボイラ
　supercritical state　超臨界圧状態
supercurrent　超電導電流
superelasticity alloy　超弾性合金
superficial　形表面上の, 見かけの
superfinishing　超仕上げ
superfluidity　超流動
superheat　動過熱する　図過熱
　superheat ratio　過熱度
superheated steam　過熱蒸気
superheater　スーパヒータ, 過熱器
　superheater equipment　過熱装置
　superheater tube　過熱[器]管
superheating　過熱
superhigh pressure boiler　超高圧ボイラ
superimpose　動載せる, 重ね合わせる
superintendent engineer　監督技師
superior　形すぐれた, 優勢な, 上位の　図すぐれた人, 上司
superplastic alloy　超塑性合金
superposition　重ね合わせ
supersaturated steam　過飽和蒸気
supersaturation　過飽和
superscript　形上付きの
　図上付き文字
supersonic　形超音速の
　supersonic aircraft　超音速機
　supersonic flow　超音速流れ
　supersonic nozzle　超音速ノズル
　supersonic speed　超音速
　supersonic wave　超音波
supersynchronous motor　超同期電動機
supervision　監視, 監督
supervisory　形監視する, 監督の, 管理の
　supervisory control　監視制御

supervisory equipment　監視装置
supervisory lamp　監視ランプ
supervisory program　監視プログラム
supervisory signal　監視信号
supplement　動補う，増補する
supplementary　形補足の，付録の
supplementary air　補助空気
supplementary angle　補角
supplementary [feed] valve　補給弁
supplementary units　補助単位
supplied air respirator　送気マスク
supplier　供給元
supply　動供給する，提供する，補充する　名供給，給水，補給
supply air　給気
supply and demand balance　需給バランス
supply cable　電源ケーブル
supply capability　供給[能]力，供給電力
supply circuit　電源回路
supply line　供給ライン
supply port　供給口
supply switch　給電スイッチ
supply transfer panel　電源配電盤
supply ventilating fan　給気通風機
supply voltage　電源電圧，供給電圧
supplying　供給
support　動～を支える
　名サポート，支点，支持，支持物
support power　供給パワー
　動電源(動力)を供給する
supporting point　支点
suppose　動仮定する，推測する
suppress　動抑える，静める
suppressed weir　全幅せき
suppression　抑制
suppression ratio　抑圧比，抑制比
supreme　形最高の，究極の
sure　形確かな，確実な，確信する，きっと～する　副確かに，本当に
　熟 be sure to ～：きっと～する
surface　名[表]面，界面，表層，外面，水面　形表面の，外面の　動表面を平らに仕上げる，浮上する

surface active agent　界面活性剤
surface activity　界面活性，表面活性
surface area　表面積
surface barrier　表面障壁
surface blow off valve　水面吹出し弁
surface boiling　表面沸騰
surface boundary　界面
surface charge density　表面電荷密度
surface coat[ing]　表面被覆
surface combustion　表面燃焼
surface condenser　表面復水器
surface corrosion　表面腐食
surface coverage　吸着率，表面被覆率
surface current　表面電流，表面流
surface defect　表面きず
surface density　表面密度
surface diffusion　表面拡散
surface eddy　表面渦
surface effect　表面効果
surface energy　表面エネルギ，界面エネルギ
surface fatigue　表面疲労
surface finishing　表面仕上げ
surface force　表面力
surface friction　表面摩擦
surface gauge　トースカン，表面ゲージ
surface grinder　平面研削盤
surface hardening　表面硬化
surface heat exchanger　表面式熱交換器
surface heat transmission　表面伝熱
surface ignition　表面点火，表面着火
surface inspection　表面検査
surface leakage　表面漏れ
surface lubricant　粘着防止剤
surface of discontinuity　不連続面
surface of revolution　回転面
surface of rupture　すべり面，破壊面
surface passivation　表面安定化，表面不活性化
surface phase　界面層
surface plate　定盤
surface porosity　表面多孔性
surface pressure　表面圧力

surface ratio 面積比
surface recombination 表面再結合
surface resistivity 表面抵抗率
surface roughness 表面粗さ
surface state 表面状態, 表面[エネルギ]準位
surface symbol 表面記号【図面で表示する表面粗さの記号】
surface table 定盤
surface tension 表面張力
surface tension modifier 界面活性剤
surface texture 表面組織
surface traction 表面力, 面力
surface treated steel 表面処理鋼
surface treatment 表面処理
surface wave 表面波
surface waviness 表面うねり
surfacing 表面仕上げ
surge 動波打つ, 激しく変動する 名サージ, うねり, 急上昇
 surge current サージ電流【電気回路の開閉時などに回路に流れるパルス状の大電流】
 surge line サージング線
 surge pressure サージ圧力
 surge protective device 避雷装置
 surge tank サージタンク【水量・水圧の急変を抑えるための調整タンク】
 surge voltage サージ電圧
surging サージング, 脈動
 surging tank サージタンク, 調圧水槽
surplus 形余りの, 過剰の
 surplus power 余剰電力
surprise 動驚かす, びっくりさせる
surrender 動引き渡す, 放棄する, 降参する
surround 動囲む, 取り巻く
surrounded by ～で囲ま(巻か)れる
surrounding 形周囲の, 付近の 名周囲, 環境
surveillance 監視, 見張り, 監督
 surveillance camera 監視カメラ
 surveillance equipment 監視装置
survey 名サーベイ, 検査, 測量 動検査する, 測量する
susceptance サセプタンス【アドミタンスの虚数部分】
susceptibility 磁化率, 帯磁率
susceptible 形影響を受けやすい
suspect 動推測する, 疑う, 怪しむ 形疑わしい, 怪しい
suspend 動吊るす, 浮遊させる, 中断する, 延期する
suspended particulate matter 浮遊粒子状物質
suspended solid [substance] 浮遊物質
suspension サスペンション, 懸架装置, 吊るすこと, 浮遊[状態], 浮遊物, 懸濁[液]
 suspension clamp 懸垂クランプ
 suspension insulator 懸垂碍子
sustain 動保持する, 耐える, 維持する, 持続させる
sustained interruption 持続停電
sustained oscillation 持続振動
sustained radiation 持続放電
swage スエージ, タップ
 swage block はちの巣〈金敷〉
swash 動はねかける, 激しくぶつける
 swash bulkhead 制水板【タンカなどでタンク内の液体載荷の動揺を抑えるための隔壁のこと】
 swash plate 回転斜板
 swash plate cam 斜板カム
 swash plate type [axial] piston pump 斜板式ピストンポンプ
sway 動揺らす(れる)
sweep 動掃除する, 一掃する 名掃除, 掃引
 sweep-back angle 後退角
 sweep-back wing 後退翼
 sweep circuit 掃引回路
 sweep-forward wing 前進翼
sweeping 形全面的な, 徹底的な, 大ざっぱな 名掃除, 一掃
swell 動膨張する, 増大する 名膨張, 増加, うねり
swelling スエリング, 膨張, 膨出, 隆起, 膨潤

swing 動 [揺れ]動く，振動する，動揺する 名 スイング，振動，動揺，旋回，揺れ，振幅
  swing angle　相差角
  swing check valve　スイング逆止め弁
  swing frame　つり下げ台
  swing joint pipe　関節管
swinging buoy　回頭浮標
swinging link　揺りリンク
swinging motion　揺動，振り子のような運動
swirl 名 スワール，旋回，渦流 動 渦巻く，旋回する
  swirl atomizer　渦巻型噴射弁
  swirl chamber　渦室，渦流室
  swirl chamber type combustion chamber　渦流室式燃焼室
  swirl flow　旋回流
  swirl spring　渦巻ばね，ぜんまい
  swirl vane　旋回羽根
swirler　スワラ，旋回器
switch 名 スイッチ，開閉器 動 変える，切り替える，交換する
  switch-board　配電盤
  switch-box　スイッチ箱，配電箱
  switch-cock　切換コック
  switch-gear　開閉装置
  switch-indicator　スイッチ標識
  switch off ～　動 ～を停止する
  switch on ～　動 ～を投入する
  switch over (from…) to ～　動 (…から) ～に切り替える
  switch-start fluorescent lamp　スタータ形蛍光ランプ
switching　スイッチング，交換，切換え，中継
  switching circuit　スイッチング回路
  switching device　スイッチング装置，開閉装置，スイッチ素子
  switching loss　スイッチング損
  switching relay　切換継電器
  switching voltage　動作過電圧
swivel 名 スイベル，回転，転向，旋回，自在継手，回り継手 動 旋回させる，回る
  swivel angle plate　自在定盤
  swivel base　旋回台
  swivel bearing　自在軸受
  swivel block　自在軸受，スイベル滑車
  swivel joint　スイベル継手
  swivel table　自在テーブル
  swivel vice　回り万力
swiveling angular table　万能テーブル
syllabus　概要，要旨，講義要目，時間割
  syllabus planning　授業計画
symbol　シンボル，記号，符号，象徴
  symbol for dimensioning　寸法補助記号
  symbol for element　元素記号
  symbol of materials　材料記号
  symbol of weld　溶接記号
  symbol string　記号列
symbolic 形 符号の，記号的な，象徴的な
  symbolic address　記号アドレス
  symbolic logic　記号論理学
  symbolic method　記号法
symbolize 動 符号で表す，象徴する
symmetric 形 相対的な，対称の，つり合った
  symmetric coordinates　対称座標
  symmetric group　対称群
  symmetric matrix　対称行列
  symmetric rotor　対称回転子
  symmetric source　対称電源
symmetrical 形 相対的な，対称の，つり合った
  symmetrical compound　対称化合物
  symmetrical deflection　対称偏向
  symmetrical wave　対称波
symmetry　対称，つり合い
  symmetry axis　対称軸
  symmetry symbol　対称記号
symptom　徴候，しるし
synchro 名 シンクロ 形 シンクロの，同調の
synchronism　同期，同時性
  synchronism indicator　同期検定器
synchronization　同期化，同調化

synchronize 動同期する，同時性をもつ
synchronized operation 同期運転
synchronizer シンクロナイザ，同期装置
synchronizing 同期調整
synchronizing current 同期化電流
synchronizing device 同期検定器
synchronizing frequency 同期周波数
synchronizing impulse 同期インパルス
synchronizing lamp 同期検定灯
synchronizing power 同期化力
synchronizing pulse 同期パルス
synchronizing relay 同期継電器
synchronizing signal 同期信号
synchronizing torque 同期化トルク
synchronous 形同期[式]の
synchronous angular speed 同期角速度
synchronous belt タイミングベルト，歯付きベルト
synchronous circuit 同期回路
synchronous commutator 同期整流器(子)
synchronous converter 同期変流機
synchronous detector 同期検波器
synchronous generator 同期発電機
synchronous impedance 同期インピーダンス
synchronous inductance 同期インダクタンス
synchronous induction motor 誘導同期電動機
synchronous machine 同期機
synchronous measurement 同期計測
synchronous motor 同期電動機
synchronous operation 同期運転
synchronous pulley 歯付きプーリ
synchronous reactance 同期リアクタンス
synchronous rectifier 同期整流
synchronous rotation 同期回転
synchronous speed 同期速度
synchronous watt 同期ワット
synchroscope シンクロスコープ，同期検定器
synthesis 合成
synthesis gas 合成ガス
synthesizer シンセサイザ，合成器
synthetic 形合成した，人造の
synthetic fiber 合成繊維
synthetic fuel 合成燃料
synthetic insulating oil 合成絶縁油
synthetic oil 合成油
synthetic resin 合成樹脂
synthetic rubber 合成ゴム
syphon 同 siphon
syringe 水差し，油差し
system システム，装置，系，組織，体系，系統，方式
関 control system：制御系
system capacity 系統容量
system configuration 系統構成
system control システム制御
system design システム設計
system deviation 制御偏差
system diagram 系統図
system efficiency システム効率
system engineering システム工学
system equation システム方程式
system of absolute units 絶対単位系
system of fits はめあい方式
system of practical units 実用単位系
system of unit 単位系
system oil システム油，循環油
system operation 系統操作，系統切換え
system stability 系統安定性
system switching 系統操作
system voltage 系統電圧
systematic code 組織符号
systematic error 系統誤差【原因が明らかで，補正可能な誤差のこと】
systematic[al] 形組織的な，系統的な，規則正しい
systematize 動組織化する，体系化する，系統立てる

T-t

T connection T結線
T joint T継手
T pipe T字管　㊂ tee fitting
T section T形断面
T square T定規
tab タブ
table テーブル，表，台，机　動表を作る
  table top 形テーブル状の
tablet 錠剤，かたまり
  関 a tablet of soap：石鹸1個
tabulate 動表にする
tabulation 表作成
tachogenerator タコゼネレータ
tachograph 記録回転計
tachometer タコメータ，速度計，回転計，回転速度計
  tachometer genarator タコゼネレータ，速度計用発電機
tack 動びょうで留める，付け加える，進路を変える
  tack welding 仮付け溶接
tacking 仮付け
tackle テークル，滑車【滑車装置】　動取り組む
tactical diameter 旋回径
tactile sensor 触覚センサ
tag 名札　動札を付ける
tail テール，尾部
  tail block テール滑車
  tail lamp 船尾灯
  tail pipe テールパイプ，吸込み管
  tail rod 先棒【ピストン頭部に固定されたピストンの案内棒】
  tail shaft プロペラ軸
  tail spindle 心押し軸
  tail stock 心押し台
  tail valve 漏気弁【逃し弁の一種】
taint 動汚れる，腐敗する
take 動取る，受ける，持つ，要する，〜をする，利用する，占める
  関 take care：注意する
  take A apart 動Aを分解する
  take a nut 動ナットを取り付ける
  take in 動取り込む(入れる)
  take into account 〜 動〜を考慮に入れる
  take over 動引き継ぐ，連れていく
  take place 動起こる，行われる
  take time 動時間をかける
  take up 動吸収する，占める，持ち上げる
  take-up motion 巻取り装置
  take-up unit テークアップ装置【ベルトなどに張りを与える装置】
talk 動話す
tallow タロー，牛脂
tandem 名タンデム【串形】　形縦に並んだ，前後に並んだ
  tandem articulated type reduction gear タンデムアーティキュレーテッド形減速装置
  tandem compound turbine 串形タービン
  tandem drive タンデム駆動【機械を直列に配置して駆動する】
  tandem engine タンデム機関，串形機関
  tandem propeller タンデムプロペラ
tangent タンジェント，正接，接線
  tangent cam 接線カム
  tangent key 接線キー
tangential 形接線の，接する
  tangential acceleration 接線加速度
  tangential cam 接線カム
  tangential fan 横流ファン
  tangential force 接線力
  tangential load 接線荷重
  tangential stress 接線応力
  tangential velocity 接線速度
tank タンク，水槽，油槽
  tank survey タンク検査
  tank top タンクトップ，タンク頂板，内部船底板
tanker [vessel] タンカ，油槽船
tantalum タンタル　㊉ Ta

tap 名栓, 蛇口, コック, タップ【雌ねじを切る工具】
動軽くたたく
　tap bolt　ねじ込みボルト, 押えねじ
　tap holder　タップホルダ【タップを保持するもの】
　tap volt　タップ電圧
tape 名テープ
動テープを貼る, テープで留める
　tape line　巻尺
　tape measure　巻尺
taper 名テーパ, 傾斜, 勾配【円すい状に先が細くなった形状】
動次第に細くなる
　taper attachment　テーパ装置
　taper bolt　テーパボルト, 勾配付きボルト
　taper gauge　テーパゲージ, 勾配ゲージ, 傾斜計
　taper key　テーパキー, 勾配キー
　taper liner　テーパライナ
　taper matching　テーパ整合
　taper pin　テーパピン, 勾配キー
　taper reamer　テーパリーマ【テーパ穴を仕上げるのに用いるリーマ】
　taper roller　テーパころ
　taper roller bearing　テーパころ軸受
　taper thread　テーパねじ
　taper turning　テーパ削り
　taper washer　テーパワッシャ【座金の一種】
tapered blade　テーパ羽根, 勾配付き翼
tapered bore　テーパ穴
tapered drum　テーパドラム
tapered pad thrust bearing　傾斜パッドスラスト軸受
tapered pin　テーパピン
tapped hole　ねじ穴
tappet　タペット【カムの運動を弁に伝える棒】
　tappet clearance　タペットすきま
　tappet guide　タペット案内
　tappet rod　タペット棒
tapping　タッピング, ねじ立て
　tapping attachment　タップ保持器

tar　タール【防食剤に使用】
tarpaulin　防水シート
task　仕事
Taylor series expansion　テイラー級数展開
Taylor vortex　テイラー渦
TDC (Top Dead Center)　上死点
teach 動教える
teak　チーク【船材になる落葉樹】
tear 名亀裂, 損傷 動裂く, 破壊する
technical 形工業技術の, 技術上の
　technical analysis　工業分析
　technical atmosphere　工業気圧
　technical work　工業仕事
technician　技術者, 専門家
technique　技法, 技術, 技能, 手法
technological 形技術的な
　technological development　技術開発
　technological innovation　技術革新
technology　技術
tee joint　T継手【T形の管継手】
teeth　歯【tooth の複数形】
teflon　テフロン【耐熱性のフッ素樹脂】
telecontrol　遠隔制御
　telecontrol board　遠隔制御盤
　telecontrol room　遠隔制御室
　telecontrol unit　遠隔監視制御装置
telegraph　テレグラフ, 電信, 電信機
　telegraph logger　テレグラフロガ
telemeter　遠隔計測器
telemotor　遠隔舵取り機
telescopic 形伸縮自在の, はめ込み式の, 入れ子式の
　telescopic arm　伸縮アーム
　telescopic joint　伸縮継手
　telescopic tube　テレスコピック管, 抜差し管, 入れ子管
telescoping gauge　入れ子ゲージ【調節可能内側測定ゲージ】
tell 動話す, 知らせる
telltale　表示器
tellurium　テルル 記 Te
telpher　テルファ, 電力運搬装置
temper 名焼戻し

〜を調節する，適度に抑える
　temper bend test　加熱曲げ試験
　temper brittleness　焼戻し脆性，焼戻し脆さ
　temper color　焼戻し色【この色によって焼戻し温度が推定できる】
　temper crack　焼戻し割れ
　temper hardening　焼戻し硬化
　temper rolling　調質圧延
temperature　温度，気温，体温
　temperature alarm　過熱警報
　temperature change　温度変化
　temperature coefficient　温度係数
　temperature color scale　色温度目盛
　temperature compensated　温度補償式
　temperature compensation　温度補償
　temperature conductivity　温度伝導度
　temperature control valve　温度調節弁
　temperature correction　温度修正
　temperature difference　温度差
　temperature distribution　温度分布
　temperature efficiency　温度効率
　temperature entropy diagram　T-S線図，温度-エントロピ線図
　temperature factor　温度係数
　temperature gradient　温度勾配
　temperature limit　限界耐熱値
　temperature of combustion　燃焼温度
　temperature of combustion gas　燃焼ガス温度
　temperature rating　温度定格
　temperature ratio　温度比
　temperature regulating valve　温度調整弁
　temperature regulator　温度調節器
　temperature rise　温度上昇
　temperature rise coefficient　温度上昇係数
　temperature rise ratio　温度上昇比
　temperature scale　温度目盛
　temperature sensing element　感温体
　temperature setting　温度設定
　temperature sensor　温度センサ
tempered　形調節された，焼戻した，鍛えた，強化した
　tempered air　予熱空気
　tempered glass　強化ガラス
　tempered steel　焼戻し鋼
tempering　テンパリング，焼戻し【焼入れ材に粘り強さを与えるための熱処理】
　tempering crack　焼戻し割れ
tempil　テンピル，測温材
template　テンプレート，型板
temporary　形一時的な，仮の
　temporary hardness　一次硬度
ten nines　テンナインズ，99.99999999％
tenacity　粘着力，粘り強さ，強靭性
tend to 〜　動〜する傾向がある，〜しがちである
tendency　傾向，癖
tense 〜　動〜を緊張させる
tensile　形引張りの，張力の
　tensile crack　引張りき裂
　tensile creep　引張りクリープ
　tensile elongation　引張り伸び
　tensile failure　引張り破壊【垂直張力による破壊】
　tensile force　張力，引張り力
　tensile fracture　引張り破壊
　tensile load　引張り荷重
　tensile modulus　縦弾性係数
　tensile rupture　引張り破壊
　tensile shear test　引張りせん断試験
　tensile strain　引張りひずみ
　tensile strength　引張り強さ，引張り強度，抗張力
　tensile stress　引張り応力
　tensile test　引張り試験
　tensile test specimen　引張り試験片
　tensile yield stress　引張り降伏応力
tensimeter　引張り計，張力計
tensiometer　張力計，伸び計
tension　テンション，引張力，張力，張り，応力，伸張，電圧
　関 in tension：張力／load in tension：引張荷重
　tension axis　張力軸
　tension bar　テンションバー

tension bolt　テンションボルト,タイロッド,引張ボルト
tension crack　破断クラック
tension dynamometer　引張り動力計
tension fracture　破断
tension fuse　電圧フューズ
tension gauge　引張り計
tension load　引張り荷重
tension meter　張力計
tension plate　押えばね板
tension pulley　張り車
tension rod　突張り棒
tension set　永久伸び
tension side　引張り側
tension spring　引張りばね
tension test　引張り試験,抗張試験
tension test specimen　引張り試験片
tension wheel　張り車
tension winch　張索ウインチ
tension wrench　トルクレンチ
tentative　形試みの,仮の
tentative specification　仮仕様書
term　項,用語,期間　関 in terms of ～：～の点で
terminal　名ターミナル,端子,端末,電極　形末端の,最終の
terminal assembly　端子板
terminal block　端子盤
terminal board　端子盤
terminal box　端子箱
terminal lever　最終レバー
terminal marking　端子記号
terminal pressure　終圧
terminal settling velocity　最終沈降速度,終端沈降速度
terminal velocity　終末(端)速度
terminal voltage　端子電圧
terminate　動終了する,終わる,終結する
termination　終端,終了
tertiary　形第三の,第三次の
tertiary air　三次空気
tertiary circuit　三次回路
tertiary winding　三次巻線
tesla　テスラ【磁束密度の計量単位：T】
test　名テスト,試験,検査　動試験する,検査する
test circuit　試験回路
test cock　試験コック,検水コック
test current　試験電流
test hammer　テストハンマ
test load　試験荷重
test paper　試験紙
test piece　試験片
test pressure　試験圧力
test results　試験結果
test specimen　試験片
test tank　検油タンク
test tube　試験管
test voltage　試験電圧
tester　テスタ,回路計,試験器
testing machine　試験機,材料試験機
tetrahedron　四面体
textile belt　布ベルト
texture　組織[集合],質感,生地　関 surface texture：表面組織
than　前/接～よりも,～に比べて　関 more than：～以上,～より大きい／no more than ～：たったの～,～しか／no less than ～：～(ほど)も
thanks to ～　前～のお陰で,～のせいで
the first law of thermodynamics　熱力学第一法則
the first (third) angle system　第一(三)角法
the International System of Units　SI,国際単位系
the second law of thermodynamics　熱力学第二法則
then　副その時,それから,その上
theorem　定理
theoretical　形理論[上]の
theoretical air　理論空気量
theoretical air fuel ratio　理論空燃比
theoretical amount of air　理論空気量
theoretical amount of oxygen　理論酸素量
theoretical combustion temperature　理論燃焼温度
theoretical cycle　理論サイクル
theoretical density　理論密度

theoretical efficiency 理論効率
theoretical error 理論誤差
theoretical mixture ratio 理論混合比
theoretical output 理論出力
theoretical power 理論動力
theoretical pump head 理論揚程
theoretical quantity of combustion gas 理論燃焼ガス量
theoretical relieving capacity 理論流出量
theoretical suction head 理論吸込み揚程
theoretical thermal efficiency 理論熱効率
theoretical throat 理論のど厚
theoretical value 理論値
theoretical volumetric efficiency 理論体積効率
theoretical work 理論仕事
theoretically 副理論的に
theorize 動理論づける
theory 理論, 学説
theory of elasticity 弾性学
theory of plasticity 塑性学
theory of relativity 相対性理論
theory of structures 構造力学
therblig analysis サーブリッグ分析
there 副そこに(で)
there are (is) ~ ~がある, ~である
thereby 副そのために, それにより
therefore 副それ故に, その結果, 従って
therein 副その中に, その点に(で)
thermal 形熱の, 温度の
thermal accumulator 蓄熱器
thermal ammeter 熱電流計
thermal analysis 熱分析, 熱解析
thermal balance 熱バランス, 熱平衡, 熱勘定
thermal boundary condition 熱境界条件
thermal boundary layer 温度境界層
thermal capacity 熱容量
thermal conductance 熱コンダクタンス
thermal conduction 熱伝導
thermal conductivity 熱伝導率
thermal contact resistance 接触熱抵抗
thermal contraction 熱収縮
thermal convection 熱対流
thermal decomposition (cracking) 熱分解
thermal deformation 熱変形
thermal degradation 熱劣化
thermal destruction 熱破壊
thermal diffusivity 温度伝導率
thermal dissociation 熱解離
thermal efficiency 熱効率
thermal effusion 熱遷移
thermal electromotive force 熱起電力
thermal electron 熱電子
thermal emittance ratio 熱ふく射率
thermal energy 熱エネルギ
thermal engineering 熱工学
thermal equilibrium 熱平衡
thermal equivalent 熱当量
thermal equivalent of work 仕事の熱当量
thermal expansion 熱膨張
関 coefficient of thermal expansion：熱膨張係数
thermal expansion coefficient 熱膨張率
thermal expansion valve 熱膨張弁
thermal fatigue 熱疲労
thermal flux 熱流束
thermal gradient 熱勾配
thermal insulating material 保温材
thermal insulation 断熱, 保温
thermal insulation performance 断熱性能
thermal load 熱負荷
thermal paper 感熱紙
thermal plasticity 熱可塑性
thermal power 熱出力
thermal property 熱的性質
thermal protection material 熱防護材
thermal radiation 熱ふく射, 熱放射
thermal ratio 温度比
thermal receiver 排気だめ

thermal refining　調質
thermal relay　サーマルリレー，熱動継電器
thermal resistance　熱抵抗
thermal resistivity　熱抵抗率
thermal sensor　温度センサ
thermal shield　熱遮蔽
thermal shock　熱衝撃
thermal shock fracture　熱衝撃破壊
thermal shock test　熱衝撃試験
thermal spraying　溶射
thermal storage　蓄熱
thermal strain　熱ひずみ
thermal stress　熱応力
thermal switch　温度スイッチ
thermal transpiration　熱遷移
thermal type over current relay　熱動過電流継電器
thermal unit　熱量単位
thermalelectricity　熱電気
thermet　サーメット【切削工具用合金の一種】
thermion　熱電子【高温の金属から放出される電子】
thermionic　形熱電子の
thermistor　サーミスタ【熱に敏感な抵抗体】
thermit　テルミット【アルミニウムと酸化鉄の粉末の混合物】
thermo-　接頭熱の，熱電気の
thermoalloy　サーモアロイ【ニッケル銅合金】
thermoammeter　熱電流計
thermobulb　サーモバルブ
thermochemical equation　熱化学式
thermocouple　サーモカップル，熱電対
thermodynamic　形熱力学の，熱力学的な
　thermodynamic cycle　熱力学的サイクル
　thermodynamic efficiency　熱力学的効率
　thermodynamic equilibrium　熱力学平衡
　thermodynamic function　状態量，熱力学関数
　thermodynamic potential　熱力学ポテンシャル
　thermodynamic property　熱特性
　thermodynamic system　熱力学系
　thermodynamic temperature　熱力学温度，絶対温度
thermodynamics　熱力学
　関 first law of thermodynamics：熱力学の第1法則
thermoelasticity　熱弾性
thermoelectric　形熱電の
　thermoelectric cooling　熱電冷却，電子冷却
　thermoelectric couple　熱電対
　thermoelectric current　熱電流
　thermoelectric effect　熱電効果
　thermoelectric generator　熱電発電機
　thermoelectric potential　熱電位
　thermoelectric pyrometer　熱電[対]高温計
　thermoelectric refrigeration　熱電冷凍，電子冷凍
　thermoelectric series　熱電荷
　thermoelectric thermometer　熱電温度計【熱起電力で測温する温度計】
　thermoelectric type instrument　熱電型計器
thermoelectricity　熱電気
thermoelectrometer　熱電流計
thermoelectromotive force　熱起電力
thermoelectron　熱電子
thermoelectronic refrigeration　熱電冷凍
thermoforming　熱成形
thermogalvanometer　熱電検流計
thermogram　サーモグラム【赤外線センサを用いた温度分布図】
thermograph　サーモグラフ，記録温度計
thermography　サーモグラフィ【物体の表面温度を測定する方法】
thermohygrostat　恒温恒湿器
thermolabel　サーモラベル【温度計の一種】

thermomechanical treatment　加工熱処理

thermometer　温度計

thermopaint　サーモペイント，示温塗料【変色により温度が判るペイント】

thermophysical property　熱物性値

thermopile　サーモパイル，熱電対列

thermoplastic　形熱可塑性の　名熱可塑性樹脂
　thermoplastic resin　熱可塑性樹脂

thermoregulator　温度調節器

thermosetting　形熱硬化性の

thermosiphon　熱サイホン

thermostat　サーモスタット，恒温器，自動調温装置

thermostatic　形温度自動調整の
　thermostatic control　サーモスタット制御
　thermostatic expansion valve　温度自動膨張弁
　thermostatic pyrometer　熱電高温計
　thermostatic type　サーモスタット形

thermotank　サーモタンク

thermovalve　サーモバルブ

thermovoltmeter　熱電圧計

these　形このような　代これら【「this」の複数形】

Thevenin's theorem　テブナンの定理

thick　形厚い，厚みのある　副厚く
　thick cylinder　厚肉円筒
　thick film　厚膜
　thick film lubrication　流体潤滑

thicken　動厚くなる(する)

thickener　増ちょう剤，増粘剤

thickness　厚さ，太さ
　thickness chord ratio　厚弦比
　thickness gauge　厚み計，すきまゲージ

thimble　シンブル，はめ輪

thin　形薄い，乏しい，細い　動薄くなる，希薄になる
　thin cylinder　薄肉円筒
　thin film　薄膜
　thin film lubrication　薄膜潤滑
　thin layer　形薄い層の　名薄層
　thin layer approximation　薄層近似
　thin plate orifice　薄板オリフィス
　thin walled　形薄肉の

thing　物体

think　動考える，思う

thinner　「thin」の比較級

thionic　形硫黄の

third　形第三の
　third angle projection (system)　第三角法
　third law of motion　作用・反作用の法則，運動の第3法則

thirdly　副第三に

thorium　トリウム　記Th

thorough　形徹底的な，完全な

thoroughly　副徹底的に，すっかり，完全に

though　～　接～だけれども，たとえ～でも　関 even though ～：たとえ～であっても，～であるのに

thought　「think」の過去・過去分詞

thousand　名千，無数　形千の，無数の
　thousand fold　形千倍の

THP (Thrust HorsePower)　スラスト馬力

thread　名ねじ山　動ねじ山をつける，糸を通す
　thread gauge　ねじ山ゲージ
　thread micrometer　ねじマイクロメータ

threaded rod　ねじ棒

threaded screw hole　ねじ穴

threading　ねじ切り

three　名3　形3の
　three bladed propeller　三枚羽根プロペラ
　three cylinder two stroke engine　3気筒二サイクル機関
　three dimension　三次元
　three phase　三相
　three phase alternating current　三相交流
　three phase induction motor　三相誘導電動機
　three phase synchronous generator

三相同期発電機
three phase transformer 三相変圧器
three way 形三方向の, 三種の
three way catalyst 三元触媒
three way cock 三方コック
three way switch 三路スイッチ
three way valve 三方弁
three wire system 三線式
threshold 閾(しきい)値, 限界, 臨界
throat スロート, のど, のど部, 絞り
throat bolt のどボルト
throat depth のど厚さ
throat pressure のど圧
throttle 名スロットル, 絞り, 絞り弁 動窒息させる, 絞る, 減速する
throttle calorimeter 絞り熱量計
throttle control device 燃料制御装置
throttle governing 絞り調速
throttle nozzle 絞りノズル
throttle (throttling) valve スロットルバルブ, 絞り弁
throttling スロットリング, 絞り
throttling calorimeter 絞り熱量計
through 前〜を通して, 〜を通って, 〜によって, 〜を通じて
形通しの, 通り抜けられる
副通り抜けて, 貫いて, 通しで
through bolt 通しボルト
throughout 〜 前〜至る所に, 〜の間じゅう
throughput 処理能力, 処理量
throw 動投げる, 投入する, 噴出する, 振る, 向ける, 浴びせる, 注ぐ 名行程, 落差, 到達距離
throw velocity 噴出速度
thrust スラスト, 側圧, 推力, 軸方向に作用する外力
動強く押す, 突く, 押し進む
thrust augmenter 推力増強装置
thrust ball bearing スラスト玉軸受
thrust bearing (block) スラスト軸受, 推力軸受
thrust coefficient スラスト定数, スラスト係数
thrust collar スラストカラー, スラストつば

thrust collar bearing スラストつば軸受
thrust constant スラスト定数
thrust face 背面, 羽根の腹
thrust factor スラスト係数
thrust force 推(進)力
Thrust HorsePower THP, スラスト馬力
thrust load coefficient スラスト荷重係数
thrust mass ratio スラスト質量比
thrust output スラスト出力
thrust pad スラスト受け
thrust reduction factor スラスト減少係数
thrust roller bearing スラストころ軸受
thrust shaft スラスト軸, 推力軸
thrust shoe スラストシュー
thrust spacer スラストスペーサ
thrust washer スラストワッシャ
thruster スラスタ【推力を発生するもの】
thumb 親指
thumb nut つまみナット
thumb screw 蝶ねじ
thus 副このように, こうして
thyristor サイリスタ, シリコン制御整流素子
tide mark 貝殻模様, 潮標
tie 名引張り棒, 結び 動縛る, 結ぶ
tie bar 控え棒
tie bolt タイボルト
tie plate 帯板
tie rod タイロッド, 引張りボルト
TIG (Tungsten [electrode] Inert Gas) welding TIG 溶接
tight 形きつい, 窮屈な, きちんと合った
tight fit 密閉
tight rolling 密着圧延
tight seal 密封(閉)
tight weld 耐密溶接
tighten 動締める, ぴんと張る, 強化する, [規則などを]厳重にする
tightening 締付け, 引締め

tightening pulley　張り車
tightening torque　締付けトルク
tightly　副しっかりと，ぴったりと
till ~　前／接~まで
tiller　チラー，舵柄(だへい)
tilt　動傾く　名傾斜，傾き
tilting box　傾転箱
tilting pad bearing　ティルティングパッド軸受
time　時，時間，回，倍
　関 three times a day：日に3回／4 times 2 is 8：2×4＝8／at times：たまに，時々／all the time：常に，絶えず，いつでも／at a time：一度に／at all times：いつも，常に／at any one time：常時，どの時点においても／at the same time：同時に，その一方で
　time chart　タイムチャート
　time constant　時定数
　time delay relay　緩動(時延)継電器
　time keeping　保持
　time lag　遅れ
　time lag of first order　一次遅れ
　time lag of higher order　高次遅れ
　time lag relay　限時継電器，緩動継電器
　time limit relay　限時継電器
　time relay　定時継電器
timer　タイマ　同 time switch
timing　タイミング
　timing belt　タイミングベルト，歯付き伝動ベルト
　timing device　噴射時期調整装置【燃料の噴射時期を調整する】
　timing gear　タイミングギア，弁軸調時歯車【給・排気弁の作動時期を調整するカム軸を駆動する歯車】
　timing motor　時限電動機
tin　名錫(すず)　記 Sn
　動錫めっきをする，缶詰にする
　tin foil　錫はく
　tin plate　ブリキ，ブリキ板
tinning　錫めっき
tinny　形(とても)小さい

tip　チップ，火口，先端，頂点
　動先をつける
　tip circle　歯先円
　tip clearance　翼端すきま
　tip clearance leakage loss　翼端漏れ損[失]
　tip leakage　翼端漏れ
　tip speed　翼端速度
　tip vortex　翼端うず
tiresome　形うんざりする
titanium　チタン　記 Ti
titration　滴定
to　前~の方へ，~へ，~まで，~に，~に対して，~に属して，~のために，~に合わせて
　to begin with　副まず第一に
toast　動こんがり焼く，火であぶる
toe-in　トーイン【自動車の前輪を八字形に狭くすること】
together　副ともに，互いに，全体として，同時に，協力して
　together with ~　前~とともに，~に加えて
toggle　トグル，留め金，ひじ金
　toggle joint　トグル装置，倍力装置【小さい力で大きな力を出す装置】
　toggle switch　トグルスイッチ【指先で操作する小形スイッチ】
tolerability　許容度
tolerance　公差，許容，許容範囲，許容度，許容値，耐用，耐性，余裕
　tolerance limit　許容限度，許容限界
tolerate　動我慢する，耐える
toluene　トルエン【溶剤の一種】
tombac　トンバック【黄銅の一種】
ton　トン【重量の単位】
　ton of refrigeration　冷凍トン
tone　名トーン，音，音調
　動音調を整える，調子をつける
tongs　やっとこ
tongue　舌【舌状の回り止め】，つまみ
tongued washer　舌付きワッシャ【回り止めに使用する】
tonnage　トン数，積量
too　副…もまた，同様に，あまり

にも，…すぎる，非常に，とても
too…to 〜　あまりに…なので〜できない
too much　過剰(度)に
tool　名工具，道具，バイト
動道具で細工する
tool angle　刃先角
tool box　工具箱
tool carbon steel　工具用炭素鋼
tool gauge　刃物ゲージ
tool mark　ツールマーク，工具きず
tool steel　工具鋼
tooth　名歯
動歯をつける，目立てをする
tooth bearing　歯当たり
tooth bearing center　歯面中心
tooth belt　歯形ベルト
tooth clutch　歯付きクラッチ
tooth crest　歯先面
tooth depth　歯たけ
tooth flank　歯元の面
tooth form (profile)　歯形
tooth gear　歯車
tooth pitch gauge　歯形ピッチゲージ
tooth profile curve　歯形曲線
tooth profile factor　歯形係数
tooth root stress　歯元応力
tooth space　歯みぞ
tooth surface　歯面
tooth thickness　歯厚
toothed　形歯付き，のこぎり状の
toothed armature　スロット付き電機子
toothed factor　歯車係数
toothed lock washer　歯付き座金
toothed wheel　歯車　同 gear
top　名トップ，頂点，頭部，こま
形一番上の
動頂上に達する，越える
top clearance　上部すきま
top coat　上塗り
Top Dead Center　TDC，上死点
top disc　トップディスク
top half　上半分
top heavy　名トップヘビー
形不安定な，頭でっかちの
top ring　トップリング
top speed　最高速度
top view　平面図
torch lamp　トーチランプ
torque　トルク，回転力
torque arm　トルクアーム
torque coefficient　トルク係数
torque constant　トルク定数
torque converter　トルクコンバータ【流体変速装置】
torque limiter　トルクリミッタ，トルク伝動制限器
torque meter　トルク計
torque motor　トルクモータ
torque wrench　トルクレンチ【トルクを測定できるレンチ】
Torr　トール【圧力の単位】
Torricelli's theorem　トリチェリの定理
torshear type bolt　トルシア形ボルト【高力ボルトの一種】
torsion　ねじり，ねじれ
torsion angle　ねじれ角
torsion axis　ねじり軸
torsion bar　ねじり棒，トーションバー【ねじりに対する復原力を持つ金属棒】
torsion coil spring　ねじりコイルばね
torsion couple　ねじり偶力
torsion damper　ねじり振動止め
torsion dynamometer　ねじり動力計
torsion galvanometer　ねじり検流計
torsion meter　トーションメータ，ねじり動力計
torsion moment　ねじりモーメント
torsion spring　ねじりばね
torsion test　ねじり試験
　関 torsion tester：ねじり試験機
torsion vibrograph　ねじれ振動計
torsional　形ねじりの
torsional buckling　ねじれ座屈
torsional couple　ねじり偶力
torsional criticals　ねじり危険回転
torsional deflection　ねじりたわみ
torsional deformation　ねじり変形

torsional dynamometer　ねじり動力計
torsional effect　ねじれ効果
torsional flexibility　ねじりたわみ性
torsional load　ねじり荷重
torsional moment　ねじりモーメント
torsional oscillation　ねじり振動
torsional resisting moment　ねじり抵抗モーメント
torsional rigidity　ねじり剛性【ねじり荷重に対する変形抵抗】
torsional spring constant　回転ばね定数
torsional strength　ねじり強さ
torsional stress　ねじり応力
torsional vibration　ねじり振動
torsionally　副ねじれて
toss　動投げる, 揺れる, ゆする
total　名トータル, 合計, 総計　形全体の, 総～, 全～　動合計する
total acid number　全酸価
total alkali　全アルカリ
total alkalinity　全アルカリ度
total amount　総量
total amplitude　全振幅
total analysis　完全分析, 全分析
total applied voltage　全印加電圧
total base number　全アルカリ価
total calorific value　総発熱量
total cross section　全断面積
total deflection　全たわみ
total differential equation　全微分方程式
total discharge head　吐出し全ヘッド
total drag　全抵抗
total dynamic head　全動圧
total efficiency　全効率, 総合効率
total elongation　全伸び, 破断伸び
total energy　全エネルギ
total equivalent brake power　相当軸出力
total face width　全歯幅
total hardness　全硬度, 総硬度
total head　全ヘッド, 全落差, 全水頭, 全揚程
total heat　エンタルピ, 全熱量

total heat entropy diagram　エンタルピ・エントロピ線図
total heating value　全発熱量
total ionization　全電離, 全電離度
total length　全長
total load　全荷重, 全負荷
total loss　全損, 全損失
total moisture content　全水分
total operating hour　総運転時間
total performance　総合性能
total pressure　全圧, 全圧力, 総圧
total pressure loss　全圧損失
total pressure vacuum gauge　全圧計
total pump head　全揚程
total resistance　全抵抗
total solid　全固形物, 全固形分
total strain　全ひずみ
total strain energy　全ひずみエネルギ
total strain range　全ひずみ幅
total suction head　吸込み全ヘッド
total sum　合計, 総和, 累計
total thermal resistance　総括熱抵抗
total weight　総重量, 全重量
total width　全幅
totality　総数, 全体, 総計
totalize　動合計する, 要約する
totally　副全く, 完全に
totally enclosed type　全閉形
touch　動触れる
tough　形タフな, 強靭な, 粘り強い, 堅い
tough hardening steel　強靭鋼
tough pitch copper　タフピッチ銅, 精銅
tough property　靭性
toughened glass　強化ガラス
toughness　粘り強さ, 靭性
tow　動牽引する
toward　～　前～の方へ
toxic　形有毒の
toxic gas　有毒ガス
trace　名跡, 微量, (配)線　動突きとめる, 解明する
trace of oxygen　微量の酸素
traceability　トレーサビリティ

traced drawing 原図, 元図
tracer トレーサ【追跡するもの】
tracing トレーシング, 複写, 透写
tracking 追跡
  tracking control　トラッキング制御, 追従制御
traction トラクション, 引張り, 牽引
traditional 形 従来の, 伝統的な
traditionary 従来から, 伝統的に
tragic 形 悲劇の, 悲惨な, 痛ましい
trail 動 引きずる, 跡をつける
trailer トレーラ, 付随車
trailing edge 後縁
train (長い)列
training 訓練, 練習
  training ship　練習船
trajectory 軌道, 軌跡
  trajectory of principal stress　主応力線
transaction 相互作用
transceiver トランシーバ
transducer 変換器, 振動子, 伝送器
transfer 動 移す, 移動する, 伝える, 変える　名 伝達, 移送, 伝導, 輸送, 移動, 転送
  関 heat transfer：伝熱
  transfer calipers　写しパス
  transfer efficiency　伝達効率
  transfer function　伝達関数
  transfer lag　伝達遅れ
  transfer pump　移送ポンプ
transference of heat 熱伝播
transform 動 変換する, 変形させる, 変圧する　名 変換
transformation 変態, 変換, 転移
  関 A_1 transformation：A_1 変態
  transformation loss　変圧損
  transformation point (temperature)　変態点
  transformation ratio　変圧比, 変性比
transformer トランス, 変圧器, 変成器
  transformer oil　変圧器油
transgranular fracture 粒内破壊
transient 形 過渡的な, 一時的な
  transient current　過渡電流
  transient deviation　過渡偏差
  transient phenomenon　過渡現象
  transient response　過渡応答
  transient state　過渡状態
  transient state vibration　非定常振動
  transient vibration　過渡振動
transistor トランジスタ
  transistor ignition system　トランジスタ点火装置
transit 名 輸送, 通過, トランシット【望遠鏡のついた測量器】　動 通過する, 横切る
transition 遷移, 転移
  transition boiling　遷移沸騰
  transition element　遷移元素
  transition fit　中間ばめ
  transition layer　遷移層
  transition metal　遷移金属
  transition point　転移点
  transition region　遷移域
  transition temperature　遷移温度, 転移温度, 転移点
translate 動 翻訳する
translation 移動, 並進, 翻訳
  translation cam　直動カム
translational motion 並進運動
transmissibility 伝達率
transmission トランスミッション, 伝達, 透過, 送信, 送電, 伝送, 伝導
  transmission dynamometer　伝動動力計
  transmission efficiency　伝達効率
  transmission gear　伝動装置
  transmission loss　伝送損失, 透過損失, 送電損
  transmission shaft　伝動軸
  transmission system　伝達システム
  transmission zone　遷移領域
transmissivity 透過率
transmissometer 透過率計
transmit 動 伝達する, 透過する
transmittance 透過率, 透過度
transmitter 伝送器, 送信機, 発信機
transmittivity 透過率
transpiration cooling 吹出し冷却

transport 名輸送, 移動 動輸送する, 運ぶ
transportation 輸送
transpose 動入れ替える
transversal 名横断線 形横断の
　transversal section　横断面
　transversal strain　横ひずみ
　transversal stress　横応力
transverse 形横[向き]の, 横断した, 横軸の 名横, 横軸, 横断
　transverse arc of transmission　接触弧
　transverse bulkhead　横隔壁
　transverse direction　横方向
　transverse impact　曲げ衝撃
　transverse load　横荷重
　transverse metacenter　横メタセンタ
　transverse moment　横揺れ
　transverse section　横断面
　transverse stability　横復原力
　transverse strain　横ひずみ
　transverse strength　横強度
　transverse stress　横応力
　transverse vibration (oscillation)　横振動
　transverse wave　横波
transversely 副横に, 横切って
trap 名トラップ【排水装置の一種】動せきとめる, 溜まる, 捕らえる
trapezoid　台形 同 trapezium
trapezoidal 形台形の
　trapezoidal rule　台形法則
　trapezoidal thread　台形ねじ
trapping efficiency　給気効率
traval centrifuge　トラバル遠心分離機
travel 名行程, 移動, 運動 動移動する, 伝わる
traveler　すべり環
traveling cable　移動ケーブル
traveling load　移動荷重
traverse 動横切る, 交差する 名横行, 横断, 障害, 横送り装置
　traverse feed　横送り
tray　トレイ【トレイとは浅い皿の意】, 脱気器のじゃま板
tread　トレッド【左右車輪の中心距離】, 踏み面
treat　動処理する, 扱う
treated oil　洗浄油
treated steel　処理鋼
　関 surface treated steel：表面処理鋼
treater　処理器
treating oil　処理油
treatment　トリートメント, 処理, 処置
trebling　補強板
tremendous 形巨大な, ものすごい
trend　傾向
　trend analysis　(動向, 傾向)分析
　trend recorder　多点記録計
tri- 接頭「3の」の意
Triac　トライアック【交流電流の制御に用いる半導体素子】
trial　試運転
　trial and error　試行錯誤
　trial speed　試運転速度
triangle　三角形
　triangle of force　力の三角形
　triangle square　三角定規
triangular 形三角の, 三角形の
　triangular arrangement　千鳥形配列
　triangular cylinder　三角柱
　triangular thread　三角ねじ
　triangular weir　三角せき
triaxial stress　三軸応力
triboelectricity　摩擦電気
tribology　摩擦学, 潤滑
tribometer　摩擦計
trichlene　トリクレン【脱脂用洗浄剤】
trigger　トリガ, 引き金
　trigger circuit　トリガ回路
　trigger point　起動点
　trigger stop　トリガストップ【自動停止装置のこと】
trigonometric function　三角関数
trim 名トリム, 縦傾斜 動刈り込む, [船の]つり合いを取る
trimetal　トリメタル, 三層メタル
trimmer　微調整用可変素子
trip　トリップ, 引き外し, 遮断 動トリップする
　trip coil　引き外しコイル

trip gear 引き外し装置
trip meter 走行距離計
triple 形3倍の，三重の 名3倍
 triple point 三重点
 triple thread 三条ねじ
 triple valve 三動弁
tripoli トリポリ【研磨剤の一種】
tripping 引き外し
 tripping current 引き外し電流
tritium トリチウム，三重水素【水素の同位元素の一つ】
trochoid トロコイド
 trochoid pump トロコイドポンプ【歯車ポンプの一種】
trolley トロリ，集電器
troostite トルースタイト【鋼の組織の一つ】
tropometer 捻転角度計
trouble トラブル，故障，不調 動心配させる，迷惑をかける
 trouble free operation 円滑な運転
 trouble shooter 機器の故障に対応する人
troublesome 形面倒な
trowel こて
truck 台車
true 形真の，真実の，本当の，当てはまる，本物の
 関 be true of ～：当てはまる
 true strain of fracture 真破断ひずみ
 true stress 真応力
 true stress of fracture 真破断応力
 true tensile strength 真の引張り強さ
 true value 真値
truly 副実に，本当に，真に
trunk トランク，通風筒
 trunk piston engine トランクピストン機関，筒形ピストン機関
 trunk piston rudder gear トランクピストン式舵取り装置
trunking ケーシング，電信中継回線
trunnion トラニオン，耳軸
 trunnion mounting cylinder トラニオン形シリンダ
truss トラス【骨組構造物の一種】
 truss structure トラス構造

try 動試みる，やってみる
 try square 直角定規
T-S (temperature entropy) diagram T-S 線図，温度-エントロピ線図
tube チューブ，パイプ，管，水管
 tube arrangement 管配置
 tube bank 管群
 tube blower 管すす吹き
 tube brush 管ブラシ
 tube cutter 管切り
 tube expander 管拡げ器【熱交換器の管を管板に取り付け密着させる工具】
 tube plate 管板
 tube plug プラグ
 tube seam 管継目
 tube stopper 管栓
 tube type combustor 筒形燃焼室【ガスタービンの燃焼装置】
 tube wall 管壁
tubular 形管の，管状の，多くの管からなる
 tubular boiler 煙管ボイラ
 tubular exchanger 管形熱交換器
 tubular radiator 管型放熱器
tubulous boiler 水管ボイラ
tufftriding タフトライド，軟窒化【液体浸窒の一種】
tug 動強く引っ張る
tumbler switch タンブラスイッチ【上下または左右に倒すスイッチ】
tune 動同調させる，調整する
tuner チューナ，同調器
tungsten タングステン 記 W
 Tungsten electrode Inert Gas welding TIG 溶接
 tungsten steel タングステン鋼
tuning 同調
 tuning coil 同調コイル
tunnel トンネル
 tunnel diode トンネルダイオード
 tunnel effect トンネル効果
 tunnel shaft 中間軸
 tunnel shaft bearing 中間軸受
 tunnel well トンネルビルジ溜り
turbidimeter 濁り度計，混濁計，

濁度計
turbidity 濁度，濁り度
turbine タービン【羽車を回転し動力を伝える機械】
  turbine blade　タービン羽根，タービン翼
  turbine casing　タービン車室
  turbine current meter　羽根車式流速計
  turbine disc　タービン円板，タービン翼車
  turbine driven　形タービン駆動式の
  turbine driven feed pump　タービン駆動給水ポンプ
  turbine drum　タービン胴
  turbine efficiency　タービン効率
  turbine flowmeter　タービン流量計
  turbine generator　タービン発電機
  turbine nozzle　タービンノズル
  turbine oil　タービン油
  turbine output　タービン出力
  turbine powered　形ターボ駆動の
  turbine pump　タービンポンプ
  turbine rotor　タービンロータ，タービン翼車
  turbine rotor blade　タービン動翼
  turbine shaft　タービン軸，タービン車軸
  turbine stage　タービン段
  turbine stator cascade　タービン静翼列
  turbine wheel　タービン羽根車
turbo- ターボ【タービンの意】
  turbo-blower　ターボ送風機
  turbo-charged engine　過給機関
  turbo-charger　ターボチャージャ，排気タービン過給機【排気を利用したタービンでシリンダ内に空気を強制的に圧送する】
  turbo-charging　ターボ過給，排気タービン過給
  turbo-compressor　ターボ圧縮機，遠心圧縮機
  turbo electric drive (propulsion)　タービン電気推進
  turbo-fan　ターボファン，ターボ送風機
  turbo-generator　ターボ発電機
  turbo-jet engine　ターボジェットエンジン
  turbo-machinery　ターボ機械
  turbo-prop engine　ターボプロップエンジン
  turbo-refrigerating machine　ターボ冷凍機，遠心冷凍機
  turbo-separator　遠心式分離器，遠心式気水分離器
  turbo-supercharger　ターボ過給機
turbulence 乱流，乱れ
  turbulence stimulation　乱流発生装置
  turbulence viscosity　乱流粘度
turbulent 形乱流の
  turbulent boundary layer　乱流境界層
  turbulent burner　渦巻バーナ
  turbulent diffusion　乱流拡散
  turbulent factor　乱流係数
  turbulent flame　乱流火炎
  turbulent flow　乱流
  turbulent friction coefficient　乱流管摩擦係数
  turbulent kinetic energy　乱れエネルギ
  turbulent layer　乱流層
  turbulent motion　乱れ運動
  turbulent viscosity　乱流粘度
turn 動回転する・させる，曲がる，ひっくり返す，向きを変える
  名回転，変化，順番
  熟 in turn：順番に，交替で，次々
  turn down　動小さくする，抑える，拒否する，低くする
  turn-down ratio　ターンダウン比，絞り比
  turn off　動（スイッチを）切る
  turn over　動始動する・させる，回転する・させる，引き継ぐ
  turn-over　ターンオーバ，回転，交替，転換，移動
  turn ratio　巻数比
turnbuckle ターンバックル【支持棒などの長さ調整器具】
turned bolt 仕上げボルト

turning 回転，旋回
- turning angle　転向角，回転角
- turning circle　旋回圏
- turning force (effort)　回転力
- turning gear　ターニング装置，回転装置
- turning moment　回転モーメント，トルク
- turning motion　回転運動
- turning motor　ターニングモータ
- turning pair　回り対偶
- turning speed　旋回速力
- turning trial　旋回試験
- turning wheel　回転輪

twice 副2回，2度

twin 名ツイン，双晶
形ツインの，一対の
- twin cam engine　ツインカムエンジン
- twin core cable　二芯ケーブル
- twin longitudinal bulkhead　二列縦隔壁
- twin rudder　二枚舵
- twin screw vessel　二軸船
- twin turbine　双流タービン
- twin vortex　双子渦

twirl 動くるくる回す，振り回す

twist 動ねじる，より合わせる
名ツイスト，ねじれ，ねじり

twisted blade ねじれ羽根

twisting load ねじり荷重

twisting moment ねじりモーメント

twisting stress ねじり応力

twisting test ねじり試験

twisting vibration ねじり振動

two 名2　形2つの

- two color water gauge　二色水面計
- two cycle diesel engine　二サイクルディーゼル機関
- two dimensional　形二次元の
- two dimensional flow　二次元流
- two dot chain line　二点鎖線
- two element system　2要素式
- two liquid manometer　二液マノメータ
- two or more　2つ以上
- two pass flow　2回流
- two phase　二相
- two phase flow　二相流
- two phase servomotor　二相サーボモータ
- two position action　二位置動作
- two speed gear　二段変速装置
- two stage air compressor　二段空気圧縮機
- two stage combustion　二段燃焼
- two step relay　二段動作継電器
- two [stroke] cycle engine　二サイクル機関
- two throw crank shaft　二連クランク軸
- two way cock　二方コック
- two wire system　二線式

type 名タイプ，型，型式，種類
動タイプする，分類する
- type A fire door　甲種防火戸
- type of equipment　機種
- type of ignition　点火様式
- type ship　基準船

typical 形代表的な，典型的な，特有の，標準的な

U-u

U bend　Uベンド
U trap　Uトラップ
U tube　U字管
 U tube manometer　U字管マノメータ，U字管圧力計
U type expansion bend　U形伸縮管継手
ullage　アレージ【タンクの液面より上の部分の空間のこと】
 ullage table　アレージ表
ultimate　形究極の，これ以上細分化できない，最大の，最高の
 ultimate analysis　元素分析
 ultimate load　破壊荷重
 ultimate moment　極限モーメント
 ultimate set　極限残留ひずみ
 ultimate strength　極限強さ
ultra-　接頭「超〜」「過〜」の意
ultraaudible sound　超音波
ultrahigh pressure　超高圧
ultrahigh strength steel　超高張力鋼，超高力鋼，超強度鋼
ultrahigh vacuum　超高真空
ultramarine blue　群青
ultraoptimeter　ウルトラオプチメータ【コンパレータの一種】
ultraprecision machining　超精密加工
ultrashort wave　極超短波
ultrasonic　形超音波の
 ultrasonic bonding　超音波圧着法
 ultrasonic cleaning　超音波洗浄
 ultrasonic detection of defects　超音波探傷
 ultrasonic detector　超音波探知器
 ultrasonic flaw detecting test　超音波探傷検査
 ultrasonic flowmeter　超音波流量計
 ultrasonic hardness meter　超音波硬さ計
 ultrasonic inspection　超音波探傷法
 ultrasonic machining　超音波加工
 ultrasonic thicknessmeter　超音波厚み計
 ultrasonic type liquid level gauge　超音波液面計
 ultrasonic vibration　超音波振動
 ultrasonic wave　超音波
Ultra Violet　名UV，紫外線　形紫外[線]の
 ultraviolet ray　紫外線
 ultraviolet sterilizer　紫外線殺菌装置
ultrawave　超音波
un-　接頭「反対」「否定」の意
unattended　無人の
unavailable energy　無効エネルギ
unavoidable　形避けられない
unavoidably　副止むを得ず
unbalance　名アンバランス，不均衡，不つり合い　動不つり合いにする
 unbalance load　不つり合い負荷
 unbalance moment　不つり合いモーメント
unbalanced couple　不つり合い偶力
unbalanced pressure　不つり合いの圧力
unburned　形未燃焼の
 unburned combustibles　未燃分
 unburned fuel　未燃焼燃料
 unburned fuel loss　未燃損失
 unburned gas　未燃焼ガス
unburnt　形未燃焼の
 unburnt gas　未燃焼ガス
 unburnt hydrocarbon　未燃炭化水素
unchangeable　形不変の
uncharged engine　無過給機関
uncoil　動伸ばす，ほどく
unconformity　不整合
unconstrained　形不限定の
 unconstrained chain　不拘束連鎖
uncontrollable　制御できない
uncooled　非(無)冷却
uncover　動おおいを取る，開放する
undamaged　形損傷を受けていない
under 〜　前〜の下に，〜のもとに，〜を受けて，〜に基づいて
under-　接頭「不足して」「少なす

ぎて」の意
undercompound generator 不足複巻発電機
undercooled steam 過飽和蒸気
undercooling 過冷却
　関 dgree of undercooling：過冷却度
underestimate 動低く評価する
underexpansion 不足膨張
undergo 動〜を体験する
underlie 動〜の下に横たわる，〜の基礎となる
underline 動下線を引く
　名アンダーライン，下線
underlying 形基本的な
underneath 副下に　前すぐ下に
underside 下方，内面，底面
undersize 形小型の
underslung 形吊り下げ式の，重心が低い
understand 動理解する
understanding 理解
understood 「understand」の過去・過去分詞
understress 過小応力
undertake 動引き受ける，了解する
undervoltage relay 不足電圧継電器
undesirable 形望ましくない
undetected 検出できない
undo 動元通りにする，ゆるめる
unelastic 形非弾性の，弾力のない
unequal 形不均一な
uneven 形不規則な
unexplained 動説明のつかない
unfinished 形未完成の
unglazed 形素焼きの
uniaxial stress 一軸応力
unidirectional valve 一方向弁
unified screw thread ユニファイねじ
uniflow 形流れが一方向の
　uniflow engine ユニフロー機関，単流機関
　uniflow scavenging ユニフロー掃気，単流掃気式
uniflux condenser 整流復水器
uniform 形一様な，等しい，一定の，均一の　名ユニフォーム

動一様にする，同一にする
uniform acceleration 等加速度
uniform circular motion 等速円運動
uniform elongation 一様伸び
uniform flow 一様流，等流
uniform load 等分布荷重，均一負荷
uniform motion 等速運動
uniform pitch 均一ピッチ
uniform pitch propeller 一定ピッチプロペラ
uniform strength 均一強さ
uniformity 一様，均等
uniformly 副一様に，均一に
　uniformly distributed load 等分布荷重
unify 動統一する，1つにする
uninformed 形知識のない，十分な情報を持たない
union 結合，ユニオン継手
　union joint ユニオン継手
　union nut ユニオンナット
unipolar ユニポーラ，単極性
　unipolar transistor ユニポーラトランジスタ
unit ユニット，単位，装置，構成部分
　関 basic unit：基本単位
　unit air charge 吸気率
　unit area 単位面積
　unit cell 単位格子
　unit circle 単位円
　unit cooler ユニットクーラ
　unit cross section area 単位断面積
　unit lattice 単位格子
　unit length 単位長さ
　unit pump 単筒形噴射ポンプ
　unit vector 単位ベクトル
　unit volume 単位体積
unite 動結合する，一体にする
universal 形万能の，普遍的な，一般的な
　名ユニバーサル，万能，普遍
　universal angle block 万能定盤
　universal bellows expansion joint ユニバーサル式ベローズ形伸縮管継手
　universal coupling 自在継手

universal gas constant　一般気体定数
universal gravitation　万有引力
universal joint　自在[軸]継手
universal motor　交直両用電動機
universal testing machine　万能材料試験機
universal vice　万能万力
unknown　形未知の　名未知数
unknown constant　未知定数
unknown quantity　未知数, 未知量
unless　接もし〜でなければ, 〜でない限り, 〜以外は
unlike 〜　前〜と違って
unlikely　形疑わしい, 可能性が低い
unload　動取り外す, 荷を降ろす
unloader　アンローダ, 負荷軽減装置【圧縮機の負荷軽減装置】
unloading [pressure control] valve　アンロード弁
unlock　動鍵を開ける
unmanned operation　無人運転
unnecessory　形不必要な
unrefined　形未精製の, 未加工の
unrelated　形関係のない, 無関係な
unsafe　形危険な
unstability　不安定性
unstable　形不安定な
unstable condition　不安定状態
unstable equilibrium　不安定つり合い
unstationary state　非定常状態
unsteady　形不安定な, 変動する
unsteady flow　非定常流
unsteady state　非定常状態
unsuitable　形不向きの, 不適切な
unsupportable　形支持されない, 支えられない, 耐えられない
until 〜　前〜まで　接〜するまで
unused　形未使用の
unusual　形異常な
unusully　副異常に, 非常に
unveil　動覆いをとる, 明らかにする
unwary　形不注意な
unwater　動排水する
unwind　動ほどける, ゆるめる
up and down　副上下に

up to 〜　前〜まで
up-to-date　形最新[式]の
uphold　動支える, 持ち上げる
uplift　動持ち上げる
upon 〜　前〜の上に　同on
upper　形上の, 上位の, 上級の
upper chamber　上部室
upper dead center　上死点
upper explosion limit　爆発上限
upper face　上面
upper limit　上限
upper yield point　上降伏点
upright　名垂直材　形直立した, 縦長の　副まっすぐに
uprighting moment　復原モーメント
upset　動ひっくり返す　名転倒
upset butt welding　アプセット突合わせ溶接
upside down　副逆さまに
upstroke　上り行程
uptake　アプテーク, 排気管, 煙道, 煙路, 通風管【ボイラ炉から煙突への煙道】
upward　副上方へ, 上向きに　形上向きの
uranium　ウラン　記U
urea　尿素
urethane rubber　ウレタンゴム
usability test　使用性能試験
use　動用いる, 消費する　名使用, 用途, 利用
　関in use：使用中
useful　形有効な, 有用な
useful horsepower　有効馬力, スラスト馬力
useful life　耐用年数, 有効寿命
useful load　有効荷重, 有効負荷
useful thread　有効ねじ部
useful work　有効仕事
usefulness　有用性
useless　形役に立たない, 無用な
user　ユーザ, 利用者
usual　形一般的な
usually　副通常, 普通は, 一般に
utensil　器具
utility　有用, 実用性

utilization 利用,活用
utilize 動利用する

UV(**Ultra-Violet**)[**ray**] 紫外線

V-v

V belt　Vベルト
　V belt drive　Vベルト駆動
V block　Vブロック
V connection　V結線
V notch weir　三角堰
V packing　Vパッキン【V形断面のリング状パッキン】
V pulley　Vプーリ
V ring　Vパッキン
V type engine　V型エンジン
V-V connection　V-V結線
V welding　矢はず継手
vacuum　真空
　㊝ degree of vacuum：真空度
　vacuum annealing　真空焼鈍し
　vacuum augmentor　真空増進器
　vacuum breaker　バキュームブレーカ，真空破壊器，真空破壊弁【流体の供給ラインに生ずる真空を除いて逆流を防ぐ装置】
　Vacuum Circuit Breaker　VCB，真空遮断器
　vacuum cleaner　掃除機
　vacuum desiccator　真空乾燥器，減圧デシケータ
　　vacuum discharge　真空放電
　vacuum distillation　真空蒸留
　vacuum dryer　真空乾燥機
　vacuum ejector　真空エジェクタ
　vacuum evaporation　真空蒸着
　vacuum fan　真空ファン
　vacuum filter　真空ろ過器
　vacuum forming　真空成形
　vacuum gauge　真空計
　vacuum gauge pressure　真空ゲージ圧力
　vacuum impregnation　真空含浸
　vacuum ion carburizing　真空イオン浸炭法
　vacuum loss　真空損失
　vacuum oxygen decarburization　真空酸素脱炭法
　vacuum plating　真空めっき
　vacuum plating of aluminum　アルミニウム蒸着
　vacuum pump　真空ポンプ
　vacuum refrigerating machine　真空冷凍機
valence　原子価
　valence electron　価電子
valid　㊡有効な，妥当な
valuable　㊡貴重な，重要な
value　㊅値，価値
　㊌評価する，価値判断をする
　value added　付加価値
　value of impact energy　衝撃値
valve　弁，バルブ
　valve action　弁作用
　valve body　弁体
　valve bonnet　弁帽
　valve cam　弁カム
　valve casing　弁箱　㊝ valve box
　valve chest　弁室
　valve clearance (gap, lash)　弁すきま【タペットと弁棒または揺れ腕と弁棒とのすきま】
　valve closing time　弁閉じ時期
　valve connector　バルブ継手
　valve controlled port scavenging　弁調節穴掃気【二サイクルエンジンの掃気法の一種】
　valve diagram　弁線図
　valve disc　弁体，弁ディスク
　valve discharge coefficient　弁流量係数
　valve face　弁フェース
　valve gear　弁装置，動弁装置
　valve guard　弁押え
　valve guide　弁案内
　valve handle　バルブハンドル
　valve head　弁頭，弁がさ
　valve-in-head scavenging　頭弁掃気
　valve inlet pressure　一次側圧力
　valve lever　バルブレバー，弁てこ
　valve lift　バルブリフト，弁揚程
　valve loop　弁ループ【軸流タービ

ンのノズルの配置】
valve mechanism 弁機構
valve motion 弁装置
valve opening time 弁開き時期
valve operating gear 弁駆動装置, 吸排気動弁装置
valve outlet pressure 二次側圧力
valve overlap 弁オーバラップ
valve plate 弁板, 弁盤
valve port 弁孔
valve positioner バルブポジショナ【ダイアフラム弁の制御装置】
valve push rod 弁突き棒
valve reseater 弁座削正器
valve rocker 弁の揺りてこ
valve rod (spindle, stem) 弁棒
valve rod guide 弁棒案内
valve rotator バルブローテータ
valve scavenging 弁掃気
valve seat 弁座
valve seat angle 弁座角
valve setting 弁調整
valve spindle 弁棒
valve spring 弁ばね
valve spring retainer 弁ばね受け
valve steel 弁用鋼
valve sticking 弁固着
valve timing バルブタイミング, 弁開閉時期
valve tip 弁棒端
valve travel 弁行程
valve yoke 弁枠
vanadium バナジウム 記 V
vanadium attack バナジウムアタック
vanadium steel バナジウム鋼
vane 羽根, 翼
vane motor ベーンモータ
vane pump ベーンポンプ
vane wheel flow meter 羽根車流量計
vaned diffuser 羽根付きディフューザ
vaneless diffuser 羽根なしディフューザ
vanish thread 不完全ねじ
vanishing バニシング【潤滑剤の変色で生じる現象】

vanishing point of stability 復原力減失角
vapor 蒸気 同 steam
vapor column 蒸気柱
vapor compression refrigerating machine 蒸気圧縮冷凍機
vapor density 蒸気密度
vapor deposition 蒸着
vapor explosion 蒸気爆発
vapor film 蒸気膜
vapor lock ベーパロック【液中に蒸気が発生し不具合を生じること】
vapor phase 気相
vapor phase cracking 気相分解
vapor plating 気相めっき
vapor pocket ベーパポケット, 蒸気空間
vapor pressure 蒸気圧
vapor proof 防湿, 防露
vapor velocity 蒸気速度
vaporization 蒸発, 気化, 揮発
vaporize 動蒸発する・させる
vaporizer 蒸発器, 気化器, 霧吹き[器] 同 evaporator
vaporizing chamber 蒸発室
vaporizing combustion chamber 気化式燃焼器
vapour 同 vapor
varactor 可変容量ダイオード
variable 形可変[式]の
名変数, 可変, 量
variable acceleration 変加速度
variable area flowmeter 面積式流量計
variable capacitor 可変コンデンサ, バリコン【可変蓄電器】
variable capacity 可変容量
variable capacity motor 可変容積形モータ
variable condenser 可変コンデンサ, バリコン
variable displacement (delivery) pump 可変容量形ポンプ
variable expansion valve 加減膨張弁
variable geometry 可変静翼
variable inductance 可変誘導

variable load　変動荷重
variable pitch　可変ピッチ
variable pitch propeller　可変ピッチプロペラ
variable pressure operation　変圧運転
variable rate spring　非線形ばね
variable resistance　可変抵抗
variable resistor　可変抵抗器
variable speed　可変速度, 無段変速
variable speed gear　変速装置
variable speed motor　可変速モータ
variable speed pump　可変速ポンプ
variable stroke pump　可変行程ポンプ
variable value control　追値制御
variable voltage control　可変電圧制御

variance　相違, 不一致, 変化, 変動, 分散
variation　バリエーション, 変動, 変異, 変化, 偏差
variation of tolerance　寸法差
varied　形多様な, 変えられた
varied flow　不等流
variety　多様性
関 a variety of：様々
various　形様々な, 多くの
various impurity　様々な不純物
variously　副様々に, いろいろに
varistor　バリスタ【抵抗器の一種】
varnish　ワニス
varnished cambric insulated cable　ワニスカンブリック絶縁ケーブル
vary　動変える・わる, 変化する・させる, 異なる
varying　形変化する
varying duty　変負荷連続使用
varying pitch　変動ピッチ
varying pitch propeller　可変ピッチプロペラ
varying speed motor　変速度電動機
varying stress　変動応力
vaseline　ワセリン【固形油の一種】
vast　形膨大な, 広大な, 巨大な
vector　ベクトル【大きさ・方向を持つ量】　関 scalar：スカラ

vector control　ベクトル制御
vector diagram　ベクトル図
vector locus　ベクトル軌跡
vector mean velocity　ベクトル平均速度
vector product　ベクトル積
vector quantity　ベクトル量
vegetable oil　植物油
vehicle　車両, 伝達手段, 媒体
veil　動ベールで覆う, 隠す
velocimeter　速度計, 流速計
velocity　速度　同 speed
velocity boundary layer　速度境界層
velocity coefficient　速度係数
velocity compound turbine　速度複式タービン
velocity compounded stage　速度複式段
velocity curve　速度曲線
velocity diagram　速度線図, 速度三角形
velocity distribution　速度分布
velocity drop ratio　速度降下率
velocity error　速度偏差
velocity fluctuation　速度変動
velocity gradient　速度勾配
velocity head　速度ヘッド, 速度水頭
velocity modulation　速度変調
velocity of light　光速度
velocity of rotation　回転速度
velocity pickup　速度ピックアップ
velocity potential　速度ポテンシャル
velocity pressure　動圧
velocity pressure compound turbine　速度圧力複式タービン
velocity ratio　速度比
velocity triangle　速度三角形, 速度線図
vena contracta　縮流, くびれ
venetian door　よろい戸
vent　ベント, 通風穴, 排気, 脱気孔, ガス抜き, 通気, 通気孔
vent condenser　ベントコンデンサ
vent hole　ガス抜き穴
vent sleeve　通風筒

vent valve　逃し弁
ventilate　動換気する
ventilation　換気，通風，換気装置
　ventilation fan　換気ファン
　ventilation trunk　通風トランク
ventilator　ベンチレータ，通風機，通気筒，換気装置
venting　ガス抜き
venture　動危険にさらす，思い切って進む　名ベンチャ，冒険
Venturi meter　ベンチュリ計【差圧流量計の一種】
Venturi scrubber　ベンチュリスクラッバ【気体中の粉じんを捕捉除去する装置】
Venturi tube　ベンチュリ管
verdigris　緑青【緑色の有毒さび】
verification　確認，検証
verify　動立証する，確かめる
vernier　バーニア，副尺
　vernier calipers　ノギス
versatility　多様性
vertex　頂点
vertical　鉛直，たて形，垂直線(面)　形垂直の，縦の
　vertical angle　対頂角
　vertical axis　縦軸
　vertical boiler　立てボイラ
　vertical engine　立て型機関
　vertical force　垂直力
　vertical incidence　垂直入射
　vertical line　垂線，垂直線
　vertical motion　上下運動
　vertical pipe　立ち管
　vertical plane　垂直面
　vertical pump　縦軸ポンプ
　vertical section　縦断面
　vertical shaft　縦軸
very　副非常に，大変，とても　形まさにその
　very Large Scale Integration　超 LSI
　very low temperature　極低温
vessel　船，容器，入れ物
via　前経由で，によって
vibrate　動振動する，揺れる
vibration　振動，変動

vibration control　制振，振動制御
vibration damper　振動ダンパ，振動止め
vibration isolation　防振
vibration level meter　振動レベル計
vibration of normal mode　固有振動
vibration proof fitting　防振継手
vibration proof rubber　防振ゴム
vibration system　振動系
vibrational motion　振動運動
vibrator　バイブレータ，振動器
vibrograph　振動[記録]計
vibroisolating material　防振材
vibroisolating rubber　防振ゴム
vibrometer (vibroscope)　振動計，振動指示計
vice　万力　関 vice
　vise versa　副逆に，逆もまた同じ
vicinity　近辺，付近，周辺
Vickers hardness　ビッカース硬さ
Vickers hardness tester　ビッカース硬さ試験機
video　ビデオ【映像または映像信号のこと】
view　動見る，眺める，観察する　関 with a view to ～ ing：～する目的で，～するつもりで
vigilant　形慎重な，用心深い
vinyl chloride resin　塩化ビニル樹脂
vinyl pipe (tube)　ビニル管
violate　動破る，犯す，違反する
virtual　形仮の，虚の，事実上の
　virtual mass　見掛けの質量
　virtual moment of inertia　見掛け慣性モーメント
　virtual reality　バーチャルリアリティ，仮想現実
virtually　副事実上，～も同然で
virus　ウイルス　関 computer virus：コンピュータウイルス
vis inertiae　惰性，惰力
viscometer　粘度計
visco-plasticity　粘塑性
viscosimeter　粘度計
viscosity　粘性，粘度，粘性率
　viscosity blending chart　混合粘度

図表
viscosity controller 粘度調節器
viscosity conversion table 粘度変換表
Viscosity Index VI, 粘度指数
viscosity index improver 粘度指数向上剤
viscosity law 粘性法則
viscosity ratio 粘度比
viscosity temperature chart 粘度温度線図
viscous 形粘性のある, 粘着性の
viscous boundary layer 粘性境界層
viscous damping coefficient 粘性減衰係数
viscous flow 粘性流
viscous fluid 粘性流体
viscous friction 粘性摩擦
viscous frictional damping 粘性摩擦減衰
viscous lubrication 粘性潤滑
viscous resistance 粘性抵抗
viscous stress 粘性応力
viscous sublayer 粘性底層
vise 万力
visibility 視界, 視程
visible 形目に見える, 明白な
visible flame 可視炎
visible indicator 視覚表示器
visible light 可視光
visible outline 外形線
visible ray 可視[光]線
visibly 副明らかに, 目に見えて
vision light のぞき窓
visual 形視覚の
visual angle 視角
visual check (inspection) 肉眼検査, 外観検査, 目視検査
visual gauge ビジュアルゲージ
visual observation 目視観測
visual persistence 残像
visual sensor 視覚センサ
visual signal 視覚信号
visualization 可視化
visually 副視覚的に
vital 形不可欠な, 極めて重要な

Viton バイトン【フッ素ゴム】
voice coil motor ボイスコイルモータ
voice recorder ボイスレコーダ
void 形空の 名ボイド, 空隙
void fraction ボイド率
void space 空所
Voith Schneider propeller フォイトシュナイダプロペラ【櫓の原理を応用した特殊プロペラ】
volatile 形揮発性の
volatile matter (constituent) 揮発分
volatile memory 揮発性メモリ
volatile oil 揮発油, 精油
volatile solvent 揮発性溶剤
volatility 揮発性, 揮発度
volatilization 揮発
volatilize 動揮発する
volt ボルト【電圧の単位：V】
volt ampere ボルトアンペア, 皮相電力【皮相電力の単位：VA】
voltage 電圧, 電位差
voltage adjustment 電圧調整
voltage amplifier 電圧増幅器
voltage build-up rate 電圧上昇率
voltage coil 電圧コイル
voltage control 電圧調整
voltage detector 検電器
voltage drop 電圧降下
voltage gain 電圧利得
voltage ratio 電圧比
voltage reducing device 電撃防止装置
voltage regulation 電圧変動率, 電圧調整
voltage regulation diode 定電圧ダイオード
voltage regulator 電圧調整器
voltage relay 電圧継電器
voltage stabilizer 定電圧回路, 電圧安定器
voltage transformation 変圧
voltaic cell ボルタ電池
voltmeter 電圧計
voltmeter-ammeter 電圧電流計
volume ボリューム, 音量, 体積, 容積, 量, 容量

volume booster　ボリュームブースタ【減圧弁の一種】
volume change　体積変化
volume coefficient of expansion　体膨張係数
volume control　ボリュームコントロール，音量調節
volume dilatometer　体積膨張計
volume displacement type supercharger　容積型過給機
volume distortion　音量ひずみ
volume efficiency　容積効率
volume expansion　体膨張, 体積膨張
volume flow [rate]　体積流量
volume flow meter　体積流量計
volume modulus　体積弾性係数
volume of combustion　燃焼室容積
volume of displacement　排水容積
volume ratio　容積比
volumeter　体積計，容積計
volumetric　形容積測定の
　volumetric coefficient　容積係数
　volumetric displacement　行程容積，押しのけ容積，容積排気量
　volumetric efficiency　体積効率，容積効率，掃気効率
　volumetric flow [rate]　体積流量
　volumetric flow meter　容積式流量計
　volumetric strain　体積ひずみ
　volumetric thermal expansion coefficient　体膨張係数
volute　名ボリュート，渦巻，渦巻室　形渦巻形の
　volute casing　渦形室, 渦巻室, 渦室
　volute pump　ボリュートポンプ，渦巻ポンプ【渦形室のある渦巻ポンプ】
　volute spring　竹の子ばね
vortex　渦，渦式，渦巻，旋回流
　vortex chamber　渦室
　vortex generator　渦発生器
　vortex motion　渦運動
　vortex pump　渦ポンプ
　vortex sound　渦音
　vortex street　渦列
　vortex theory　渦理論
　vortex type combustor　渦型燃焼器
vorticity　渦度
Vulcan gear　フルカンギア
Vulcan hydraulic transmission gear　フルカン液体伝導装置
vulcanite　バルカナイト，エボナイト，硬質ゴム
vulnerable　形脆弱な，損傷されやすい

W-w

W ワット【電力の単位：W】
W-type engine W形機関，W形発動機
wafer ウエハ【ICの基板】，薄片
wait 動待つ，待機する
wake 伴流，後流，航跡
 関 effective wake：有効伴流
 wake coefficient (factor) 伴流係数
 wake percentage 伴流率
 wake stream 流線伴流
wall 壁，内壁，仕切り壁
 wall box 軸受壁わく
 wall deslagger デスラッガ【スートブロワの一種】
WAN (Wide Area Network) ワン，広域通信網
Wankel engine ワンケルエンジン【ロータリエンジンの一種】
want 動〜が欲しい，望む
Ward Leonard system ワード・レオナードシステム【直流電動機の速度制御方式】
warily 副用心して，油断なく
warm 動暖める 形温かい，熱い
 warm up 動暖める
 warm-up 暖機，暖機運転
 warm-up steam valve 暖機蒸気弁
 warm working 温間加工
warming ウォーミング，暖機，加温
 warming device 暖機装置【大型機関を始動させる前に，局部過熱による破損を防ぐため予め均一加熱する装置】
 warming-up 暖機，暖機運転
warmly 副温かに，熱心に，熱烈に
warn 動警告する
warning 警報，警戒，注意
 warning sign 警報表示
 warning system 警報装置
warp 動[板などを]そらせる，ひずませる，曲げる
 名そり，ゆがみ，曲がり，たて糸
warpage 曲がり，焼結ひずみ，そり【冷却のむらによって起き，遅く冷えた側がへこみ，早く冷えた側が膨らむ】
warped blade ねじれ翼
warping ゆがみ
 warping drum 網巻胴
 warping function ワーピング関数，ねじりの関数
 warping winch 網巻ウインチ
warrant 動保障する，正当化する
warranty period 保証期間
Warren girder ワーレンけた
wash 動洗う，洗浄する 名洗浄，打ち寄せる波，洗剤，洗浄液
 wash board 波よけ板
 wash bulkhead 制水隔壁
 wash cloth ウエス，タオル
 wash coat 薄め塗装
 wash plate 制水板
washback ウォッシュバック
washer ワッシャ，座金，洗浄器，洗濯機
 washer based cap nut 座付き袋ナット
washing 洗浄，洗濯
 washing machine 洗浄機，洗濯機
wastage 損(消)耗
waste 動廃棄する，処分する，消耗する，浪費する，無駄にする 名ウエス，廃物，ぼろ布，廃棄物，くず 形廃物の
 waste cloth ウエス
 waste disposal 廃棄物処理
 waste gas 廃ガス，排ガス
 waste heat boiler 廃熱ボイラ
 waste heat utilization 廃熱利用
 waste liquid (liquor) 廃液
 waste oil 廃油
 waste pipe ドレン管，排水管
 waste substance 廃棄物
 waste valve 逃し弁
 waste water 廃水，下水，汚水
wastefully 副不経済に，むだに
waster 不良品

watch 動見張る，監視する
名ワッチ，当直，監視，警備，時計
  watch officer　当直士官
  watch spring　ぜんまい
watchfully　用心深く
watchkeeper　当直員
watchkeeping　見廻り，当直
water　名水，海
動水をまく，給水する
  water absorbing power　吸水力
  water annealing　水なまし
  water ballast　水バラスト
  water brake　水ブレーキ
  water brake dynamometer　水ブレーキ動力計
  water circulation　水循環
  water column　水柱
  water consumption　水消費量
  water content　水分
  water cooled　水冷式の
  water cooled bearing　水冷軸受
  water cooled engine　水冷機関
  water cooled valve　水冷弁
  water cooling　水冷
  water cylinder　水シリンダ
  water droplet　水滴
  water drum　水ドラム
  water examination　水質検査
  water flow　水流，流水量
  water gas　水性ガス
  water gauge　水位計，ガラス水面計
  water hammer[ing]　ウォータハンマ，水撃[作用]
  water hardening　水焼入れ【冷却に水を用いて行う焼入れ】
  water head　水頭
  water horsepowwer　水馬力，水動力【ポンプによってなされる有効仕事】
  water injection　水噴射
  water jacket　水ジャケット
  water jet　ウォータジェット，水噴流
  water jet cutting　水ジェット切断
  water jet ejector　ウォータジェットエジェクタ
  water leg　みずあし【空気除去装置の一種】
  water [level] gauge　水位計，水面計
  water level indicator　水位指示器
  water line　水線，喫水線
  water main　給水主管
  water measuring tube　検水管
  water meter　水量計
  water paint　水性ペイント
  water permeability　誘水性
  water pipe　水管
  water plane　水線面
  water pollution　水質汚濁
  water power　水力，水動力
  water pressure　水圧
  water proof　耐水，防水
  water pump　水ポンプ【流体機械の一種】
  water purifier　浄水器
  water purifying plant　浄水装置
  water quenching　水焼入れ，水中急冷
  water regulating valve　制水弁
  water resisting property　耐水性
  water rheostat　水抵抗器
  water saturation　水飽和
  water screen　水冷壁
  water seal　水シール，水封じ
  water seal packing　水封パッキン
  water seal pump　水封じポンプ
  water seal type safety apparatus　水封式安全器
  water sealing　水封じ
  water separator　水分離器
  water service plan　配水管図
  water side　水側
  water sight　検水器
  water softener　軟水器
  water softening plant　軟水装置
  water soluble　形水溶性の
  water soluble fluid　水溶性工作油
  water space　水部〈ボイラ〉
  water supply　給水
  water supply pipe　給水[主]管
  water supply pump　給水ポンプ
  water supply tank　給水タンク

water temperature　水温
water test　水圧試験，水密試験
water trap　ドレントラップ
water tube　水管
water tube boiler　水管ボイラ
　関 cylindrical boiler：丸ボイラ
water tube cleaner　水管掃除機
water tube wall　水冷炉壁
water vapor　水蒸気
water vapor pressure　水蒸気圧
water wall　水冷壁
waterproof　形耐水の，防水の
　名防水，耐水　動防水する
waterproof belt　耐水ベルト
waterproof canvas (cloth)　防水布
waterproof coat　防水コート
watertight　名水密
　形水密の，防水の，耐水の
watertight bulkhead　水密隔壁
watertight compartment　水密区画［室］
watertight door　水密戸，水密扉
watertight joint　水密継手
watertight subdivision　水密区画
watertight test　水密試験
watertight work　水密工事
watt　ワット【電力の単位：W】
watt hour　ワット時，Wh
watt hour meter　電力量計，積算電力計
wattless　形無効電力の
wattless component　無効分
wattless current　無効電流
wattless power　無効電力
wattmeter　電力計
wave　名波，電波，音波，波動
　動揺れる，波打つ，ひるがえる
wave absorber　電波吸収体
wave band　周波数帯
wave crest　波頂
wave detector　検波器
wave director　導波器
wave drag　造波抵抗，造波抗力
wave dynamometer　波力計
wave energy　波動エネルギ
wave equation　波動方程式

wave form (profile, shape)　波形
wave front　波面，波先
wave function　波動関数
wave height　波高
wave length　波長
wave making resistance　造波抵抗
wave mechanics　波動力学
wave meter　波長計，周波計，電波計
wave motion　波動
wave motion equation　波動方程式
wave number　波数
wave period　波周期
wave power　波力
wave surface　波面
wave train　波列，波連
wave velocity　波速
wave wake　波伴流
wave washer　波形座金
wave winding　波巻【電機子巻線法の一つ】
waved　形波形の，波状の
waviness　うねり
wavy flow　波状流
wax　名ワックス，蝋（ろう）
　動ろうを塗る，ワックスで磨く
wax coating　ろう質膜
waxing　ろう引き
way　方法，すべり面，案内面，みぞ，通路，方向，手段，点
　関 in a way：いわば，ある意味では／in some ways：ある点では／in such a way as to：〜するように，〜するような方法で
weak　形弱い　反 strong：強い
weak acid　弱酸
weak base　弱塩基
weak electrolyte　弱電解質
weak spring card　弱ばね線図【内燃機関の筒内圧力線図の一種】
weak spring indicator　弱ばねインジケータ
weaken　動弱くする，弱まる
weakly acidic resin　弱酸性樹脂
weakly basic resin　弱塩基性樹脂
weakness　弱点，欠点

wear 名摩耗, 摩損, 消耗 動摩耗する, 摩損する
 wear allowance 摩耗しろ
 wear and abrasion resistance 耐摩耗性
 wear and tear 衰耗, 摩損, 損傷, 損耗, 老朽化, 摩耗と劣化
 wear limit 摩耗限界, 摩耗限度
 wear off (down, out) 摩耗する, 〜をすり減らす
 wear out failure 摩耗故障
 wear rate 摩耗率
 wear resistance 耐摩耗性
 wear resisting steel 耐摩耗鋼
wearable computer 装着型コンピュータ
wearing 形消耗させる, すり減る, 着るための
 wearing coat 表面被覆
 wearing plate 摩耗板
 wearing ring ウエアリング, カバーライナ
weather 名気象, 天候 動[荒天などに]耐える, [危機などを]乗り越える
 weather deck 露天(暴露)甲板
 weather proof steel 耐候性鋼材
 weather resistance 耐候性
 weather strip 気密材
weathering 風化作用
weatherometer ウェザロメータ, 耐候試験機
weatherproof 形耐候性の
web ウェブ, 腹部, クモの巣
Weber ウェーバー【磁極の強さの単位:Wb】
 Weber number ウェーバー数
wedge 名ウェッジ, くさび 動くさびを打ち込む
 wedge block gauge ウェッジブロックゲージ【くさび形をしたブロックゲージ】
 wedge friction wheel みぞ付き摩擦車
 wedge-shaped 形くさび形の, V字形の

weekly 形毎週の, 週一回の 副毎週, 週一回
Weibull coefficient ワイブル係数
weigh 動計る, 測る, 圧する
 weigh shaft 逆転軸
weighing 計量, 秤量
 weighing machine 秤(はかり)
weight 名ウエイト, 重さ, 重量[物], おもり 動重くする, 重みを加える
 weight and capacity 重量と容積
 weight and measures 度量衡
 weight efficiency 重量効率
 weight flow coefficient 流量係数
 weight per equivalent 当量
 weight percent[age] 重量百分率, 重量パーセント
 weight saving 軽量化
 weight ton 重量トン
weighted average (mean) 加重平均, 重み付き平均
weight[ing] flow 重量流量
weight[ing] function 重み関数, 荷重関数
weighting table 台盤, 台ばかり
weightless condition 無重力状態
weightlessness 無重力
weir 堰(せき)
Weir's pump ウエアースポンプ
weld 名溶接, 鍛接, 溶着部 動溶接する
 weld crack 溶接割れ
 weld decay 溶接腐食
 weld metal 溶着金属
 weld nugget ウエルドナゲット【溶接部に生じる溶融部分】
 weld penetration 溶込み
 weld zone 溶接部
weldability 溶接性, 鍛接性
welded joint 溶接継手
welded joint pipe (tube) 溶接管
welder 溶接機
welding 溶接, 鍛接, 溶着, 融着
 welding arc voltage アーク電圧
 welding current 溶接電流
 welding deformation 溶接変形
 welding equipment 溶接装置

welding flame 溶接炎
welding flux 溶接フラックス
welding ground アース，アース線
welding heat 溶接熱
welding jig 溶接用ジグ
welding leads 溶接電らん，溶接ケーブル
welding machine 溶接機
welding press 溶接プレス
welding robot 溶接ロボット
welding rod 溶接棒
welding source 溶接電源
welding strain 溶接ひずみ
welding stress 溶接応力
welding symbol 溶接記号
welding tip 溶接火口
welding torch 溶接吹管
welding transformer 溶接変圧器
weldment 溶接物，鍛接，溶接部
well 井戸 副うまく，よく，十分に 関 A as well as B：A も B も，B だけでなく A も／as well：その上，なお，〜した方が良い
well-balanced 釣り合いの良い
Weston cell ウエストン電池
wet 名湿気，水分 形ぬれた，湿った，湿気のある，湿式の 動ぬらす，湿らす
wet adiabatic curve 湿潤断熱線
wet air 湿り空気
wet and dry bulb hygrometer 乾湿球湿度計
wet bulb temperature 湿球温度 関 dry bulb temperature：乾球温度
wet combustion gas 湿り燃焼ガス
wet compression 湿り圧縮
wet dock 係船ドック
wet gas 湿性ガス
wet gas meter 湿式ガスメータ
wet liner 湿式ライナ
wet process 湿式
wet quenching 湿式消火
wet saturated steam 湿り飽和蒸気
wet steam (vapor) 湿り蒸気
wet sump lubrication 湿式潤滑
wet type combustor 湿式燃焼器

wet type dust collector 湿式集塵機
wetness [**fraction**] 湿り度
wettability ぬれ性
wetted area ぬれ面積
wetted perimeter ぬれぶち
wetted surface area 浸水表面積
wetting agent 湿潤剤
wetting power 湿潤力
wetting property ぬれ性
what 代何，〜するもの（こと）
whatever 代〜するのは何でも，何が〜でも
Wheatstone bridge ホイートストンブリッジ【電気抵抗の測定装置】
wheel ホイール，車輪，大歯車 動車で運ぶ，車輪を付ける
wheel and axle 輪軸
wheel barrow 手押し車，一輪車
wheel center 輪心
wheel house ブリッジ，操舵室，船橋
wheel hub 車輪ボス
wheel rotor ホイールロータ
wheel stage 羽根車段〈タービン〉
wheel train 歯車列
wheel tread 踏み面
when 〜 接〜する時
whenever 〜 接〜する時はいつでも
where 副どこに（で） 接するところに（の）
whereas 〜 接一方では〜，ところが〜
whether 〜 接〜かどうか，〜であろうとなかろうと
which 関代〜するところの，そしてそれは 関 in which：その中で
while 〜 接〜する間に，一方では〜，〜であるが
whippability 泡立ち性
whirl 動回転する，渦巻く，くるくる回す 名回転，旋回
whirling 名ふれ回り 形回転する，旋回する
whirling arm 回転腕
whirling shaft 側軸

whirling speed　ふれ回り速度，危険速度
whirlpool　渦，渦巻
　whirlpool chamber　渦室
whisker　ホイスカー，髭(ひげ)結晶，針電極
white　名白　形白い
　white brittleness　白熱脆性
　white cast iron　白鋳鉄
　white copper　白銅
　white graphite　ホワイトグラファイト【固体潤滑剤の一つ】
　white heart malleable cast iron　白心可鍛鋳鉄
　white heat　白熱
　white metal　ホワイトメタル【すずまたは鉛を主成分とする軸受合金】
　white pig iron　白銑
　white spot　白点
　white zinc paint　白亜鉛ペイント
Whitworth screw thread　ウイットねじ
whole　形全体の，すべての，全～
　whole depth　全歯たけ
　whole number　整数
wick　ろうそく(ランプ)の芯，灯心
　wick-feed oiler　毛糸給油器
　wick-lubricator　灯心注油器
wide　形広い，広範囲にわたる
　Wide Area Network　WAN，ワン，広域通信網
　関LAN：Local Area Network
　wide blade　幅広羽根〈タービン〉
　wide liner　幅広ライナ
　wide range grease　ワイドレンジグリース【高速転がり軸受用グリース】
widely　副広く
widespread　形普及した，広く行きわたった
width　広さ，幅
　width across abutment faces　組立幅〈軸受〉
　width across corner　対角距離
　width across flat　二面幅
　width of tooth　刃幅
　width series　幅系列
　width series number　幅[系列]記号
　width variation　幅不同
will　助～だろう，よく～する，～でしょう，～するもんだ
winch　ウインチ【巻上げ機械】
wind　名風　動巻く，巻き付ける
　wind direction　風向
　wind energy　風力エネルギ
　wind force (power)　風力
　wind furnace　風炉
　wind load　風荷重
　wind mill　風車
　wind power generation　風力発電
　wind pressure　風圧
　wind resistance　空気抵抗
　wind shield　風よけ，風防，前窓
　wind tunnel (channel)　風洞
　wind-up torque　ワインドアップトルク【ねじりモーメントの一種】
　wind vane　風向計
windage　風損，風圧面
　windage loss　風損
　windage resistance　造風抵抗
winder　巻取機，ワインダ
winding　巻線，巻上げ
　winding diagram　巻線図
　winding drum (barrel)　巻き胴
　winding engine　巻上げ機関
　winding number　巻数
　winding pitch　巻線ピッチ
　winding ratio　巻線比
　winding rope　巻網
　winding wire　巻線
windlass　ウインドラス，揚錨機，巻上げ機
wing　翼，羽根
　wing blade　W羽根〈タービン〉
　wing bolt　蝶ボルト
　wing furnace　側炉
　wing loading　翼面荷重
　wing nut　蝶ナット
　wing pump　ウイングポンプ，羽根ポンプ【容積変化によって送水するポンプ】

winterization

wing screw 蝶ねじ
wing screw propeller 外側スクリュプロペラ
wing shaft 側軸
wing tank 舷側タンク
wing tip 翼端
winterization 防寒
wipe 動拭く
wiper ring ワイパリング, スクレーパリング
wire 名ワイヤ, 電線, 針金 動電線を引く, 接続する, 針金で結ぶ
wire bench 線引機
wire braid hose ワイヤブレードホース【高圧用ホース】
wire brush ワイヤブラシ
wire buff ワイヤバフ
wire cut electric spark machine ワイヤカット放電加工機
wire drawing 線引き, 絞り作用
wire drawing machine 線引機
wire electrical discharge machine ワイヤ放電加工機
wire fillet 針布
wire gauge 針金ゲージ【針金などの径を測るゲージ】
wire gauze 金網
wire harness ワイヤハーネス【内部配線の総称】
wire resistance strain gauge 抵抗線ひずみ計
wire rod 線材
wire rope 鋼索, ワイヤロープ
wire sling ワイヤスリング
wire spring 線ばね
wire strain gauge 抵抗線ひずみ計用ゲージ
wire wound 巻線
wireless 名ワイヤレス, 無線装置 形無線の
wireless compass 無線コンパス
wireless control 無線操縦
wireless installation 無線装置
wireless office (station) 無線局
wiring 配線
wiring diagram 配線図

~ **wise** 形/副 ~のような(に), ~の方向の(に)
関 film-wise:膜状の(に)/clockwise:時計回りの(に)
wish 動 ~を望む, ~したいと思う
with 前 ~と一緒に, ~を伴う, ~した状態で
with all ~ 前 ~があるのに, ~があるので
with regard to ~ 前 ~に関しては
withdraw 動引き抜く
wither 動しぼむ, 枯れる
withhold 動差し控える, 与えずにおく
within ~ 前 ~の内部に(で), ~の範囲内で, ~以内で
without ~ 前 ~なしに(で), ~のない, ~しないで, もし~がなかったら
without regard for (to) ~ 前 ~にかまわず, ~を無視して
withstand 動耐える, 抵抗する
wolfram タングステン 記 W
wood 木材
wood meal 木粉
wood plug 木栓
wooden 形木製の
wooden pattern 木型
wooden plug 木栓
woodruff key 半月キー
Wood's metal ウッドメタル, ウッド合金【ヒューズなどに用いられる低融点合金】
Woodward viscosimeter ウッドワード粘度計
wore/worn 動「wear」の過去, 過去分詞
word 語, 単語, 言葉 関 in other words:言い換えれば, つまり
work 名仕事【仕事の単位:joule】, 加工[物], 工作[物], 仕事量, 動作, 作業
動働く, 作動する, 作業する
work bench 作業台, 工作台
work carry 回し金〈旋盤〉
work[ing] done factor 作動率

work function　仕事関数
work hardening　加工硬化
work head　工作主軸台，工作物台，ワークヘッド，主軸台
work holder　工作物受台，工作物保持台，ワークホルダ
work material　被削材，工作物
work of compression　圧縮仕事
work of fracture　破壊仕事
work of rupture　破断仕事
work piece　工作物，被溶接物，被加工物，細工物
work place　作業現場
work plan　作業計画
work ratio　仕事比
work rest　ワークレスト，支持刃
work retainer　ワークホルダ
work sampling　ワークサンプリング
work separator　工作物分離装置
work shop　工作室
work station　ワークステーション
work supervisor　作業責任者
work support blade　支持刃
work table　テーブル，ワークテーブル

workability　加工性，加工度
workable days　可働日数
worker　作業者，労働者，職工
working　形働く，運転する，工作の，仕上げの　名作用，働き，加工，稼働，動作，使用，工作，作業
　working accuracy　工作精度
　working angle　作用角
　working area　作業域，動作範囲，作業場
　working atmosphere　作業環境
　working condition　使用条件
　working current　使用電流
　working cylinder　作動シリンダ
　working depth　有効歯たけ，作用高さ
　working diagram　施工図，工作図
　working distance　作動距離
　working drawing　工作図，製作図
　working face　当たり面，作用面
　working factor　作動率，稼働率
　working fit　動きばめ
　working fluid　動作流体，作動流体
　working gas　作動ガス，動作ガス
　working gauge　工作ゲージ
　working life　耐用年数
　working light　作業灯
　working limit　加工限度
　working load　使用荷重
　working mixture　動作混合ガス
　working model　実用模型
　working pitch circle　かみあいピッチ円
　working pressure　使用圧力，動作圧力，作動圧
　working principle　動作原理
　working quenching　鍛造焼入れ
　working rake angle　作用すくい角
　working relief angle　作用逃げ角
　working speed　動作速度
　working strength　作用強さ
　working stress　使用応力
　working stroke　動作行程，動力行程，膨張行程
　working substance　作動物質
　working surface　作動面
　working tension　作用張力
　working voltage　使用電圧
workload　仕事量，仕事負荷
workmanship　工作
workshop　工場
　workshop practice　工場実習，機械工作法
world wide　形全世界の
worm　ウォーム【食違い軸歯車の一種で歯数の少ない歯車をウォーム，これと噛み合う歯数の多い歯車をウォームホイール，両歯車の対をウォームギアという】
　worm brake　ウォームブレーキ
　worm drive　ウォーム駆動
　worm gear[ing]　ウォームギア，ウォーム歯車装置
　worm shaft　ウォーム軸
　worm wheel　ウォームホイール
worn　動「wear」の過去分詞【wear-wore-worn】

worse 形「bad」の比較級，より悪い，より劣った
worst 形「bad」の最上級，最も悪い，一番ひどい
worth 価値
Worthington pump ウォーシントンポンプ
wound 動「wind」の過去・過去分詞，傷つける 名傷
  wound core 巻鉄心
  wound field 巻線磁界
  wound rotor 巻線形回転子
  wound rotor induction motor 巻線形誘導電動機
  wound rotor type motor 巻線形電動機
woven lining ウーブンライニング【摩擦用ライニング】
wow and flutter ワウフラッタ【周波数が変動すること】
wrap 動包む

wrapper plate 側板〈ボイラ〉
wreck 動破壊する，難破させる
wrench 名スパナ，レンチ 動ねじる，ひねる
wringing リンギング，こすりあわせ
wrist pin リストピン，ピストンピン
wrong 形誤った，故障した，正常でない 動不当な取り扱いをする，誤解する
  wrong operation 誤操作
  wrong way alarm transmitter 誤作動警報発信器
  wrong with 動〜にとって良くない
wrongly 副間違って，不正に
wrought 形鍛えられた，精製した
  wrought aluminum alloy 展伸用アルミニウム合金
  wrought iron 鍛鉄，錬鉄
  wrought steel 錬鋼，鍛鋼，鍛練鋼
wye branch Y字管

X-x

X-axis　X 軸
xenon　キセノン【希ガス元素の一種】㊝ Xe
 xenon arc lamp　キセノンランプ
xerography　ゼログラフィ【静電写真法】
X-ray　X 線
 X-ray computed tomography　X 線 CT
 X-ray defect inspection　X 線探傷法
 X-ray diffraction　X 線回折
 X-ray examination　X 線検査
 X-ray inspection　X 線検査
 X-ray non-destructive testing　X 線探傷法
 X-ray photograph　X 線写真
 X-ray plant　X 線装置
 X-ray transmission method　X 線透過法
X-type engine　X 形機関, X 形発動機
X-Y plotter　XY プロッタ
X-Y recorder　XY 記録計
Xylene　キシレン【無色で有毒なベンゼン誘導体】
X-Y-Z axial measuring　三次元測定器

Y-y

Y alloy Y合金【アルミニウムの合金の一種】
Y bend Y継手, 二又, Y形管
Y connection Y結線
Y type wrench Y形レンチ
Y-Δ connection Y-Δ結線
YAG (Yttrium Aluminum Garnet) laser YAGレーザ, ヤグレーザ
yard ヤード, 操車場, 置き場, 構内, 作業場
  yard number 造船番号
  yard plan ヤードプラン
  yard-pound system ヤードポンド法
yarn ヤーン, 撚糸, 小索
yaw 動左右に揺れる 名船首揺れ
  yaw angle ヨー角, 片揺れ角
  yaw heel ヨーヒール【揺れの一種】
  yaw meter 片揺れ計
yawing ヨーイング, 横揺れ, 片揺れ, 偏揺れ, 船首揺れ
  yawing axis 片揺れ軸
  yawing moment 片揺れモーメント
yawl ヨール, 雑用艇

yellow flag 検疫旗
yield 降伏, 歩留まり
  yield load 降伏荷重
  yield locus 破壊包絡線
  yield point 降伏点
  yield point load 降伏点荷重
  yield ratio 降伏比
  yield strength 降伏強さ, 耐力
  yield stress 降伏応力
  yield temperature 作動温度
yielding 降伏
  yielding point 降伏点
  yielding rubber 緩衝ゴム
  yielding valve 弱め弁
yoke ヨーク, 継鉄, 外枠
  yoke cam ヨークカム【駆動カムの一種】
  yoke end Uリンク
  yoke joint 枠継手
  yoke pin 十字軸
Young's modulus ヤング率, ヤング係数, 縦弾性係数
yttrium イットリウム 記Yt

Z-z

z bar　Z 形材, Z 形鋼
z steel　Z 形鋼
Z transfer function　Z 伝達関数
Zener　ツェナー
　Zener breakdown　ツェナー降伏
　Zener diode　ツェナーダイオード, 定電圧ダイオード
　Zener effect　ツェナー効果
　Zener voltage　ツェナー電圧
zenith　天頂, 頂点
Zenith carburetor　ゼニス気化器
zeolite　ゼオライト【硬水の軟化やイオン交換体・触媒などに使用】
Zeolly turbine　ツェリータービン, 圧力複式衝動タービン
zero　名ゼロ, 零点, 零位　形ゼロの
　zero adjustment　零点調整
　zero bevel gear　ゼロベベルギア
　Zero Defects movement　ZD 運動, 無欠点運動
　zero gravity　無重力
　zero input response　零入力応答
　zero method　零位法【測定法の一種】
　zero offset　原点オフセット
　zero point　零点
　zero synchronization　ゼロ同調
Zeuner valve diagram　ツォイナー弁線図【蒸気往復機関におけるすべり弁の弁線図】
ziglo penetrant method　ジグロ探傷法
zigzag　名ジグザグ形, Z 字形　形ジグザグ形の, Z 字形の　副ジグザグ形に, Z 字形に
　zigzag fastening　千鳥締め
　zigzag intermittent weld　千鳥溶接
　zigzag riveted joint　千鳥リベット継手
　zigzag rule　折れ尺
　zigzag type　千鳥形
zinc　亜鉛　記Zn
　Zinc Alloy for Stamping　ZAS, ザス【亜鉛合金の一種】
　zinc base bearing metal　亜鉛基軸受合金
　zinc bath　亜鉛バス
　zinc casting　亜鉛鋳物
　zinc oxide　酸化亜鉛
　zinc plate　亜鉛板
　zinc plating　亜鉛めっき
　zinc protector　保護亜鉛
　zinc sheet　亜鉛板
　zinc slab　亜鉛板, 亜鉛片
zincing　亜鉛めっき
zircaloy　ジルカロイ【ジルコニウム合金の一種】
zirconium　ジルコニウム　記Zr
zone　帯, 地帯
　zone of contact　接触領域
zoom lens　ズームレンズ, 可変焦点レンズ

ISBN978-4-303-30120-0

英和 舶用機関用語辞典

2008年4月10日　初版発行	ⓒ 2008
2025年7月20日　4版発行	

編　者　商船高専機関英語研究会　　　　　　　　　　　検印省略
発行者　岡田雄希
発行所　海文堂出版株式会社
　　　　本　社　東京都文京区水道2-5-4（〒112-0005）
　　　　　　　　電話　03(3815)3291代　FAX 03(3815)3953
　　　　　　　　https://www.kaibundo.jp/
　　　　支　社　神戸市中央区元町通3-5-10（〒650-0022）
日本書籍出版協会会員・自然科学書協会会員

PRINTED IN JAPAN　　　　　　　　　　　　印刷　ディグ／製本　ブロケード

JCOPY ＜出版者著作権管理機構　委託出版物＞

本書の無断複製は著作権法上での例外を除き禁じられています。複製される場合は，そのつど事前に，出版者著作権管理機構（電話 03-5244-5088，FAX 03-5244-5089, e-mail: info@jcopy.or.jp）の許諾を得てください。